Biochar Applications in Agriculture and Environment Management

Jay Shankar Singh • Chhatarpal Singh
Editors

Biochar Applications in Agriculture and Environment Management

Springer

Editors
Jay Shankar Singh
Department of Environmental Microbiology
Babasaheb Bhimrao Ambedkar University
Lucknow, Uttar Pradesh, India

Chhatarpal Singh
Department of Environmental Microbiology
Babasaheb Bhimrao Ambedkar University
Lucknow, Uttar Pradesh, India

ISBN 978-3-030-40996-8 ISBN 978-3-030-40997-5 (eBook)
https://doi.org/10.1007/978-3-030-40997-5

This Springer imprint is published by the registered company Springer Nature Switzerland AG.
The registered company address is: Gewerbestrasse 11, 6330 Cham, Switzerland

Foreword

There has been an increased pressure on the meagre available land resources due to rapid urbanization, which, in turn, has resulted in a gradual but steady deterioration in the cultivable land worldwide. In addition, the adoption of "green revolution" technologies in several countries has made agriculture chemical-intensive, which has contributed significantly to the deterioration of the health of the available agricultural soils. The cultivable agricultural land is scarce and, consequently, is a limiting factor, leaving no scope for further expansion of agriculture. The use of crop residues (CRs) generated in copious amounts annually, in management of soil fertility, reclamation and restoration of marginal lands, reducing air pollution, etc. appears as an innovative option. The management of huge amount of CRs is expected to relieve this pressure to a large extent. Burning of CRs is threatening soil health, causing air pollution, and drastically affecting the soil microbial diversity. Decomposition of paddy residues under anaerobic flooded conditions has substantially increased greenhouse gas emissions, particularly methane. Therefore, appropriate sustainable management of huge amount of CRs produced every year is the need of the hour. Appropriate strategies for management of CRs and augmented environmental awareness can show alternatives to the current practices of irrational burning of CRs.

Crop residues-based biochar has recently received significant attention of researchers with its potential to improve the soil physicochemical status and its contribution to the enrichment of beneficial soil microbiomes. Limited literature is available on impact of CRs-based biochar application on nutrient dynamics, greenhouse gas emission mitigations, crop yields, and beneficial microbial community dynamics of agroecosystems. The editors have invited experts to contribute articles in the area of sustainable soil health and environmental management. This volume will enhance our knowledge about better performance of soil microbes in soil, resulted due to CRs burning, and how these changes along with other environmental factors will affect the soil fertility and crop yields. Overall, the book ***Biochar Applications in Agriculture and Environment Management***, gives an overview of the current status of CRs management. This book also gives directions to popularize related technologies that can mitigate the decline in soil fertility of cultivable land. The chapters in the book also help us to focus on beneficial soil microbial diversity for sustainable agriculture and decrease pressure on the need for more cultivable land.

Vice-Chancellor, University of Hyderabad
India
December 28, 2019

Appa Rao Podile,

Preface

Biochar use in agricultural soil has been projected as effective options to mitigate the soil fertility health due to climate change. Its sole use may also cause salt stress in the soil as it contains large concentration of nutrient metals. Crop residues-based biochar has been recently proposed as suitable soil fertility enhancer as it improves the physicochemical structure and beneficial microbial biomass of soil. Previous studies have indicated that wood- or manure-based biochars can reduce leaching of fertilizer N in highly weathered subtropical or tropical soils using pot experiments. Previous investigations argued inconsistent effects of biochar additions on greenhouse gas emissions, attributable to many factors, such as soil and biochar types, moisture regime, nutrient availability, and potentially other unknown causes. However, few studies and books describe the impact of biochar on inorganic nutrient dynamics, greenhouse gas emission mitigation, pollution decontamination, and beneficial microbial community dynamics of cropland soils. Therefore, this book, ***Biochar Applications in Agriculture and Environment Management***, considers leading experts to contribute articles that may describe the role of biochar application on soil nutrient dynamics of disturbed agricultural fields and as soil conditioner for saline/stressed/nutrient poor soils. This book also describes the impact of biochar application on C sequestration and greenhouse gas emissions/mitigation from agricultural soil and the role of biochar in remediation of contaminated agricultural soils to evaluate the addition of biochar in dynamics of beneficial soil microbial communities and soil microbial biomass levels. However, literature on the interactive effect of both biochar and organic fertilizers is scarce. Therefore, further study on importance of organic farming and intensive C-sequestration strategies such as biochar production from agricultural wastes is important.

We are thankful to all contributors for their informative articles to this book. We are sure that this volume will be useful to students, scientists, and academicians concerned with the management of CRs and its applications in sustainable agricultural development.

Lucknow, Uttar Pradesh, India Jay Shankar Singh
Rampur, Uttar Pradesh, India Chhatarpal Singh

Contents

Contributors

Keerthika A. ICAR-Central Arid Zone Research Institute, Regional Research Station, Pali Marwar, Rajasthan, India

Tripti Agarwal Department of Agriculture and Environmental Sciences, National Institute of Food Technology Entrepreneurship and Management, Kundli, Sonipat, Haryana, India

Nijara Baruah Plant Physiology and Biochemistry Laboratory, Department of Environmental Science, Tezpur University, Tezpur, Assam, India

Palakshi Borah Plant Physiology and Biochemistry Laboratory, Department of Environmental Science, Tezpur University, Tezpur, Assam, India

Bikram Borkotoki Biswanath College of Agriculture, Assam Agricultural University, Biswanath Chariali, Assam, India

Shaon Kumar Das ICAR-National Organic farming Research Institute, Gangtok, India

Rahul Dev ICAR-Vivekananda Parvatiya Krishi Anusandhan Sansthan, Almora, Uttarakhand, India

Rachana Dubey ICAR-Research Complex for Eastern Region, Patna, Bihar, India

Ram Kishor Fagodiya Division of Irrigation and Drainage Engineering, CSSRI, Karnal, Haryana, India

Goutam Kumar Ghosh Palli-Siksha Bhavan, Visva-Bharati, Sriniketan, West Bengal, India

Balendu Shekher Giri Department of Chemical Engineering and Technology, Centre of Advanced Study, Indian Institute of Technology, Banaras Hindu University, Varanasi, India

Lina Gogoi Department of Energy, Tezpur University, Tezpur, Assam, India

Nirmali Gogoi Plant Physiology and Biochemistry Laboratory, Department of Environmental Science, Tezpur University, Tezpur, Assam, India

Mandavi Goswami Department of Bioengineering, Integral University, Lucknow, Uttar Pradesh, India

Chandan Kumar Gupta Department of Agro-meteorology and Environmental Science, Birsa Agricultural University, Ranchi, Jharkhand, India

Dipak Kumar Gupta ICAR-Central Arid Zone Research Institute, Regional Research Station, Pali Marwar, Rajasthan, India

Surabhi Hota National Bureau of soil Survey and Land Use Planning, RC, Jorhat, Assam, India

Roomesh Kumar Jena National Bureau of soil Survey and Land Use Planning, RC, Jorhat, Assam, India

P. Kannan Department of Soils and Environment, Agricultural College and Research Institute, Madurai, Tamil Nadu, India

Rupam Kataki Department of Energy, Tezpur University, Tezpur, Assam, India

Shakeel A. Khan Centre for Environmental Science and Climate Resilient Agriculture, IARI, New Delhi, India

Puja Khare Agronomy and Soil Science Division, CSIR-Central Institute of Medicinal and Aromatic Plants, Lucknow, India

Aacademy of Scientific and Innovative Research (AcSIR), Ghaziabad, India

Ki-Hyun Kim Department of Civil and Environmental Engineering, Hanyang University, Seoul, Republic of Korea

D. Krishnaveni Alagappa Government Arts and Science College, Alagappa University, Karaikudi, Tamil Nadu, India

Amit Kumar Host Plant Section, Central Muga Eri Research & Training Institute, Jorhat, Assam, India

Arun Kumar Department of Environmental Science, Guru Nanak Khalsa College, Yamuna Nagar, Haryana, India

Lal Chand Malav National Bureau of soil Survey and Land Use Planning, RC, Nagpur, Maharashtra, India

M. B. Noor Mohamed ICAR-Central Arid Zone Research Institute, Regional Research Station, Pali Marwar, Rajasthan, India

Umesh Pankaj Division of Biotechnology, CSIR-Institute of Himalayan Bioresource Technology (IHBT), Palampur, Himachal Pradesh, India

S. Ponmani Mother Terasa College of Agriculture, Pudukkottai, Tamil Nadu, India

A. G. Rajalakshmi Department of Biotechnology, SNMV College of Arts and Science, Coimbatore, India

Prasenjit Ray National Bureau of soil Survey and Land Use Planning, RC, Jorhat, Assam, India

S. K. Ray National Bureau of soil Survey and Land Use Planning, RC, Jorhat, Assam, India

Kangkan Roy Department of Chemical Engineering and Technology, Centre of Advanced Study, Indian Institute of Technology, Banaras Hindu University, Varanasi, India

Amita Shakya Department of Agriculture and Environmental Sciences, National Institute of Food Technology Entrepreneurship and Management, Kundli, Sonipat, Haryana, India

Gulshan Sharma ICAR-National Bureau of Soil Survey and Land Use Planning, Regional Center, Jorhat, Assam, India

Gulshan Kumar Sharma National Bureau of soil Survey and Land Use Planning, RC, Jorhat, Assam, India

A. K. Shukla ICAR-Central Arid Zone Research Institute, Regional Research Station, Pali Marwar, Rajasthan, India

Chhatarpal Singh Department of Environmental Microbiology, Babasaheb Bhimrao Ambedkar University, Lucknow, Uttar Pradesh, India

Jay Shankar Singh Department of Environmental Microbiology, Babasaheb Bhimrao Ambedkar University, Lucknow, Uttar Pradesh, India

Ram Sharan Singh Department of Chemical Engineering and Technology, Centre of Advanced Study, Indian Institute of Technology, Banaras Hindu University, Varanasi, India

Himanshu Tiwari Department of Chemical Engineering and Technology, Centre of Advanced Study, Indian Institute of Technology, Banaras Hindu University, Varanasi, India

Yui Fai Tsang Department of Science and Environmental Studies, The Education University of Hong Kong, Tai Po, New Territories, Hong Kong, China

Kumar Vikrant Department of Chemical Engineering and Technology, Indian Institute of Technology, Banaras Hindu University, Varanasi, India

Department of Civil and Environmental Engineering, Hanyang University, Seoul, Republic of Korea

Saowanee Wijitkosum Environmental Research Institute, Chulalongkorn University, Bangkok, Thailand

Krishna Kumar Yadav Institute of Environment and Development Studies, Bundelkhand University, Jhansi, India

Vineet Yadav Agronomy and Soil Science Division, CSIR-Central Institute of Medicinal and Aromatic Plants, Lucknow, India
Aacademy of Scientific and Innovative Research (AcSIR), Ghaziabad, India
Jawaharlal Nehru Customs House, Central Revenue Control Laboratory, New Delhi, India

Chapter 1
Applying Rice Husk Biochar to Revitalise Saline Sodic Soil in Khorat Plateau Area – A Case Study for Food Security Purposes

Saowanee Wijitkosum

Abstract Salt-affected soils occur in the areas where excess dissolved mineral salts accumulate in the root zone that crop yields are adversely affected from the salts released by weathering of rock or those initially present in the soil-forming materials. In addition, evaporation and transpiration processes, due to high temperatures and droughts, can cause salt movement with capillary action inducing its accumulation in surface soil. Excess amounts of salts cause adverse effects on the physical and chemical properties of soil, microbiological processes and food security. Biochar produced from rice husk (RH) under the pyrolysis condition (400–500 °C) from a retort designed to produce laboratory quality biochar that is easy for farmers to use in order to promote self- sustaining biochar production. This study aimed to explore the use of rice husk biochar as a soil amendment in order to solve the salt-affected soil problems. The study area was Bung O sub- district, Kham Thale So district where the critical of salt-affected soil and drought area in Nakhon Ratchasima, Thailand. The results indicated that adding RH biochar with organic fertilizer into soil can improve both physical and chemical properties in every parameter. Particularly, the soil became less alkalinity. The results also showed an increased ion exchange capacity, higher amount of major and minor soil nutrients and the reduced amount of all sodium in the soil in every parameter. This included the absorption rate of sodium in the soil, the conductance of the soil, all of the amount of sodium in the soil, and the increasing amount of exchangeable magnesium and the amount of exchangeable calcium. The two elements contained positive ions which could replace the sodium ions in the salty soil making the soil less salty.

Keywords Salt-affected soil · Saline sodic soil · Biochar · Rice husk · Food security purposes

---

S. Wijitkosum (✉)
Environmental Research Institute, Chulalongkorn University, Bangkok, Thailand
e-mail: Saowanee.w@chula.ac.th

J. S. Singh, C. Singh (eds.), *Biochar Applications in Agriculture and Environment Management*, https://doi.org/10.1007/978-3-030-40997-5_1

1.1 Introduction

Salt-affected soils become one of the global issues that adversely affect soil resources in various areas (Martinez-Beltran and Manzur 2005). Even though saline soils occur naturally on Earth's crust (Rengasamy 2006) human activities and inappropriate land use have contributed to the acceleration of the saline soil problems both in terms of its severity and an expansion of the affected areas (Shahbaz and Ashraf 2013; Brinck and Frost 2009; Martinez-Beltran and Manzur 2005; Metternicht and Zinck 2003). The human activities included overgrazing, the use of chemical fertilizers and deforestation (Lakhdar et al. 2009). Climate change and soil degradation problems contribute to an expansion of soil salinity. The adverse impact in the next 25 years is expected to be 30% loss of land (Wang et al. 2003). For agricultural countries, these problems possess significant threats to agriculture, land use and food security especially in rural farming areas. High salt concentration negatively affects soil microbial activities as well as soil chemical and physical properties causing a decline in soil productivity (Amini et al. 2015; Wong et al. 2009; Yuwaniyama 2004). Moreover, salt- affected soil also causes physiological drought and its expansion to nearby areas also negatively affects the ecosystems and resource utilization. Soil salinization is complex and highly interrelated with specific conditions in the area.

In general, salt- affected soil mostly occurs in arid and semiarid regions of the world. Despite its location in the tropics, Thailand has long been suffering from salt- affected soil due to physical characteristics of the location. In recent years, the problems have become more severe and the surrounding environment and the people's quality of life are affected. The north-eastern part of Thailand (14° 14′ to 18° 27′ N and 101° 0′ and 105° 35′ E) is a vast area covering 1700,000 sq km, of which 17% is salt affected soil area with different levels of concentration. The majority of the areas are located in Nakhon Ratchasima province, especially in Kham Thale So district where the impacts are severe that patches of salt stains are visible on the surface (Wijitkosum 2018). Isaan Catchment Hydrogeological and Agricultural model was used to predict that, without an appropriate management plan for salt-affected soil in the Kham Thale So area, the severity of the problems and the distribution of salt-affected soil areas would increase by 21% in the next 30 years (Yuwaniyama 2004).

There are many possible solutions to mitigate soil salinity problems including physical, chemical and biological methods. The biological solutions are such as using salt tolerant plants (Grover et al. 2003; Manchanda and Garg 2008; Schubert et al. 2009), cover cropping, planting soil enhancement plants, shifting the crop calendars (Venkateswarlu and Shanker 2009) or mixing the saline soil with soil amelioration substances such as cow manure or chemical fertilizers (Qadir and Oster 2002). Using soil amelioration substances, either natural or chemical, is the most popular method among farmers. This method allows the soil amelioration and the agricultural processes to proceed simultaneously. However, using chemical substance for soil amelioration is a costly method and may lead to long-term effects on the ecosystem (Qadir and Oster 2002). Therefore, using environmental friendly soil amelioration substance is a more sustainable choice as it is beneficial to the ecosystem as well as farmers'quality of life (Venkateswarlu and Shanker 2009; Chaganti et al. 2015; Sun et al. 2016).

Biochar, a carbonaceous organic material, is a soil ameliorating substance that can be obtained from various types of biomass. The types of biomass or raw materials and the pyrolysis process play significant roles in the final products as the biochar's characteristics are highly depending on the types of raw materials and the pyrolysis processes it was obtained from (Sriburi and Wijitkosum 2016; Lehmann and Joseph 2009; Sohi et al. 2009; Yamato et al. 2006). This exclusive characteristic allows biochar researchers to design and produce the substance to best suit specific agricultural purposes. Furthermore, biochar has been widely researched and accepted as an efficient soil amendment substance that improves soil quality in various conditions including acid soils (Martinsen et al. 2015; Obia et al. 2015; Slavich et al. 2013; Yuan et al. 2011; Chan et al. 2007; Lehmann et al. 2003; Glaser et al. 2002), hard soils (Herath et al. 2013; Uzoma et al. 2011; Laird et al. 2010), infertile soils (Wijitkosum and Kallayasiri 2015; Rajakovich et al. 2012; Laird et al. 2010), sandy soil (Uzoma et al. 2011), alkaline soil (Sun et al. 2016; Abrishamkesh et al. 2015; Wu et al. 2014) as well as preventing soil erosion (Wang and Xu 2013; Chaganti et al. 2015). However, studies on salt-affected soils are limited to laboratory research or in column study (ex. Schultz et al. 2017; Chaganti et al. 2015; Wu et al. 2014) but not in field research. There is some research found that biochar is like organic matter that can effectively reduce the absorption of crops to Na^+ and reduce the salinity stress to crops (Thomas et al. 2013, Yang et al. 2014). There are a limited number of reports describing the effects of biochar on rice grown in saline-sodic paddy soil. Moreover, biochar research in Thailand and the effects of biochar on salted soil is still scant. This study aims to explore the use of biochar as a soil amendment in order to solve the salt- affected soil problems. Biochar used in this study was obtained from a retort designed to produce laboratory quality biochar that is easy for farmers to use in order to promote self-sustaining biochar production.

1.2 Salt Layers Underneath the Khorat Plateau: The Source of Salt Affected Soil in Nakhon Ratchasima Province

Beneath the Khorat Plateau lies multilayers of various clastic sedimentary rocks in the Khorat group including conglomerate, sandstone, siltstone, shale and mudstone (Division of Mineral Resources Conservation and Management 2015). As for Maha Sarakham formation, the layers of rock salt underneath the soil surface contributes directly to the saline soil on the top surface. The formation covers approximately 34.18% of the whole region (Department of Mineral Resource 1982; Thai- Australia Tung Kula Ronghai Project 1983). Moreover, the depths of rock salts differ from one area to another. In some areas, the depths can exceed hundreds of meters or only

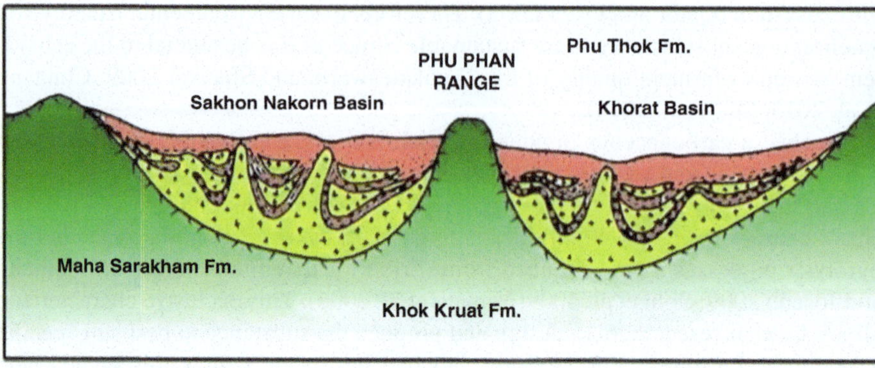

Fig. 1.1 Salt-affected areas in the north-eastern part of the country and its natural geographical cause. (Source Suwanich 1986)
A model showing the elevated Phu Phan Range that divides north-eastern part of Thailand into Sakhon Nakorn Basin and Khorat Basin resulting curved underground salt layers. These salt layers emerged near the surface causing salt-affected soils in the region

up to 20–25 m deep in Nakhon Ratchasima. The rock salts located closer to the surface affect evaporation and brings saline solution up to the surface through the capillary force (Fig. 1.1).

The dissolution of rock salts and its dispersion in groundwater are the main causes of salt-affected soil problems in Nakhon Ratchasima. Moreover, necessary shortcuts and catalysts are crucial for the saline soil to widely spread over the plateau. Salt domes act as shortcuts that allow the source of salt to emerge closer to the surface while faults act as pathways to facilitate the upward movement of saline groundwater (Fig. 1.1). Being dissolved by the groundwater, capillary force in the formerly salt domes acts as a catalyst to pull the brine to the surface against the gravity. At this point, water evaporates from the soils and only salt remains. This process happens continuously in dry weather conditions where the groundwater table is less than four feet and the top soil is sandy (Division of Mineral Resources Conservation and Management 2015).

Research indicated that the altitude of the area influences the level of ground water that contributed to the distribution of salt-affected soil in the area. The concentration of salinity was varied depending on soil horizon (Kovda et al. 1973) and seasons (Blaylock 1994). The level of saltiness was varied greatly (Leksungnoen 2006) and was influenced by the saltiness of ground water (Japan International Cooperation Agency 1991). Moreover, heavy deforestation in the past and land use changes had direct impacts on the distribution of salt- affected soil (Williamson et al. 1989). In monsoon seasons, the salt stains washed off and retained in the soil. However, in dry season or in draught period, the water retained in the soil will evaporate and make the salt stains visible on the surface (Yuvaniyama 2003; Manchanda and Garg 2008). In this relatively dry area the farmers faced severe water scarcity problems during the dry season and lost a considerable amount of production land due to soil salinization.

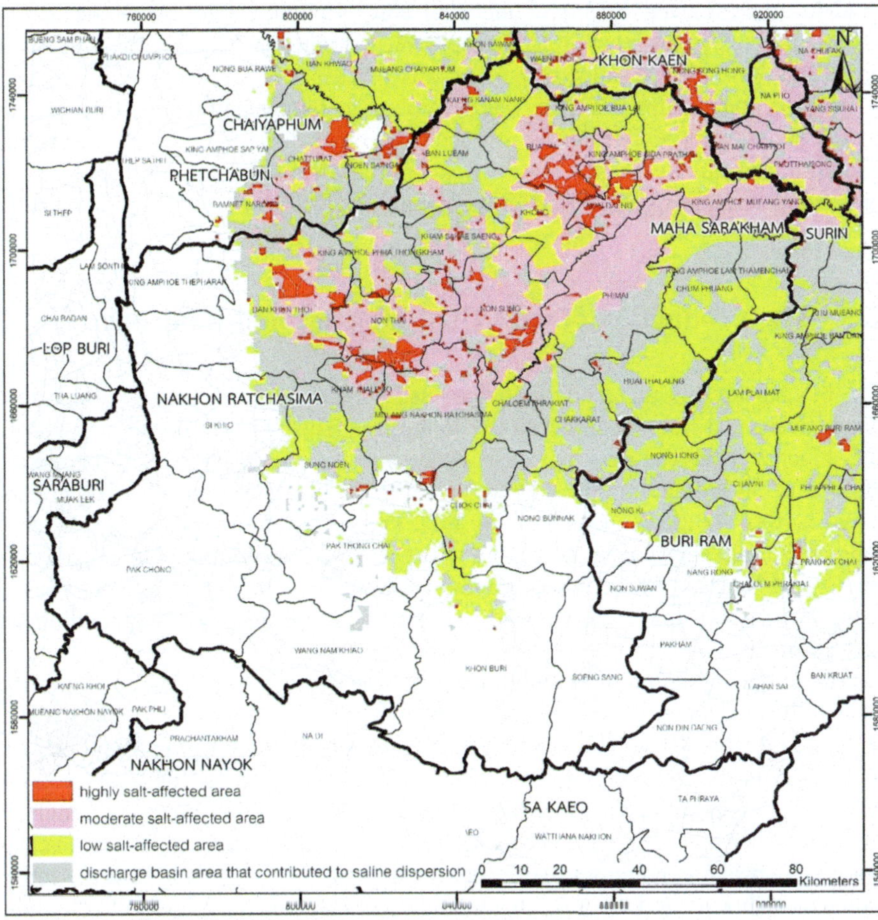

Fig. 1.2 A map showing saline-affected areas in Nakhon Ratchasima province

Nakhon Ratchasima province covers an area of approximately 20,493.96 sq km. 92.13% of Nakhon Ratchasima is land area and only 38.86% of the area is not affected by salt deposits. The saline soil area comprises lowland area (53.33%), high land area (46.66%) and salt farms (0.02%). Four different concentration levels of salt were found scattered in the lowland area: low level (less than 1% of salt stains), moderate level (1–10%), high level (10–50%) and heavy level (more than 50%). The majority of the area was affected at a low level (30.33%), followed by a moderate level (20.42%), a high level (1.61%) and a heavy level (0.97%) (Fig. 1.2). However, Nakhon Ratchasima possesses the largest salt-affected areas and suffers the most from the problems in comparison to the whole north-eastern region. The salt- affected area in Nakhon Ratchasima covers 4809.58 sq km and 5866.48 sq km is prone to saline distribution. Of all affected area, 2.33% suffered salinity at an extreme, 3.85% at a high level, 49.02% at a moderate level and 44.79% suffered a mild level. Moreover, 3.75% of the total area is uncultivable. Moderate to fair saline- affected areas, where rice was cultivated, were 15.63% and discharge basin area that contributed to saline dispersion was 17.19%.

Nakhon Ratchasima province was severely affected by salinization that patches of salt stains were visible on the surface. The rapid dispersion of saline soil had compromised quality of life for many people and caused many problems to the province's economy, society and environment.

1.3 The Production of Rice Husk Biochar and Its Properties

The biochar used in this study was rice husk biochar (RH biochar) obtained from paddy husks that were locally available in the area. The paddy husks were produced into biochar using the slow pyrolysis process in a 4 × 200 litters Retort for Rice Husk Biochar Production (patent number: 1601001281) at a controlled temperature between 400–500 degree Celsius (Fig. 1.3). The temperature was at an appropriate range for producing biochar according to the Food and Agriculture Organization of the United Nations (FAO 2009). The retort is cost- efficient and can be built easily using locally available materials and in addition can use locally available biomass as feedstock.

The biochar sampling method used in this study was adapted from the Standardized Product Definition and Product Testing Guidelines for Biochar that is used in Soil (IBI 2014). The samples were randomly selected from ground biochar and analyzed for specific characteristics including surface area, total pore volume and average pore diameter using the Brunauer- Emmett- Teller method of analysis of the nitrogen adsorption isotherms. Moreover, pH (pH meter with 1:2 (v/v) char: water), electrical conductivity (EC; EC meter with 1:5 (v/v) char: water), cation exchange capacity (CEC; leaching method), organic matter (OM; Walkley and Black method), total carbon (total C; Shimadzu TOC Tcvh), total nitrogen (total N; Kjeldahl method), phosphorus (P_2O_5; Vanadomolybdophosphoric acid colorimetric method), potassium (K_2O; AAS) and water holding capacity (WHC) were analysed. The carbon (C), hydrogen (H), nitrogen (N) and sulfur (S) atom contents (wt.%) were measured using a Carbon, Hydrogen Nitrogen and Sulfur/ Oxygen Analyzer (Leco CHN628 model).

According to the physical and chemical analysis of RH biochar, the results indicated a specific surface area of 41.43 m^2/g and total pore volume of 0.034 cm^3/g. The results were higher than those of the feedstock (rice husk) that had 2.06 m^2/g of specific surface area and 0.0038 cm^3/g of total pore volume. The majority of RH biochar composites were cellulose, hemicellulose and lignin. Moreover, the RH biochar had many pores in the structure (Fig. 1.4 bottom left and bottom right) which showed an average pore diameter of 32.73 Å. The X-ray elemental mapping with a magnification of 1000× revealed that biochar contained various elements on its surface. The proportion of oxygen was the highest followed by silicon, phosphorus, potassium and magnesium, respectively. The main elements on the surface of

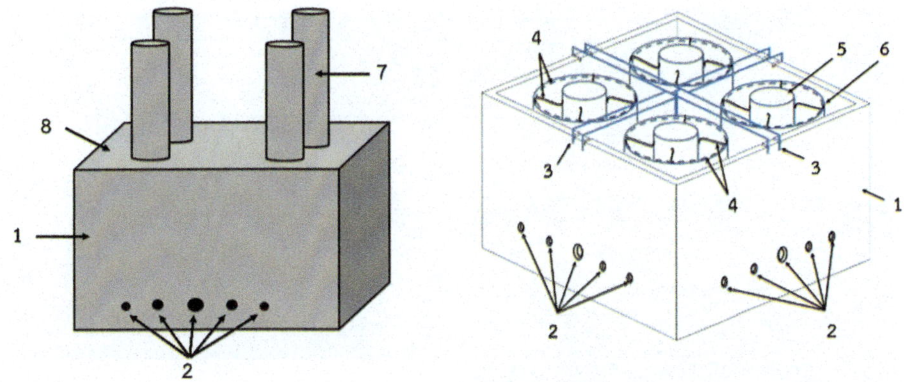

a. Model of the rice husk biochar retort (outside)

b. Model of the rice husk biochar retort (inside)

1 Outside concrete retort

2 Air intake holes

3 Outside removable steel lid support rebar

4 Heat exchanger support

5 Heat exchangers

6 200 litter steel drum

7 Smoke chimney

8 Outside removable steel lid

Fig. 1.3 The 4 × 200 litters Retort for Rice Husk Biochar Production (patent number: 1601001281) at a controlled temperature

RH biochar were oxygen (44%) and silica (20%). Nwajiaku et al. (2018a) and Mahmoud et al. (2011) found that silicon increased with an increase in pyrolysis temperature which was a result of the change in the form of silica relative to the time of charring. It is reported that the amorphous form of silica changed into crystalline form when heat was applied (Todkar et al. 2016; Parry and Smithson 1964; Nwajiaku et al. 2018b).

This study concluded that the two elements on the surface were in the form of silicon dioxide (SiO_2), which was consistent with a study conducted by Nwajiaku et al. (2018a), Mahmoud et al. (2011) and Mansaray and Ghaly (1997) who studied elements of rice husk in oxide forms. The study indicated that SiO_2 was the major element, accounting for more than 90%, along with other elements existing in oxide forms such as K_2O, P_2O_5 and MgO. The result was in accordance with the analysis conducted by Scanning Electron Microscope and Energy Dispersive X-ray Spectrometer (SEM-EDS (IT300)) that nutrients such as potassium, phosphorus and magnesium were crucial for crop growth. Therefore, this study concluded that RH biochar was an effective accelerator of crop growth.

RH biochar had weak alkalinity (pH 7.9), EC of 0.35 dS/m and CEC of 17.34 cmol/kg. The results from Elemental Analyzer (CHNS) indicated that RH biochar's composites comprised 45.68 wt% C, 0.93 wt% N and 2.22 wt% H. Its H/C molar was 0.27 and C/N molar was 49.50, which reflected the stability of the biochar (IBI

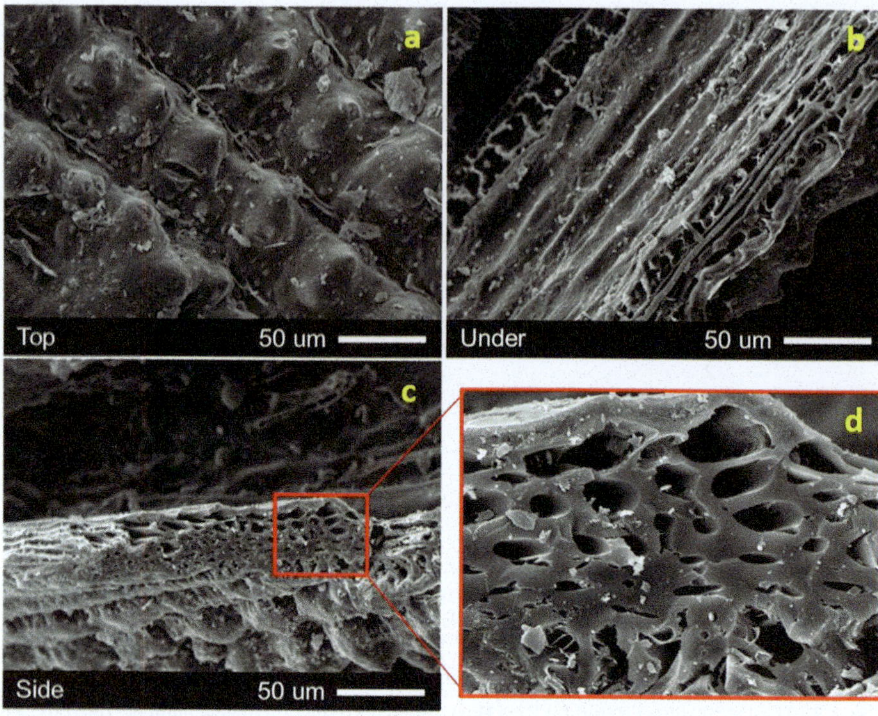

Fig. 1.4 Physical characteristics of rice husk biochar from multipoint BET method and scanning electron microscopy (SEM) at a magnification of 500× (**a** and **b**) and the porous structure of rice husk biochar (**c** and **d**)

2015; Downie et al. 2009; Lehmann et al. 2009). This property made the carbon in RH biochar very stable (Kim et al. 2012; Sohi et al. 2009; Downie et al. 2009), created a highly porous carbon structure (Chen et al. 2012; Sohi et al. 2009) which made it suitable for use as a soil amendment and soil carbon storage. In terms of its nutritional value, RH biochar had macro-nutrients: total N content of 0.51%, with Total P_2O_5 of 0.29% and K_2O of 1.02 wt.%. Moreover, RH biochar also had a high OM (13.06 wt%), which contributed to an increased OM level in the soil and improved the soil fertility. The pyrolysis of RH biochar at an appropriate temperature (400–500 °C) also increased the porosity on biochar's surface which led to an increased number of ions on its surface (Sohi et al. 2009; Sun et al. 2014). This study concluded that RH biochar produced from the retort had superior physical characteristic, which was consistent with a study of Mathurasa and Damrongsiri (2017) who found that pyrolysis of rice husks at 500 °C for 2 h increased its surface area, total pore volume, pore diameter, and CEC value of the biochar product. The surface of RHB had a greater fraction of silica than of the unprocessed rice husk. As a result, the RH biochar had high capacity to adsorb and retain organic carbon and non- organic matters within the soil. In addition, it also contributed to nutrient adsorption or covalent interaction on a large surface area (Schmidt and Noack 2000;

Amonette and Joseph 2009; Downie et al. 2009; Atkinson et al. 2010; Wijitkosum and Kallayasiri 2015). This helped with aeration and reduced soil density (Jones et al. 2010; Bhogal et al. 2009; Novak et al. 2009; Hati et al. 2007). Before applying RH biochar, the biochar was soaked in liquid fertilizer for three minutes and was left to dry on the grille. Liquid fertilizer was alkaline (pH 6.8), with EC of 0.82 dS/m, OM of 0.02 wt.%, total K of 0.04 wt.%, total N of 20.16 mg/ L and total P of 44.07 mg/L. RH biochar soaked in the liquid fertilizer yielded in even more alkalinity (pH 8.9), higher EC of 0.49 dS/m, higher specific surface area (98.49 m^2/g) and higher total pore volume (0.069 m^3/g). The structure of RH biochar showed an average pore diameter of 28.17 Å, with water holding capacity of 138.30%. Due to its large surface area and high porosity, biochar is able to adsorb and retain nutrients from the liquid fertilizer. This resulted in higher nutrient contents and a better structure as a soil amelioration substance. Biochar characteristics after being soaked in the liquid fertilizer contained 0.54% of total N, 0.35% of total P_2O_5, 1.10% of total K_2O and 7.78% of silicon.

1.4 Using Biochar as a Soil Amendment for Salt-Affected Soil: A Pilot Study in Kham Thale So District, Nakhon Ratchasima Province

The study area is located in Bung O sub- district, Kham Thale So district, Nakhon Ratchasima province. The study area covers approximately 203.60 sq km (14°57′ 39″ N101°56′ 51″ E). The area is inclined from the west towards the east and its altitude is 178–247 m above mean sea level. The soil is sandy soil and loamy sand of low fertility and high permeability (Fig. 1.5). The three most affected areas in Kham Thale So are Phan Dung, Nong Sruang and Bung O sub-districts which is the most affected area among the three. The study area is located in a plain area, far from natural water sources and with an annual average precipitation date of 54 days/year.

Soil analysis obtained from the study area indicated that the soil was strongly alkaline (pH 8.5–10.20) and electrical conductivity of the saturated extract (ECe) of 14.36–57.80 dS/m. The soil texture was loamy sand and silt loam with CEC of 2.40–6.20 cmol/kg. The soil had a very low level of primary macronutrients (total N of 17.50–297.50 mg/kg, available phosphorus (avail. P) of 2.00–17.00 mg/kg, and exchangeable potassium (exch. K) of 3.00–31.00 mg/kg and very low level of secondary macronutrients (exchangeable calcium (exch. Ca) of 1221.44–1279.63 mg/kg, and exchangeable magnesium (exch. Mg) of 22.50–23.22 mg/kg). The soil had very low fertility with 0.03–0.48% of OM. The results from the analysis of soil salinity parameters were: 0.03–0.54% of total sodium (total Na), 2425.50 mg/ kg of extractable sodium (extrac. Na), 283.60–4745.00 mg/kg of extractable chloride (extrac. Cl) and 34.35–1158 of Sodium Adsorption Ratio (SAR).

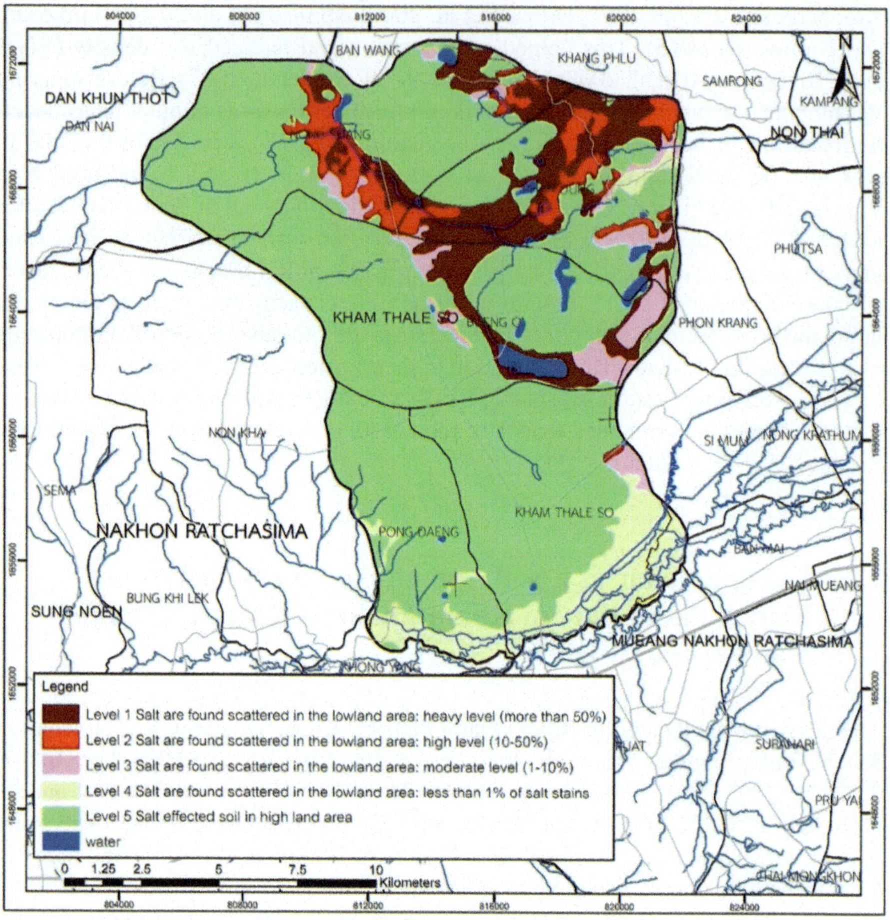

Fig. 1.5 The critical area affected by salt-affected soil in Kham Thale So, Nakhon Ratchasima

The soil properties indicated that the soil in the area was saline sodic soil. The soil analysis revealed both salinity and sodicity problems and characteristics of both types. The saline sodic soil was referred to the soil of which ECe of saturated paste extract was greater than 4 dS/m, its SAR was greater than 13, its exchangeable sodium percentage (ESP) was greater than 15 and its pH value was usually between 8.5 and 10. (U. S. Salinity Laboratory Staff 1954; FAO 1988; Richard 1954; Eynard et al. 2006; Hinrich et al. 2001; Rengasamy 2010).

Considering the ECe and SAR values to predict the quantity of salt and its impacts on plant growth, the results revealed that the soil in the pilot study area was of extremely high salinity. The soil was not suitable for regular plants but were suitable for halophytes or other plants with efficient salt-tolerance mechanisms (ECe greater than 16 dS/m) (El-Zanaty et al. 2006; Flowers et al. 1986). The high salinity made the area uncultivable and was left barren. There was no vegetative cover

Fig. 1.6 The pictures of saline sodic soils and salt stains

excepts halophytes and salt-tolerant plants (El-zanaty et al. 2006; Flower et al. 1986) such as Acacias ampliceps and Sesbania (Sesbania rostrate L.) rostrate that local farmers grew to battle soil salinity in the area.

The salinity and sodicity of the soil properties caused a declined vegetation growth (Chen et al. 2007; Mathur et al. 2006; Shi and Wang 2005; Raul et al. 2003) from salt toxicity (Bacilio et al. 2004; Munns 2005; Orcutt and Nilsen 2000), high osmotic suction (Kaymakanova et al. 2008; Sheldon et al. 2004), nutrient deficiency (Lakhdar et al. 2009; Yuwaniyama 2004), high pH, surface crust and degraded soil structure. A field survey revealed visible salt stains scattered on the soil surface and no vegetative cover was present. In relation to the findings, Wijitkosum (2018) also reported that the area in Kham Thale So district, Lam Ta Kong watershed was also critically affected by salinity, especially in the northern parts where dense salt crusts (>50%) were distributed widely on the soil surface (Fig. 1.6).

The salt-affected soil problems in Kham Thale So district were the results of groundwater flows that spread more saline water into the recharge area. This process accelerated the local flow system allowing more groundwater with rock salt solution to be brought up to the soil surface (Williamson et al. 1989; Yuwaniyama 2004). Moreover, Wijitkosum (2018) found that the study area was facing the highest risk of drought due to its location which was further from the water sources together with a very low precipitation volume of only 931 mm/year. Additionally, the salt- affected soil also caused physiological drought. The situation was at its peak during drought season when water evaporated from the soil leaving behind salt stains which appeared on the soil surface (Yuvaniyama 2003; Manchanda and Garg 2008). The expansion of saline soil to nearby areas also had an effect on ecosystems and resource utilization. These problems greatly and adversely affected agricultural areas.

1.5 Application of Rice Husk Biochar on Saline Sodic Soil in Rice Paddy Fields

In this study, soil samples in the experimental plots were collected both pre- cultivation and post- cultivation to analyse their physical and chemical characteristics including pH, soil texture, ECe, CEC, OM, total N, total Na, avail. P, exch. K, exch. Mg, exch. Ca and SAR. Soil in the experimental plots was classified as strongly alkaline loamy sand (%sand = 72, %silt = 20, %clay = 8) with a pH of 10.20 and CEC of 5.79 cmolc/kg. The salt-affected soils had a high value of ECe with 48.47 dS/m and SAR of 1158. It had a high concentration of dissolved mineral salt, exch. Mg of 22.50 mg/kg, exch. Ca of 1279.63 mg/kg and total Na of 0.329%. Moreover, the affected area had very low soil fertility with 0.103% of OM, 0.004% of total N, 14.50 mg/kg of exch. K and 3.88 mg/kg of avail. P. Based on their ECe, SAR and pH, the soil in the experimental plots was saline sodic soil. The study was a field experimental design aimed to explore the impacts of biochar as soil amelioration substance on the salt-affected areas under the influences of actual natural factors such as weather conditions, temperature and precipitation. The experimental plots located at UTM 193.051605–193.370087 and 196.0820153–200.17691. The study was carried out using randomized complete block design (RCBD). The size of each experimental plot was 3.0 m in width, 4.0 m in length and 0.3 m in depth. There were 20 experimental plots in total which were divided into five treatments, each with four replicates. The five treatment conditions included soil plus 1.25 kg/ sq m organic fertilizer for control untreated treatment (TM) representing regular farming activity. The other four treatments were soil plus 1.25 kg/sq m organic fertilizer with different amounts of added biochar at 2.0 kg/sq (T-MBR2), 3.0 kg/sq (T-MBR3), 4.0 kg/sq (T- MBR4) and 5.0 kg/sq (T-MBR5). There were two equal applications of organic fertilizer for all treatment: once at two weeks before rice planting and subsequently at booting stage.

The organic fertilizer was produced from the composting of soybean stems and its characteristics were in accordance with all the parameters the Organic Fertilizer Standard of the Thai Department of Agriculture in 2005. The characteristics of organic fertilizer were: pH 8.3, EC of 3.50 dS/m, 40.30 wt.% OM, 23.43 wt.% total organic carbon (TOC), 1.70 wt.% total N, 0.87 wt.% total P,3.54 wt.% total K and a 13.75 C/N ratio. Jasmine rice 105 (Oryza sativa L.), namely Khao Dawk Mali (KDML105) or Thai Hom Mali 105″, was planted in the experimental plots. After mixing the soils according to the given ratio, the treatments were left in the sun to dry for 14 days prior to adding water to each plot. Each plot was irrigated with 10 cm of water in depth. A day after irrigation, the rice was transplanted into the experimental plots.

1.6 Effects of RH Biochar on Physiochemical Properties of Saline Sodic Soil

After harvesting, the soil samples form all experimental plots were randomly selected in the amount of 1000 g/plot at a depth of 30 cm from soil surface. All soil samples were analyzed for physical and chemical properties. The results revealed changes in soil properties after applying RH biochar (Table 1.1). Soil pH decreased in every experimental plot to pH 9.50–9.80. The change of pH was not statistically significant in comparison with the soil before the cultivation. On the other hand, ECe was significantly decreased in all treatment relative to pre-treatment. ECe decreased in every treatment from 46.47 to 6.78–7.00 and the ECe values were higher in treatments without biochar (TM). The CEC value in all biochar- treated treatments showed statistically significant decreases in comparison with the pre-cultivation soil. The highest amount of biochar added (T-MBR5) had the highest CEC increase (8.34 cmol/kg). The soil fertility value rose in every treatment with biochar (0.180–0.260) and the result was statistically significant whereas the fertility values declined in all non- biochar treatments (0.060%). The OM value increased the most in the treatment with the highest amount of biochar. The primary macronutrients in soil underwent certain changes after the rice cultivation. In the treatment without biochar (TM), the total N value decreased to the point that it could no longer be measured. In the treatments with biochar, the total N values were maintained: 0.004% in T-MBR3 and 0.011% in T-MBR4 which was the highest increase in the total N value. The result from the T-MBR4 treatment was the only treatment that the increase of total N showed statistically significant difference (Table 1.1).

Moreover, avail. P and exch. K values increased in every treatment whereas avail. P and exch. K values were the lowest in TM treatments. In other words, there was 10.00 mg/kg of avail. P and 38.00 mg/kg of exch. K in treatment with organic fertilizer (TM). The avail. P value and exch. K value were the highest values in the T-MBR5 treatment with the highest volume of biochar (22.00 mg/kg and 191.00 mg/kg, respectively), of which the values of avail. P and exch. K increased by 5.67 and 13.17 times, respectively. As for the treatments with lowest volume of biochar (T-MBR2), the avail. P and exch. K values increased by 3.61 and 6.90 times, respectively. In the treatment without biochar (TM), the avail. P and exch. K values only

Table 1.1 Characteristics of the soil from pre- and post-treatments

Parameter	Units	Pre-treatment	Post-treatment				
			TM	T-MBR2	T-MBR3	T-MBR4	T-MBR5
pH	–	**10.20 ± 0.18**	9.80 ± 0.24^{a}	9.50 ± 0.57^{a}	9.70 ± 0.22^{a}	9.50 ± 0.51^{a}	9.60 ± 0.27^{a}
ECe	dS/m	**46.47 ± 21.97**	$7.00 \pm 0.73^{a,*}$	$6.90 \pm 0.26^{a,*}$	$6.83 \pm 0.32^{a,*}$	$6.78 \pm 0.35^{a,*}$	$6.86 \pm 0.26^{a,*}$
CEC	cmol/kg	**5.79 ± 0.76**	6.93 ± 1.07^{a}	$7.66 \pm 0.99^{a,*}$	$8.03 \pm 0.46^{a,*}$	$8.25 \pm 0.81^{a,*}$	$8.34 \pm 0.69^{a,*}$
OM	%	**0.103 ± 0.036**	0.060 ± 0.022^{b}	$0.210 \pm 0.077^{a,*}$	$0.180 \pm 0.014^{a,*}$	$0.230 \pm 0.067^{a,*}$	$0.260 \pm 0.064^{a,*}$
Total N	%	**0.004 ± 0.003**	nd	0.007 ± 0.004^{ab}	0.004 ± 0.001^{b}	$0.011 \pm 0.002^{a,*}$	0.007 ± 0.002^{ab}
Avail. P	mg/kg	**3.88 ± 1.25**	$10.00 \pm 2.83^{c,*}$	$14.00 \pm 3.27^{bc,*}$	$18.00 \pm 3.27^{ab,*}$	$19.00 \pm 4.55^{ab,*}$	$22.00 \pm 3.65^{a,*}$
Exch. K	mg/kg	**14.50 ± 2.61**	$38.00 \pm 5.89^{c,*}$	$100.0 \pm 13.37^{b,*}$	$100.00 \pm 14.24^{b,*}$	$173.00 \pm 28.61^{a,*}$	$191.00 \pm 14.63^{a,*}$
Exch. Ca	mg/kg	**1279.63 ± 341.25**	$1479.00 \pm 319.89^{a,*}$	$1446.00 \pm 327.20^{a,*}$	$1451.25 \pm 345.74^{a,*}$	$1551.75 \pm 302.50^{a,*}$	$1548.50 \pm 337.99^{a,*}$
Exch. Mg	mg/kg	**22.50 ± 5.02**	$46.00 \pm 11.05^{a,*}$	$63.00 \pm 10.61^{a,*}$	$49.00 \pm 10.58^{a,*}$	$58.00 \pm 13.19^{a,*}$	$55.00 \pm 4.08^{a,*}$
Total Na	%	**0.329 ± 0.113**	0.196 ± 0.065^{a}	$0.167 \pm 0.049^{a,*}$	$0.153 \pm 0.052^{a,*}$	$0.147 \pm 0.040^{a,*}$	$0.152 \pm 0.064^{a,*}$
Extrac. Na	mg/kg	**2425.50 ± 801.02**	1520.00 ± 271.34^{a}	$1360.00 \pm 331.96^{a,*}$	$1226.00 \pm 319.78^{a,*}$	$1163.00 \pm 228.57^{a,*}$	$1080.00 \pm 238.95^{a,*}$
SAR	–	**1158 ± 853.92**	$12.45 \pm 2.90^{a,*}$	$10.86 \pm 2.57^{a,*}$	$11.61 \pm 2.10^{a,*}$	$11.29 \pm 1.86^{a,*}$	$10.36 \pm 1.97^{a,*}$
C/N ratio	–	**2.463 ± 1.646**	nd	0.060 ± 0.019^{ab}	0.030 ± 0.012^{b}	0.080 ± 0.036^{a}	0.050 ± 0.022^{ab}

Note: Data are shown as the mean ± 1SD

$^{a, b, ab, c}$ means a significantly different betweeen treatments at the 0.05 level

*mean a differenct pre and post-soil are significantly differenct at the 0.05 level ($p < 0.05$)

increased by 2.58 and 2.62 times, respectively. Comparing the amount of avail. P and exch. K in soil, all treatments after the cultivation revealed results that were statistically significant difference in comparison with treatments prior to the cultivation.

After rice cultivation, the amount of micronutrients such as exch. Ca and exch. Mg increased in all treatment. Exch. Ca value increased in every treatment. The treatment with the highest increase of exch. Ca value was TM (1479.00 mg/kg). The exch. Mg increased in every treatment with the value of 46–63 mg/kg with was statistical significance. The amount of exch. Mg was higher in the soil after rice cultivation in the biochar treated treatments (49.00–63.00 mg/kg) than the one without biochar treated (46.00 mg/kg). The highest amount of exch. Mg was found in the treatment with the lowest volume of biochar (T-MBR2).

The parameter illustrating soil salinity revealed that certain changes occurred in every treatment (Table 1.1). After the cultivation, the total Na value decreased the most in the T-MBR4 treatment at 0.147%. The extrac. Na value and the SAR value decreased the most in the TMBR-5 at 1080 mg/kg and 10.36, respectively. The value of extrac. Na in TMBR-5 decreased by 2.25 times and the SAR value decreased by 111.78 times in comparison to the soil prior to the cultivation. However, in the TMBR-4 treatment, the amount of total Na decreased by 2.24 times in comparison to the pre-cultivated soil. As for treatments with only fertilizer, the amount of total Na, extrac. Na and SAR decreased by 1.68, 1.60 and 93.01 times, respectively, in comparison to the pre-cultivated soil. Meanwhile, treatments with the least of biochar rate- treated, the amount of total Na, extrac. Na and SAR decreased by 1.97, 1.78 and 106.63 times, respectively, in comparison to the pre-cultivated soil. Even though the soil salinity parameter in the post-cultivation treatments indicated reduced amount to total Na, extrac. Na and SAR, the treatments incorporated with biochar revealed a better decreased rate than the non-biochar incorporated treatments.

1.7 Rice Yields from the Saline Sodic Soil Area

In this study, Jasmine Rice 105 was cultivated in five experimental plots each with four replicates which totalled 20 plots. Each plot measured 12 sq m (3 × 4 sq m). Three rice saplings were placed in each of the plots (three saplings in a clump) (Fig. 1.7a, b). On average, there were eight clumps per square metre (Fig. 1.7c, d). At the time of harvesting, it was apparent that out of the five experimental fields, there were merely three treatments that produced harvestable yields, of which only one treatment yielded top quality rice crops (Fig. 1.7e, f). However, rice seedlings in other plots grew up to a vegetative stage but failed to mature or ripe which made it unable to harvest (Fig. 1.7g). Some seedlings died at an early stage and some died off soon after. The rice turned yellow and withered away (Fig. 1.7h). The three experimental plots that were successfully grown were those with biochar-treated

Fig. 1.7 The transplanting of Jasmine Rice 105 in experimental plots with saline sodic soil. (**a**) The preparation of the experimental plots. (**b**) Rice saplings. (**c**) Rice saplings after transplanting. (**d**) Rice saplings in plot T-MBR2 after transplanting. (**e**) Rice at the maturing stage in plot T-MBR2 (**f**) Yield from plot T-MBR2. (**g**) Rice saplings in plot T-MBR5 – a month after transplanting. (**h**) Dead rice in an irrigated TM plot

Table 1.2 Yields of Jasmine Rice 105

Treatments	Ears per area (ears/m²)	Numbers of ears per clump (ears/clump)	Whole grain rice yields (g/m²)	White rice yields (g/m²)
T-MBR2	100.00	12.50	142.14	80.82
T-MBR3	27.00	3.38	38.38	21.82
T-MBR4	43.33	5.42	61.60	35.02

treatments at 2.0 kg/m² (T-MBR2); 3.0 kg/m² (T-MBR3) and 4.0 kg/m² (T-MBR4), respectively.

The experimental plot with the least amount of biochar (2.0 kg/m²) yielded of jasmine rice with the maximum number of ears of paddy per area and ears of paddy per clump. The results followed by the plot with RH biochar of 4.0 kg/m² and of 3.0 kg/m², respectively. The details are provided in Table 1.2.

Even though the results from the 6-month cultivation period were not promising, there was an apparent improvement on the quality of soil. The change allowed the rice saplings to grow in the current saline sodic soil experimental plots and yielded acceptable crops. The changes that were influenced by biochar application were pH, ECe and CEC (Wijitkosum and Jiwnok 2019; Wijitkosum and Kallayasiri 2015; Yooyen et al. 2015; Laird et al. 2010). It also improved ion exchange ability (Schultz et al. 2017), better microbials activities as well as absoption ability of the roots (Akhtar et al. 2015a; Yooyen et al. 2015; Wang et al. 2014).

Biochar has an ability to adsorb salt from saline soil (Melas et al. 2017; Schultz et al. 2017; Chaganti and Crohn 2015; Akhtar et al. 2015a; Novak et al. 2009) and many studies revealed that adding salt into salt-affected soil reduced sodium ion accumulation in different parts of rice plant. Moreover, biochar addition reduced plant sodium uptake under salt stress soil by transient Na^+ binding due to its high adsorption capacity and by releasing mineral nutrients (particularly K^+, Ca^{2+}, Mg^{2+}) into the soil solution (Melas et al. 2017; Akhtar et al. 2015b). Feng et al. (2018) showed that the sodium ion accumulation significantly decreased and reduced sodium ion toxicity for better rice growth and yield formation after biochar applied in saline- sodic paddy soil. Moreover, biochar increased root biomass which accelerated plant growth (Alcívar et al. 2018; Wijitkosum and Jiwnok 2019; Yooyen et al. 2015) which was the result from biochar's ability to exclude salts from root system (Akhtar et al. 2015a). Biochar also able to induce exudation of organic compounds produced by the plant microbiota, which positively affects root development (Arjumend et al. 2015; Kammann et al. 2011). Similar results were also found by Akhtar et al. (2015a) biochar positively affects plant root growth under saline conditions, due to its high adsorption capacity, which might lead to reduce Na^+ uptake or enhanced Na^+ exclusion or both from roots. This indicated the overall improvement of the soil and its fertility as well as the reduced soil salinity enabling the rice to grow in the previously barren land. However, the improvement of saline sodic soil is a longitudinal and continuous process. Such outstanding revelation included the improvement of the salt- affected soil quality by adding soil organic carbon released from the RH biochar. The rice cultivation also required an adequate amount of water throughout the process. The sufficient amount of water aided the movement of salt from underground to the surface. Moreover, the betterment of the salt-affected soil required the right amount of time to further enhance the soil property. It contributes to a higher volume of products. This finding was in accordance with a number of research attempts indicating that in order to effectively improve soil with biochar, one needed to dedicate the right amount of time to make the biochar mechanism work as effectively as possible.

1.8 Rice Husk Biochar as an Organic Soil Amendment for Reclamation of Saline Sodic Soil

RH biochar is a highly stable substance rich in nutrients. It was produced by the slow pyrolysis of rice husk in a retort controlled temperature between 400–500 °C. The results indicated that the soil properties showed a better improcement when the soil was incorporated with both biochar and organic fertilizer than the treatment with fertilizer alone. Moreover, the improved quality of the soil enabled rice growth in saline soil within the 6-month application period. This experiment also indicated that, adding RH biochar into saline sodic soil improved all soil properties and providing better results than adding fertilizer alone. The results are shown below.

1.9 pH Value and Potential to Reduce Alkalinity

Research indicated that applying organic amendments into the salt-affected soil reduced soil pH (Wong et al. 2009; Makoi and Ndakidemi 2007). The results from this study showed that RH biochar added into soil decreased the soil pH. Even though the results did not show statistically significant difference between treatments, the biochar treated treatments yielded lower pH than the non-treated treatments (Fig. 1.8).

This was possible due to the fact that RH biochar had lower alkalinity (pH 6.8) than the soil (pH 10.20). Many studies reported that most biochar are alkaline and soil pH increases due to biochar application (Wijitkosum and Kallayasiri 2015; Martinsen et al. 2015; Obia et al. 2015; Yuan et al. 2011; Laird et al. 2010; Glaser et al. 2002). All of these studies were conducted on acidic soils with pH lower than biochar. On the other hand, studies in alkaline soil indicated that biochar application into the soil can decreases soil pH (Wijitkosum and Jiwnok 2019; Sun et al. 2016; Wu et al. 2014; Liu and Zhang 2012; Yamato et al. 2006), especially in saline sodic soils and sodic soils (Abrishamkesh et al. 2015; Liu and Zhang 2012).

High pH values of saline sodic soils and sodic soils are primarily associated with high exchangeable sodium percentage (ESP) (Shaygan et al. 2017), a reduction in soil ESP of those incorporated with biochar can be concluded as one of a possible mechanisms responsible for the decrease in soil pH (Lashari et al. 2015). Furthermore,

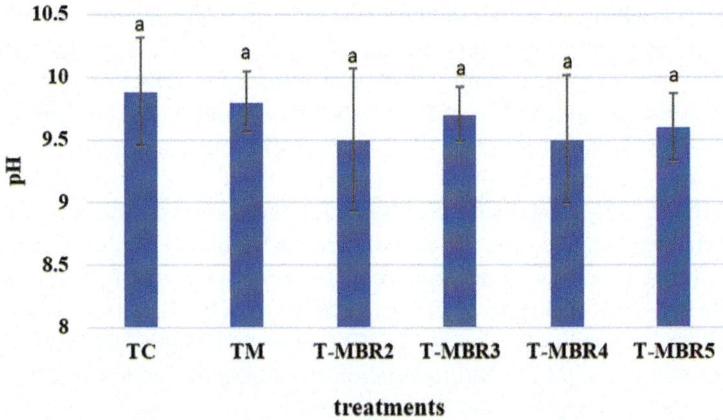

Fig. 1.8 Soil pH after the soil amelioration treatment (post-cultivation)

as for the biochar induced reduction in ESP, the initial pH of biochar may play a significant role in the pH changes of salt-affected soils (Sun et al. 2016; Amini et al. 2015; Liu and Zhang 2012). The pH of biochar depends on the types of feedstock and the pyrolysis condition as reported by Yuan et al. (2011), Chen et al. (2008), Yamato et al. (2006) and Glaser et al. (2002). The alkalinity of biochar results from their ash content releasing base cations and alkaline properties of organic functional groups (Yuan and Xu 2012). Many studies found that, not only the pH of biochar, the pH of soil also impacted the final pH of the biochar-soil mixture (Saifullah et al. 2018; Wijitkosum and Kallayasiri 2015). In this study, soil pH from all biochar- treated treatments were lower than the treatments with solely added fertilizer. The pH level in the soil changed due to its ability to exchange cations in the soil solution. Adding biochar into the soil enabled chemical activities that triggered the ion exchange which led to a higher level of soil pH (Warnock et al. 2007; Chan et al. 2008; Yuan et al. 2011) while biochar pH decreases (Amelong et al. 1997; Wijitkosum and Kallayasiri 2015). Hinsinger et al. (2003) explained that the decreased pH in biochar-treated treatments was due to the high amount of biochar's CEC which promoted the plant's uptake of cation (e.g. K^+, Ca^{2+}, and Mg^{2+}), resulting in H^+ being released from roots to balance the charges. The change of the pH occurred from oxidations between functional groups of biochar and soil solution (Cheng et al. 2006). Moreover, application of biochar also stimulates microbial activities. Biochar's highly porous structure and large surface area provides "shelter" for soil microorganisms such as microbes which live in the plants' rhizosphere and increase macro-nutrient availability, soil aeration and soil hydrology (Sriburi and Wijitkosum 2016; Hardie et al. 2014; Downie et al. 2012; Lehmann and Joseph 2009). Thus, the carbon dioxide partial pressure increases during the decomposition of organic matters in the soil and causes the development of pH-reducing conditions. Biochar effectively adjusts the pH of the soil in agricultural areas as shown in Wijitkosum and Kallayasiri (2015) which the results indicated that biochar obtained from wood scraps improved soil pH in the agricultural area from pH 6.70 to pH 7.51. The study took into account the suitable amount of biochar applied per area and the types of biomass that were made into biochar. Liu and Zhang (2012) reported that biochar produced a decreasing pH trend, which can reduce the effect of high pH on the growth and development of rice and soil nutrient availability. It was necessary for the biomass types to be appropriate for the nature of the soil as buffering capacity of the soil may hamper its ability to improve the pH (Collins 2008; Yuan and Xu 2010). It was also reported that biochar's capability for reducing salt-affected soil pH was mainly influenced by the types of raw materials (Schultz et al. 2017). Biochar was able to lower soil acidity in accordance with the amount of biochar being mixed within the soil. The ability to lower soil acidity increased in accordance with the amount of added biochar (Yuan et al. 2011; Jien and Wang 2013). However, the decreasing trends of soil pH and its relation the amount of added biochar were not conclusive and did not reveal apparent differences.

1.10 Soil CEC and Levels of Base Cations with Biochar Addition

Soil CEC was increased in all treatments but there was a significant difference between the pre and post cultivation in only the biochar-treated treatment. Moreover, CEC results in all biochar-treated treatments were higher than the treatments without biochar treated (Fig. 1.9). The results corroborated Wijitkosum and Jiwnok (2019), Wijitkosum and Kallayasiri (2015), Yooyen et al. (2015), Abdullaeva (2014) and Laird et al. (2010) which observed that CEC in various types of soil were increased after the addition of biochar from agricultural residues.

Adding RH biochar into the soil helped increase soil CEC significantly, but the significant difference on soil CEC among the biochar-treated treatments were not detected (Fig. 1.9). This result was similar to previous studies that also reported the increases in soil CEC after application different types of soil amendments such as composts (Aggelides and Londra 2000; Ouédraogo et al. 2001), biochar (Laird et al. 2010; Cheng et al. 2008; Liang et al. 2006). Furthermore, the high CEC value, increased due to biochar application, had a tendency to control the soil salinization process in agricultural lands (Liu and Zhang 2012).

The value of soil CEC increased after adding RH biochar was due to an effect of biochar's structure. Ions from the biomass that was made into biochar helped enhance cation exchange capacity within the soil (Chan et al. 2008; Lehmann et al. 2003). After the pyrolysis process, the structure of biochar was formed by aromatic compound (Schmidt and Noack 2000) and carboxyl groups were created on its surface which led to high ion exchange capacity (Joseph et al. 2009; Lehmann et al. 2007; Cheng et al. 2006). CEC is the quantification of the capacity of a material to bind positive charged ion or molecule on negatively charged surfaces like clays and soil organic matter (Brady and Weil 2008). Biochar has large surface areas and a large number of ions per area resulting in a higher rate of ion exchange within the

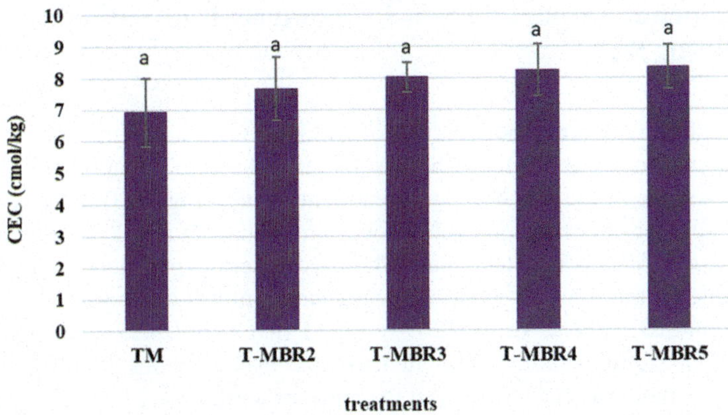

Fig. 1.9 CEC values at post treatment and post cultivation stage

soil (Liang et al. 2006). However, the amount of biochar added to the soil has an influence on the ion exchange capacity. This study concluded that the more biochar added, the better the ion exchange performance became (Fig. 1.10). The results were in accordance with Zhang et al. (2019); Chintala et al. (2014); Liang et al. (2006) in both laboratory and field researches. Chintala et al. (2014) incubated sodic soil with biochar made from corn stover (Zea mays L.) in different ratios (0, 52, 104, and 156 Mg ha^{-1}) over the period of 165 days. The results indicated that the increase in soil CEC values was significantly higher in treatments with corn stover biochar at all application rates. The CEC values increased in accordance with the increased amount of incorporated biochar. The CEC values were 14.71 for 52 Mg ha^{-1}, 17.33 for 104 Mg ha^{-1} and 19.04 for 156 cmolc kg^{-1}. As for the field research, Nigussie et al. (2012) conducted an experiment on lettuces (*Lactuca sativa*) by incorporating biochar produced from maize stalk at the rates of 0, 5 and 10 t/ha on soils (pH = 5.23; sand:silt:clay = 20%:40%:40%). The results indicated that CEC values at the post-cultivation stage were 27.22, 31.61 and 33.69 meq/100 g. However, Abdullaeva (2014) founded that the CEC in soil were not depending on the rate of biochar-treated applied. The results revealed that the amount of CEC increased in accordance with the amount of biochar added to the treatments. Moreover, the CEC amount also increased even higher the longer the biochar was incorporated within the soils (Zhang et al. 2019).

The increase of CEC is possible from the exchange of ions between soil solution and biochar through cation exchange which is a reversable process that occurs constantly. The process allows plants to utilise the nutrients continuously. Therefore, the increase in the amount of exchangeable cations in the amended soils suggested an improvement in soil fertility and nutrient retention, which may be attributed to the high specific surface area and a number of carboxylic groups of the biochar (Cheng et al. 2006; Metson 1961).

1.11 Biochar Impact on Soil ECe and SAR of Saline Sodic Soils

Even though the post-treatment results of soil ECe were not significantly different from one treatment to another, the results still showed a significant difference between the ECe of the pre- and pos- treatment results (Fig. 1.10). The decrease of ECe was apparent in one crop cycle. However, Lashari et al. (2013) indicated that the decrease of ECe was detectable in the 2-year field experiment under the combined application of biochar and poultry manure compost compared to non-treated soil. In contrast, Alcívar et al. (2018) found that soil amendment applications resulted in a significant reduction in soil ECe except in the biochar treatment. However, Usman et al. (2016) showed that biochar addition at higher rates increased ECe values due to the concentration of soluble salts in the ash.

However, the slight decreases of ECe among the treatments might be due to the high amount of salt accumulated in plants' roots. The experimental fields were high in temperature, effected by direct and strong sunlight and retained little amount of freshwater. On the other hand, saline solution in groundwater was similar to that of surface water which contributed to a high level of evapotranspiration which bring a large amount of salt to the surface. Therefore, washing away the salt from the surface was limited. The comparison between ECe results among the pre- and the post-treatments were taken from the cultivation in the experiment. Each plot was irrigated with water at 10 cm in depth (Williamson et al. 1989; Qadir and Oster 2004; Manchanda and Garg 2008; Yuvaniyama 2003). The reduction of ECe in saline sodic soil was attributed to the biochar-induced improvement in soil porosity and hydraulic conductivity that accelerated leaching of salts. ECe reduction in biochar-treated soil was be attributed to the adsorption of some soluble ions by functional groups existing at the biochar surface.

The SAR of all treatments were significantly decreased in post-cultivation, but the SAR among post- treatments were not significant different. However, the SAR of saline sodic soil in all biochar-treated treatments were lower than the fertilizer-treated treatment alone. However, Alcívar et al. (2018) indicated that soil incorporated with humic substances yielded a smaller decrease in SAR soil incorporated with humic substances alone.

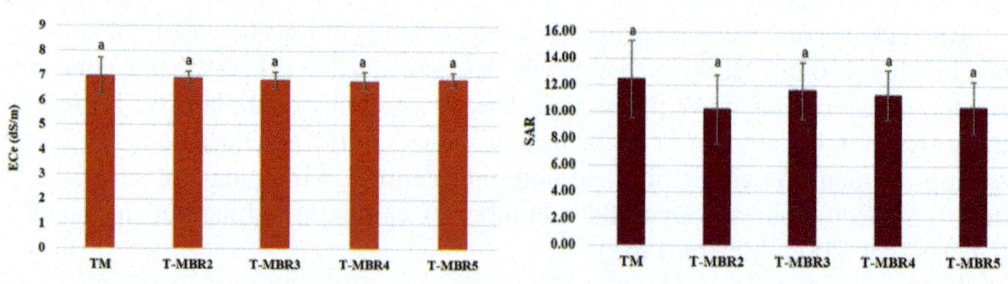

Fig. 1.10 ECe and SAR values of the post-cultivation stage

Adding RH biochar into the saline sodic soil as a soil amendment increased salt leaching which was facilitated by soil aggregation. In addition, Kim et al. (2016) discussed the changes of SAR values within the soil that there was an apparent relationship between Na^+ and Ca^{2+} proportions in the soil solution. The content of Na and Ca varied depending on biochar types. The rate and types of biochar applied into the soil are the two most important factors controlling the impact of biochar on SAR of saline sodic soils. In contrast, the result from this study found that the SAR values in saline sodic soils were not depending on the rate of biochar-treated applied (Fig. 1.10). The decreases in soil EC and SAR in this study were similar to the findings of Lashari et al. (2015), Hammer et al. (2015) and Akhtar et al. (2015a).

1.12 Effects of Biochar on Na^+, Mg^{2+}, Ca^{2+} of Saline Sodic Soil

Post-cultivation soil analysis revealed non-significant differences among treatments with the control having higher soil total Na and extrac. Na^+ than soils that received biochar amendments (Fig. 1.11). The results of this study were similar the Chaganti et al. (2015) of which the application of biochar helped reduce highest amount of extra. Na^+ by 80% in comparison to the control treatment.

This study found that adding RH biochar into the soil significantly decreased total Na and extrac. Na^+, but the significant difference on soil total Na and extrac. Na^+ of the addition of various amount of RH biochar was not detected (Fig. 1.11). The findings corroborated Lashari et al. (2013) who reported that there was a significant decrease in soil pH and salt and sodium contents with the application of a biochar. Alcívar et al. (2018) reported that adding biochar alone and biochar incorporated with humic substance significantly reduced extrac. Na^+ in soil. There are many studies (ex. Alcívar et al. 2018; Major et al. 2010; Lehmann et al. 2003) reported that the incorporation of biochar into soil successfully reduced Na^+ concentrations. Moreover, application of RH biochar into the soil can significantly increase soil $Mg^{2+}$and Ca^{2+}. The ions were being absorbed onto biochar surfaces

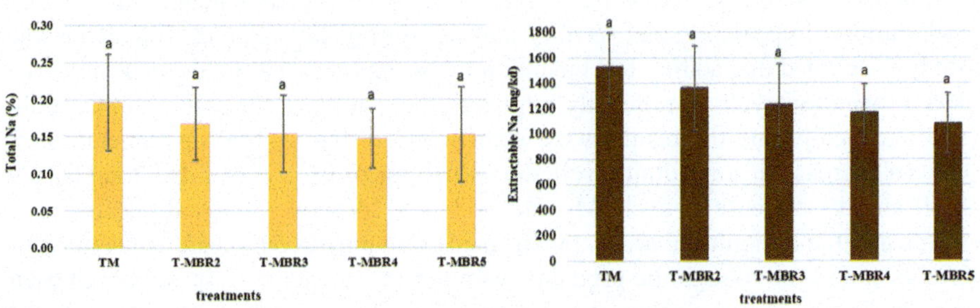

Fig. 1.11 Total Na and extractable Na at the post-cultivation stage

which increased K^+, Ca^{2+}, Mg^{2+} concentrations (Alcívar et al. 2018; Lashari et al. 2015; Hammer et al. 2015; Akhtar et al. 2015a, b; Major et al. 2010).

The reclamation of saline sodic soils requires the removal of sodium from the soil exchange sites into soil colloids by divalent cation to promote soil flocculation (Jalali 2008). This is a key point for a successful reclamation of saline–sodic soil. Soil incorporated with organic amendments, such as fertilizer and biochar, has an increased amount of soil CEC that attracts Ca^{2+} instead of Na^+ from the soil solution. The results of this research study was similar to the findings of Hammer et al. (2015), Rajakovich et al. (2012), Tsai et al. (2012) and Laird et al. (2010) who reported that biochar was rich in nutrients such as Ca^{2+} and Mg^{2+} and enhanced their availability in the soil when added as soil amendments.

This was due to the pyrolysis condition of biochar. Feedstocks that underwent the pyrolysis process contained a higher amount of P, Mg and Ca (Cao and Harris 2010). Therefore, saline sodic soils benefit from biochar application in various ways including an increased content of soil organic carbon and nutrients, especially cationic ones (eg. K^+, Ca^{2+}, Mg^{2+}), increased CEC, more stable soil structure, enhanced physical properties by balancing water contain and air porosity as well as the replacement of Na^+ from exchange sites by providing Ca^{2+} in soil solution (Zheng et al. 2018; Usman et al. 2016; Yue et al. 2016; Rajakovich et al. 2012).

Many studies also indicated that the reclamation process was a lengthy process. Lashari et al. (2013) reported no significant differences in total N between plots applied with biochar manure compost in combination with pyroligneous solution and untreated plots in the first year of the experiment. However, there was a 69% increase in total N in the second year. Biochar amendment had significant effects on soil Ca in Year 1, but this did not persist the following year. Amended soils' P, K, or Mg levels were not significantly different than those of the control soils in either year though their concentrations decreased at the end of the study. The decrease in base cation concentrations by the end of the second growing season was due to crop uptake and leaching losses in this sandy and saline-sodic soil.

1.13 Conclusion and Recommendation

The application of RH biochar into saline sodic soils has great potential for reducing soil salinity, improving soils fertility and promoting rice growth. Therefore, this study concluded that adding RH biochar into the experimental fields of Jasmine rice 105 in one crop cycle significantly improved the quality of the soil both physically and chemically and the results were statistically different. Main parameters indicated a decrease in soil salinity were the reductions of ECe, extrac. Na, total Na and SAR and the increase of exch. Mg and exch. Ca. Moreover, the soil became more fertile and the agricultural areas transformed from uncultivable bare lands into cultivable soils even though the products were not of top quality. The study indicated that treatments of soil with biochar induced changes in soil that are favourable, but long- term studies are required to monitor the extent of these effects. Moreover, one of the crucial factors for saline sodic soil amelioration was also maintaining the irrigation level in the rice fields to prevent transportation of salt to the soil surface. Therefore, using RH biochar to revitalise saline sodic soil should maintain the irrigation level in the plots.

Acknowledgments This research was supported by the "Innovation in Increasing the Organic Carbon in Soil for Sustainable Agricultural Purpose in Saline Soil Areas: 1st Year Pilot Project at the Lam Ta Kong Watershed", Ratchadaphisek Somphot Endowment Fund, Chulalongkorn University (CU-57-090-IC).

References

Abdullaeva Y (2014) Biochar effects on fertility of saline and alkaline soils (Navoiy Region, Uzbekistan). Thesis PhD in Agriculture, University of Uzbekistan, Uzbekistan

Abrishamkesh S, Gorji M, Asadi H, Bagheri-Marandi GH, Pourbabaee AA (2015) Effects of rice husk biochar application on the properties of alkaline soil and lentil growth. Plant Soil Environ 61(11):475–482

Aggelides SM, Londra PA (2000) Effects of compost produced from town wastes and sewage sludge on the physical properties of a loamy and a clay soil. Bioresour Technol 71(3):253–259

Akhtar SS, Andersen MN, Liua F (2015a) Residual effects of biochar on improving growth, physiology and yield of wheat under salt stress. Agric Water Manag 158:61–68

Akhtar SS, Andersen MN, Liu F (2015b) Biochar mitigates salinity stress in potato. J Agri Crop Sci 201:368–378

Alcívar M, Zurita-Silva A, Sandoval M, Muñoz C, Schoebitz M (2018) Reclamation of saline–sodic soils with combined amendments: impact on quinoa performance and biological soil quality. Sustainability 10(9):30–83

Amelong W, Flach KW, Zech W (1997) Climate effects on soil organic matter composition in great plain. Soil Sci Soc Am J 61:115–123

Amini S, Ghadiri H, Chen C, Marschner P (2015) Salt-affected soils, reclamation, carbon dynamics, and biochar: a review. J Soils Sediments 16(3):939–953

Amonette J, Joseph S (2009) Characteristics of biochar: micro chemical properties. In: Lehmann J, Joseph S (eds) Biochar for environmental management: science and technology. Earthscan, London, pp 33–52

Arjumend T, Abbasi MK, Rafique E (2015) Effects of lignite-derived Humic acid on some selected soil properties, growth and nutrient uptake of wheat (*Triticum Aestivum* l) grown under greenhouse conditions. Pak J Bot 47(6):2231–2238

Atkinson CJ, Fitzgerald JD, Hipps NA (2010) Potential mechanisms for achieving agricultural benefits from biochar application to temperate soils: a review. Plant Soil 337:1–18

Bacilio M, Rodriguez H, Moreno M, Hernandez J-P, Bashan Y (2004) Mitigation of salt stress in wheat seedlings by a *gfp*-tagged *Azospirillum lipoferum*. Biol Fertil Soils 40:188–193

Bhogal A, Nicholson FA, Chambers BJ (2009) Organic carbon additions: effects on soil biophysical and physico-chemical properties. Eur J Soil Sci 60:276–286

Blaylock AD (1994) Soil salinity, salt tolerance and growth potential of horticultural and landscape plants. Co-operative Extension Service, University of Wyoming, Department of Plant, Soil and Insect Sciences, College of Agriculture, Laramie, Wyoming

Brady NC, Weil RR (2008) The nature and properties of soil, 14th edn. Prentice-Hall, Upper Saddle River

Brinck E, Frost C (2009) Evaluation of amendments used to prevent sodification of irrigated fields. Appl Geochem 24(11):2113–2122

Cao X, Harris W (2010) Properties of dairy-manure-derived biochar pertinent to its potential use in remediation. Bioresour Technol 101:5222–5228

Chaganti VN, Crohn MD (2015) Evaluating the relative contribution of physiochemical and biological factors in ameliorating a saline-sodic soil amended with composts and biochar and leached with reclaimed water. Geoderma 259-260:45–55

Chaganti VN, Crohn DM, Simunek J (2015) Leaching and reclamation of a biochar and compost amended saline–sodic soil with moderate SAR reclaimed water. Agric Water Manag 158:255–265

Chan KY, Van Zwieten L, Meszaros I, Downie A, Joseph S (2007) Agronomic values of greenwaste biochar as a soil amendment. Aust J Soil Res 45(8):629

Chan KY, Van Zwieten LV, Meszaros I, Downie A, Joseph S (2008) Using poultry litter biochars as soil amendments. Aust J Soil Res 46:437–444

Chen Z, Zhou M, Newman IA, Mendham NJ, Zhang G, Shabala S (2007) Potassium and sodium relations in salinised barley tissues as a basis of differential salt tolerance. Funct Plant Biol 34(2):150

Chen B, Zhou D, Zhu L (2008) Transitional adsorption and partition of nonpolar and polar aromatic contaminants by biochars of pine needles with different pyrolytic temperatures. Environ Sci Technol 42:5137–5143

Chen Y, Yang H, Wang X, Zhang S, Chen H (2012) Biomass-based pyrolytic polygeneration system on cotton stalk pyrolysis: influence of temperature. Bioresour Technol 107:411–418

Cheng CH, Lehmann J, Thies JE, Burton SD, Engelhard MH (2006) Oxidation of black carbon by biotic and abiotic processes. Org Geochem 37:1477–1488

Cheng CH, Lehmann J, Engelhard M (2008) Natural oxidation of black carbon in soils: changes in molecular form and surface charge along a climosequence. Geochim Cosmochim Acta 72:1598–1610

Chintala R, Schumacher TE, McDonald LM, Clay DE, Malo DD, Papiernik SK, Clay SA, Julson JL (2014) Phosphorus sorption and availability from biochars and soil/biochar mixtures. Clean (Weinh) 42:626–634

Collins H (2008) Use of biochar from the pyrolysis of waste organic material as a soil amendment: laboratory and greenhouse analyses. A quarterly progress report prepared for the Biochar Project (December 2008)

Department of Mineral Resources (1982) Geologic map of Thailand. Department of Mineral Resources, Thailand

Division of Mineral Resources Conservation and Management (2015) Saline soil area development in the northeast: geological approaches. Research report. Department of Mineral Resources Conservation and Management, Bangkok

Downie A, Crosky A, Munroe P (2009) Physical properties of biochar. In: Lehmann J, Joseph S (eds) Biochar for environmental management: science and technology. Earthscan, London, pp 13–32

Downie AE, Munroe P, Cowie A, Van Zwieten L, Lau D (2012) Biochar as a geo-engineering climate solution: hazard identification and risk management. Crit Rev Environ Sci Technol 42:225–250

El-Zanaty AA, El-Nour A, El-Fouly MM (2006) Modern methods for counteracting salinity stress: a review. In: 2nd International conference on water resources and arid environments 26–29 November, 2006. Riyadh, Saudi Arabia

Eynard A, Schumacher TE, Lindstrom MJ, Malo DD, Kohl RA (2006) Effects of aggregate structure and organic C on wettability of Ustolls. Soil Till Res 88:205–216

FAO (1988) Saline soil and their management. Available from: http://www.fao.org/docrep/x5871e/x5871e04.htm [25 February 2015]

FAO (2009) The research progress of biomass pyrolysis processes. Available from: http://www.fao.org/docrep/t4470e/t4470e0a.htm [5 January 2015]

Feng DD, Zhao YJ, Zhang Y, Sun SZ, Gao JM (2018) Steam gasification of sawdust biochar influenced by chemical speciation of alkali and alkaline earth metallic species. Energies 11(1):1–16

Flowers TJ, Hajibagheri MA, Clipson NJW (1986) Halophytes. Q Rev Biol 61:313–337
Glaser B, Lehmann J, Zech W (2002) Ameliorating physical and chemical properties of highly weathered soils in the tropics with charcoal: a review. Biol Fert Soils 35:219–230
Grover A, Aggarwal PK, Kapoor A, Katiyar-Agarwal S, Agarwal M, Chandramouli A (2003) Addressing abiotic stresses in agriculture through transgenic technology. Curr Sci 84:355–367
Hammer EC, Forstreuter M, Rillig MC, Kohler J (2015) Biochar increases arbuscular mycorrhizal plant growth enhancement and ameliorates salinity stress. Appl Soil Ecol 96:114–121
Hardie M, Clothier B, Bound S, Oliver G, Close D (2014) Does biochar influence soil physical properties and soil water availability? Plant Soil 376:347–361
Hati KM, Swarup A, Dwivedi AK, Misra AK, Bandyopadhyay KK (2007) Changes in soil physical properties and organic carbon status at the topsoil horizon of a vertisol of central India after 28 years of continuous cropping, fertilization and manuring. Agric Ecosyst Environ 119:127–134
Herath HMSK, Camps-Arbestain M, Hedley M (2013) Effect of biochar on soil physical properties in two contrasting soils: an Alfisol and an Andisol. Geoderma 209–210:188–197
Hinrich LB, Brian LM, O'Connor GA (2001) Soil chemistry, 3rd edn. Wiley, New York
Hinsinger P, Plassard C, Tang C, Jaillard B (2003) Origins of root-mediated pH changes in the rhizosphere and their responses to environmental constraints: a review. Plant Soil 248:43–59
IBI (2014) Standardized product definition and product testing guidelines for biochar that is use in soil. Version 2 [online] Available from: https://www.biochar-internationalorg/wp-content/uploads/2018/04/IBI_Biochar_Standards_V2%200_final_2014.pdf [1 March 2015]
IBI (2015) Standardized product definition and product testing guidelines for biochar that is use in soil. Version 2.1 [online] Available from: https://biochar-international.org/wp-content/uploads/2019/01/IBI_Biochar_Standards_V21_Final1.pdf [1 May 2017]
Jalali M (2008) Effect of sodium and magnesium on kinetics of potassium release in some calcareous soils of western Iran. Geoderma 145(3):207–215
Japan International Cooperation Agency (JICA) (1991) The master plan study on the integrated rural development of salt-affected land in Northeast Thailand- Main Report and Appendix. Japan International Cooperation Agency (JICA), 199p
Jien S-H, Wang C-S (2013) Effects of biochar on soil properties and erosion potential in a highly weathered soil. Catena 110:225–233
Jones BEH, Haynes RJ, Phillips IR (2010) Effect of amendment of bauxite processing sand with organic materials on its chemical, physical and microbial properties. J Environ Manag 91:2281–2288
Joseph S, Peacocke C, Lehmann J, Munroe P (2009) Developing a biochar classification and test methods. In: Lehmann J, Joseph S (eds) Biochar for environmental management. Earthscan, London, pp 107–126
Kammann C, Linsel S, Goßling JW, Koyro HW (2011) Influence of biochar on drought tolerance of Chenopodium quinoa willd and on soil-plant relations. Plant Soil 345:195–210
Kaymakanova M, Stoeva N, Mincheva T (2008) Salinity and its effects on the physiological response of bean (Phaseolus Vulgaris L.). J Cent Eur Agric 9(4):749–756
Kim KH, Kim JY, Cho TS, Choi JW (2012) Influence of pyrolysis temperature on physical properties of biochar obtained from the fast pyrolysis of pitch pine (Pinus rigida). Bioresour Technol 118:158–162
Kim HS, Kim KR, Yang JE, Ok YS, Owens G, Nehls T, Wessolek G, Kim KH (2016) Effect of biochar on reclaimed tidal land soil properties and maize (*Zea mays* L) response. Chemosphere 142:153–159
Kovda VA, Hagan RM, van den Berg C (1973) Irrigation, drainage and salinity: an international source book. FAO/UNESCO, Paris
Laird D, Fleming P, Wang B, Horton R, Karlen D (2010) Biochar impact on nutrient leaching from a Midwestern agricultural soil. Geoderma 158:436–442
Lakhdar A, Rabhi M, Ghnaya T, Montemurro F, Jedidi N, Abdelly C (2009) Effectiveness of compost use in salt-affected soil. J Hazard Mate 171:29–37

Lashari MS, Liua Y, Li L, Pan W, Fu J, Pana G, Zheng J, Zheng J, Zhang X, Yu X (2013) Effects of amendment of biochar-manure compost in conjunction with pyroligneous solution on soil quality and wheat yield of a salt-stressed cropland from Central China Great Plain. Field Crop Res 144:113–118

Lashari MS, Ye Y, Ji H, Li L, Grace WK, Lu H, Zheng J, Pan G (2015) Biochar-manure compost in conjunction with pyroligneous solution alleviated maize salt stress and improved growth in a salt affected soil from Central China: a two-year field experiment. J Sci Food Agric 95:1321–1327

Lehmann J, Joseph S (2009) Biochar for environmental management: an introduction. In: Lehmann J, Joseph S (eds) Biochar for environmental management: science and technology. Earthscan, London, pp 1–13

Lehmann J, da Silva JP Jr, Steiner C, Nehls T, Zech W, Glaser B (2003) Nutrient availability and leaching in an archaeological Anthrosol and a Ferralsol of the Central Amazon basin: fertilizer, manure and charcoal amendments. Plant Soil 249:343–357

Lehmann J, Kern DC, Glaser B, Woods WI (2007) Amazonian dark earths: origin properties management. Springer, New York

Lehmann J, Czimczik C, Laird D, Sohi S (2009) Stability of biochar in soil. In: Lehmann J, Joseph S (eds) Biochar for environmental management: science and technology. Earthscan, London, pp 169–182

Leksungnoen N (2006) Some ecological characteristics and adaptations of natural vegetation in saline soil, Amphoe Khan Thale So, Changwat Nakhon Ratchasima. Dissertation, Kasetsart University, Bangkok (in Thai with English abstract)

Liang B, Lehmann J, Solomon D, Kinyangi J, Grossman J, O'Neill B, Neves EG (2006) Black carbon increases cation exchange capacity in soils. Soil Sci Soc Am J 70:1719–1730

Liu XH, Zhang XC (2012) Effect of biochar on pH of alkaline soils in the loess plateau: results from incubation experiments. Int J Agr Biol 14:745–750

Mahmoud EFA-H, Hegazy HS, Hassan NS, Naguib DM (2011) Effect of silica ions and nano silica on rice plants under salinity stress. Ecol Eng 99:282–289

Major J, Rondon M, Molina D, Riha SJ, Lehmann J (2010) Maize yield and nutrition during 4 years after biochar application to a Colombian savanna oxisol. Plant Soil 333:117–128

Makoi JHJR, Ndakidemi PA (2007) Reclamation of sodic soils in northern Tanzania, using locally available organic and inorganic resources. Afr J Biotechnol 6:1926–1931

Manchanda G, Garg N (2008) Salinity and its effects on the functional biology of legumes. Acta Physiol Plant 30:595–618

Mansaray KG, Ghaly AE (1997) Physical and thermochemical properties of rice husk. Energy Sources 19(9):989–1004

Martinez-Beltran J, Manzur CL (2005) Overview of salinity problems in the world and FAO strategies to address the problem. In: Proceedings of the International Salinity Forum, Riverside, California, pp 311–313

Martinsen V, Alling V, Nurida NL, Mulder J, Hale SE, Ritz C, Rutherford DW, Heikens A, Breedveld GD, Cornelissen G (2015) pH effects of the addition of three biochars to acidic Indonesian mineral soils. Soil Sci Plant Nutr 61:821–834

Mathur N, Singh J, Bohra S, Bohra A, Vyas A (2006) Improved productivity of mung bean by application of thiourea under arid conditions. World J Agric Sci 2:185–187

Mathurasa L, Damrongsiri S (2017) Potential of using surfactants to enhance ammonium and nitrate adsorption on rice husk and its biochar. Appl Envion Res 39:11–22

Melas GB, Ortiz O, Alacaniz JM (2017) Can biochar protect labile organic matter against mineralization in the soil? Pedosphere 27:822–831

Metson AJ (1961) Methods of chemical analysis for soil survey samples. Soil Bureau Bulletin 12, Department of Scientific, Industrial Research, New Zealand

Metternicht GI, Zinck JA (2003) Remote sensing of soil salinity: potentials and constraints. Remote Sens Environ 85:1–20

Munns R (2005) Genes and salt tolerance: bringing them together. New Phytol 167:645–663

Nigussie A, Kissi E, Misganaw M, Ambaw G (2012) Effect of biochar application on soil properties and nutrient uptake of lettuces (Lactuca sativa) grown in chromium polluted soils. Am Eurasian J Agric Environ Sci 12:369–376

Novak JM, Busscher WJ, Laird DL, Ahmedna M, Watts DW, Niandou MAS (2009) Impact of biochar amendment on fertility of a southeastern coastal plain soil. Soil Sci 174:105–112

Nwajiaku IM, Sato K, Tokunari T, Kitano S, Masunaga T (2018a) Improvement of rice husk residue silicon availability for replenishing available silicon in paddy soil. Int J Plant Soil Sci 24:1–11

Nwajiaku IM, Olanrewaju JS, Sato K, Tokunari T, Kitano T, Masunaga T (2018b) Change in nutrient composition of biochar from rice husk and sugarcane bagasse at varying pyrolytic temperatures. Int J Recycl Org Waste Agric 7:269–276

Obia A, Cornelissen G, Mulder J, Dörsch P (2015) Effect of soil pH increase by biochar on NO, N2O and N2 production during denitrification in acid soils. PLoS One 10:e0138781

Orcutt DM, Nilsen ET (2000) The physiology of plants under stress soil and biotic factors. Wiley, New York, 680 pp

Ouédraogo JT, Maheshwari V, Berner DK, St-Pierre C-A, Belzile F, Timko MP (2001) Identification of AFLP markers linked to resistance of cowpea (Vigna unguiculata L.) to Striga gesnerioides. Theor Appl Genet 102:1029–1036

Parry DW, Smithson F (1964) Types of opaline silica depositions in the leave of British Grasses. Ann Bot 28:169–185

Qadir M, Oster J (2002) Vegetative bioremediation of calcareous sodic soils: history, mechanisms, and evaluation. Irrig Sci 21(3):91–101

Qadir M, Oster J (2004) Crop and irrigation management strategies for saline-sodic soils and waters aimed at environmentally sustainable agriculture. Sci Total Environ 323(1–3):1–19

Rajakovich S, Enders A, Hanley K, Hyland C, Zimmerman AR, Lehmann J (2012) Corn growth and nitrogen nutrition after additions of biochars with varying properties to a temperate soil. Biol Fert Soils 48:271–284

Raul L, Andres O, Armado L, Bernardo M, Enrique T (2003) Response to salinity of three grain legumes for potential cultivation in arid areas (plant nutrition). Soil Sci Plant Nutr 49(3):329–336

Rengasamy P (2006) World salinization with emphasis on Australia. J Exp Bot 57(5):1017–1023

Rengasamy P (2010) Soil processes affecting crop production in salt-affected soils. Funct Plant Biol 37:613–620

Richard LA (1954) Diagnosis and improvement of saline and alkalis soils. Agriculture, 160, Handbook 60. US Department of Agriculture, Washington, DC, 160 pp

Saifullah DS, Naeem A, Rengel Z, Naidu R (2018) Biochar application for the remediation of salt-affected soils: challenges and opportunities. Sci Total Environ 625:320–335

Schmidt MWI, Noack AG (2000) Black carbon in soils and sediments: analysis, distribution, implications and current challenges. Global Biogeochem Cy 14:777–793

Schubert S, Neubert A, Schierholt A, Sümer A, Zörb C (2009) Development of salt-resistant maize hybrids: the combination of physiological strategies using conventional breeding methods. Plant Sci 177(3):196–202

Schultz E, Chatterjee A, DeSutter T, Franzan D (2017) Sodic soil reclamation potential of gypsum and biochar addition: influence on physiochemical properties and soil respiration. Commun Soil Sci Plant Anal 48:1792–1803

Shahbaz M, Ashraf M (2013) Improving salinity tolerance in cereals. Crit Rev Plant Sci 32(4):237–249

Shaygan M, Reading LP, Baumgartl T (2017) Effect of physical amendments on salt leaching characteristics for reclamation. Geoderma 292:96–110

Sheldon A, Menzies NW, So HB, Dalal R (2004) The effect of salinity on plant available water SuperSoil 2004: 3rd Australian New Zealand Soils Conference, 5–9 December 2004, University of Sydney, Australia

Shi D, Wang D (2005) Effects of various salt-alkaline mixed stresses on Aneurolepidium chinense (Trin.) Kitag. Plant Soil 271(1–2):15–26

Slavich PG, Sinclair K, Morris SG, Kimber SWL, Downie A, Van Zwieten L (2013) Contrasting effects of manure and green waste biochars on the properties of an acidic ferralsol and productivity of a subtropical pasture. Plant Soil 366:213–227

Sohi S, Lopez-Capel E, Krull E, Bol R (2009) Biochar's roles in soil and climate change: a review of research needs. CSIRO Land and Water Science Report 05/09, 64 pp

Sriburi T, Wijitkosum S (2016) Biochar amendment experiments in Thailand: practical examples. In: Bruckman VJ, Varol EA, Uzun BB, Liu J (eds) Biochar a regional supply chain approach in view of climate change mitigation, 1st edn. Cambridge University Press, Cambridge, pp 351–367

Sun Y, Gao B, Yao Y, Fang J, Zhang M, Zhou Y, Chen H, Yang L (2014) Effects of feedstock type, production method, and pyrolysis temperature on biochar and hydrochar properties. Chem Eng J 240:574–578

Sun J, He F, Shao H, Zhang Z, Xu G (2016) Effects of biochar application on Suaeda salsa growth and saline soil properties. Environ Earth Sci 75:630

Suwanich P (1986) Potash and rock salt in Thailand. Nonmetallic Minerals Bulletin No 2, Economic Geology Division, Department of Mineral Resources, Bangkok, Thailand, 339 pp

Thai-Australia Tung Kula Ronghai Project (1983) Tung Kula Ronghai Salinity Study. McGowan International PTY LTD

Thomas V, Albert JRG, Perez RT (2013) Climate-related disasters in Asia and the Pacific. ADB Economics Working Paper Series No. 358

Todkar BS, Deshmukh SM, Deorukhka OA (2016) Extraction of silica from rice husk. Int J Eng Res Dev 12:69–74

Tsai W-T, Liu S-C, Chen H-R, Chang Y-M, Tsai Y-L (2012) Textural and chemical properties of swine- manure-derived biochar pertinent to its potential use as a soil amendment. Chemosphere 89:198–203

United State Salinity Laboratory Staff (USSL) (1954) Diagnosis and Improve of saline and alkali soil [Online]. Available from: http://www.ars.usda.gov/sp2UserFiles/Place/20360500/hb60_pdf/hb60complete.pdf [26 February 2015]

Usman ARA, Al-wabe MI, Ok YS, Al-harbi A, Wahb-allah M, El-naggar AH, Ahmad M, Al-faraj A, Al- omran A (2016) Conocarpus biochar induces changes in soil nutrient availability and tomato growth under saline irrigation. Pedosphere 26:27–38

Uzoma KC, Inoue M, Andry H, Fujimaki H, Zahoor A, Nishihara E (2011) Effect of cow manure biochar on maize productivity under sandy soil condition. Soil Use Manag 27:205–212

Venkateswarlu B, Shanker AK (2009) Climate change and agriculture: adaptation and mitigation strategies. Indian J Agron 54:226–230

Wang G, Xu Z (2013) The effects of biochar on germination and growth of wheat in difference saline- alkali soil. Asian Agric Res 5:116–119

Wang W, Vinocur B, Altman A (2003) Plant responses to drought, salinity and extreme temperatures: towards genetic engineering for stress tolerance. Planta 218(1):1–14

Wang Y, Li X, Zhang B, Wu Y (2014) Meso-damage cracking characteristics analysis for rock and soil aggregate with CT test. SCIENCE CHINA Technol Sci 57(7):1361

Warnock DD, Lehmann J, Kuyper TW, Rillig MC (2007) Mycorrhizal responses to biochar in soil-concepts and mechanisms. Plant Soil 300:9–20

Wijitkosum S (2018) Fuzzy AHP for drought risk assessment in Lam Ta Kong watershed, the north-eastern region of Thailand. Soil & Water Res 13:218–225

Wijitkosum S, Jiwnok P (2019) Effect of biochar on Chinese kale and carbon storage in an agricultural area on a high rise building. AIMS Agric Food 4:177–193

Wijitkosum S, Kallayasiri W (2015) The Use of biochar to increase productivity of indigenous upland rice (*Oryza sativa* L.) and improve soil properties. Res J Pharmaceut Biol Chem Sci 6:1326–1336

Williamson DR, Peck AJ, Turner JV, Arunin S (1989) Groundwater hydrology and salinity in a valley in Northeast Thailand. In: Groundwater contamination. IAHS Publ No. 185

Wong VNL, Dalal RC, Greene RSB (2009) Carbon dynamics of sodic and saline soil following gypsum and organic material additions: a laboratory incubation. Appl Soil Ecol 41:29–40

Wu Y, Xu G, Shao HB (2014) Furfural and its biochar improve the general properties of a saline soil. Solid Earth 5:665–671

Yamato M, Okimori Y, Wibowo IF, Anshori S, Ogawa M (2006) Effects of the application of charred bark of *Acacia mangium* on the yield of maize, cowpea and peanut, and soil chemical properties in South Sumatra, Indonesia. Soil Sci Plant Nutr 52:489–495

Yang F, Zhang GL, Yang JL, Li D-C, Zhao YG, Liu F, Yang R-M, Yang F (2014) Organic matter controls of soil water retention in an alpine grassland and its significance for hydrological processes. J Hydrol 519:3086–3093

Yooyen J, Wijitkosum S, Sriburi T (2015) Increasing yield of Soybean by adding biochar. J Environ Res Dev 9:1066–1074

Yuan JH, Xu RK (2010) The amelioration effects of low temperature biochar generated from nine crop residues on an acidic Ultisol. Soil Use Manag 27:110–115

Yuan JH, Xu RK (2012) Effects of biochars generated from crop residues on acid soils from tropical and subtropical China. Soil Res 50:570–578

Yuan JH, Xu RK, Wang N, Li JY (2011) Amendment of acid soils with crop residues and biochars. Pedosphere 21:302–308

Yue Y, Yao Y, Lin Q, Li G, Zhao X (2016) The change of heavy metals fractions during hydrochar decomposition in soils amended with different municipal sewage sludge hydrochars. J Soils Sediments 17:763–770

Yuvaniyama A (2003) Handbook for saline management. Land Development Department (In Thai), Bangkok

Yuwaniyama A (2004) Manual of Saline soils management and reclamation, Technical paper of research and development of salt-affected areas group, Land Development Department, Bangkok

Zhang M, Muhammad R, Zhang L, El-desouki Z, Jiang C (2019) Biochar Induces Changes to Basic Soil Properties and Bacterial Communities of Different Soils to Varying Degrees at 25 mm Rainfall: More Effective on Acidic Soils. Front Microbiol 10:1321

Zheng H, Wang X, Chen L, Wang Z, Xia Y, Zhang Y, Wang H, Luo X, Xing B (2018) Enhanced growth of halophyte plants in biochar-amended coastal soil: roles of nutrient availability and rhizosphere microbial modulation. Plant Cell Environ 41:517–532

Chapter 2
Impact of Pyrolysis Techniques on Biochar Characteristics: Application to Soil

Vineet Yadav and Puja Khare

Abstract Society currently faces global environmental challenges of the burning of waste plants reside which demand innovative, interdisciplinary and complex solutions. In India about 500 million tons agricultural and agro-industrial residues are being generated annually in the country. A major amount of this agricultural residue farmers treats as waste, are burn in field itself. Hence, there is need to combat these problems through a sustainable management system, which will revive depletion of waste generated from agriculture itself. Conversion of this agricultural waste into biochar through pyrolysis could be a positive solution for minimizing agricultural waste. Production of biochar also offers many opportunities for enhancing soil Physico-chemical characteristics and carbon sequestration. However these characteristics alter with different factors like type and temperature of pyrolysis, biomass holding time. In the present chapter the detailed information of pyrolysis techniques and their impact on soil fertility are discussed.

Keywords Biochar · Pyrolysis · Soil · Biomass · Crop residue

V. Yadav
Agronomy and Soil Science Division, CSIR-Central Institute of Medicinal and Aromatic Plants, Lucknow, India

Aacademy of Scientific and Innovative Research (AcSIR), Ghaziabad, India

Jawaharlal Nehru Customs House, Central Revenue Control Laboratory, New Delhi, India

P. Khare (✉)
Agronomy and Soil Science Division, CSIR-Central Institute of Medicinal and Aromatic Plants, Lucknow, India

Aacademy of Scientific and Innovative Research (AcSIR), Ghaziabad, India

J. S. Singh, C. Singh (eds.), *Biochar Applications in Agriculture and Environment Management*, https://doi.org/10.1007/978-3-030-40997-5_2

2.1 Introduction

India has a huge amount of agriculture land area (159.7 million hectares i.e. 60.4%) and ranks second globally in agricultural production at $367 billion. Simultaneously, a huge quantity of unused agricultural spent produced, whose disposal is one of the major concerns, currently. The MNRE (Ministry of New and Renewable Energy) jointly with Indian Institute of Science, Bangalore evaluated (24 February 2014) that approximately 500 million tons of agro-industrial and agricultural wastes are being produced annually in the country. The major quantities of these wastes generated from agricultural sources are sugarcane bagasse, aromatic crop residue, paddy, wheat straw, and husk, etc. (Quispe et al. 2017). In which, about 50% of finds application in various industrial and agricultural purposes like animal feedstock, paper industry, and energy generation, etc. However, major amount of agricultural residues, to which farmers treat as waste, are burn in field itself (Fig. 2.1). One of the most important negative effects of burning of these wastes is CO_2 release, which is the main greenhouse gas that is generated by human activities. The progress in the agriculture through utilization of agricultural land leads to depletion of nutrient contents of soil and their organic carbon (OC) content (Yadav et al. 2018b)

According to the survey (The Economic Times: 29 January 2018), 35 million tonnes of paddy crop in 3 adjacent states (Punjab, Haryana and Western UP) are burnt in late October. In the last decades, there is much news bulletin on the drawback of crop residue burning in the field (Fig. 2.2).

Therefore, a sustainable management method is required to tackle these two agricultural problems through reutilize these agricultural waste. The conversion of

Fig. 2.1 Agricultural crop residue burning and greenhouse gases emission

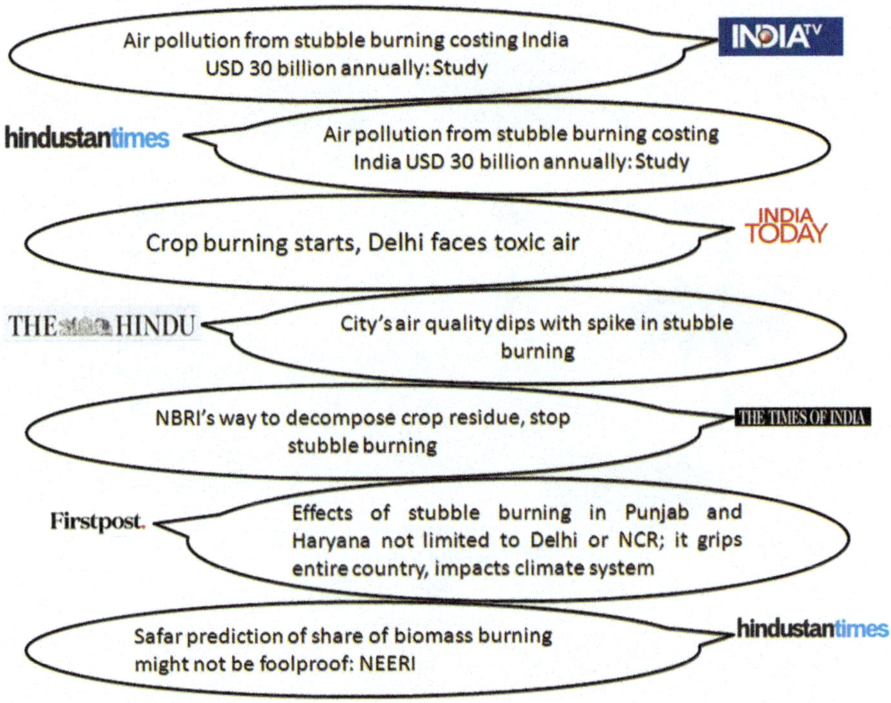

Fig. 2.2 News bulletin on agricultural crop residue burning

biomass wastes to value-added products as biochar through pyrolysis has been recommended to avoid the negative effects of the direct burning of these biomasses (Fig. 2.3). As biochar is resistant to biological decay, it can be preserved in the terrestrial systems for more longer time than the other plant residues or compost, hence; the beneficial effects of biochar are extended (Lehmann et al. 2011). Biochar has the potential to enhance yield of agricultural crop and long-term (hundreds to thousands of years) storage of carbon in soils (Lehmann et al. 2006). However, the benefits of biochar to the agricultural soil depending on a number of factors including feedstock, (Tiwari et al. 2019a; Tiwari et al. 2019b; Singh et al. 2019; Kour et al. 2019), biochar production methods and application rates (Jha et al. 2010). The feedstock is possibly an important variable for evaluating the final composition and characteristics of biochar.

2.2 Background

The beginning of biochar is associated with the ancient populations of the Amazon region, known as Terra Preta de Indio, where the dark earth was generated by the use of techniques of slash and char (Lehmann and Rondon 2006). Nowadays, bio-

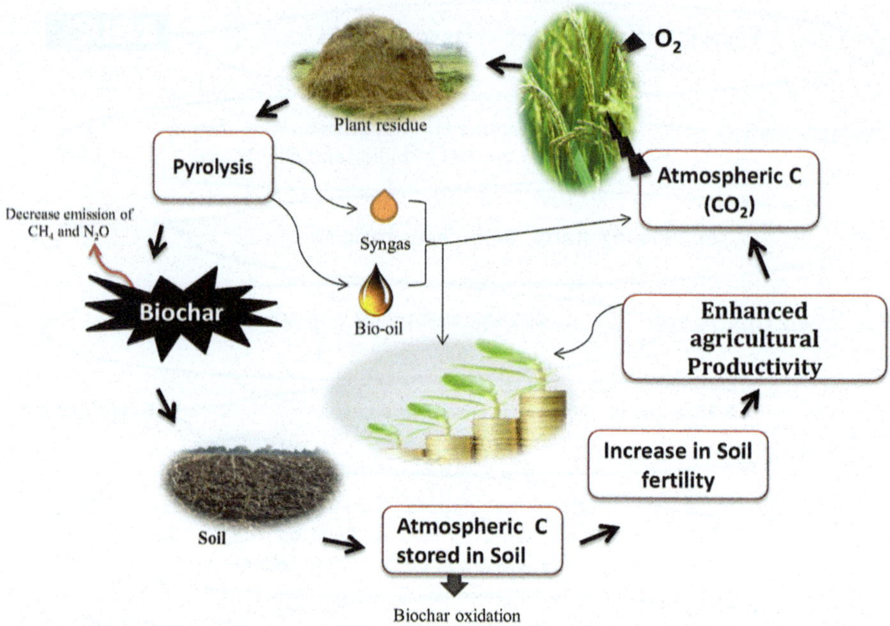

Fig. 2.3 Schematic of the utilization of agro-waste through pyrolysis

char is known to be the best amendment for the fertility of soil and its sustainability, so far many researchers and farmers are still paying attention to its hidden secrets all over the world. Biochar is a constructed scientific term. The different agencies biochar defined in a different way: The International Biochar Initiative (IBI) had shown its definition as "A solid material obtained from the thermo-chemical conversion of biomass in an oxygen-limited environment" (The International Biochar Initiative (IBI)."The porous carbonaceous solid material produced by the thermochemical conversion of feedstock in an oxygen depleted atmosphere that has physicochemical properties suitable for safe and long-term storage of carbon in the environment" Shackle et al. (2012).

2.2.1 *Production of Biochar*

The conversion of biomass through pyrolysis is a very old technology, which is still applicable for energy production and biomass conversion. The definition of pyrolysis of feedstock is always the same that is thermo-chemical conversion of feedstock material in the absence of oxygen. However, different methodologies exist for the pyrolysis of biomass, and it can be further categorized into three subcategories: fast pyrolysis, slow pyrolysis and full gasification (Fig. 2.4).

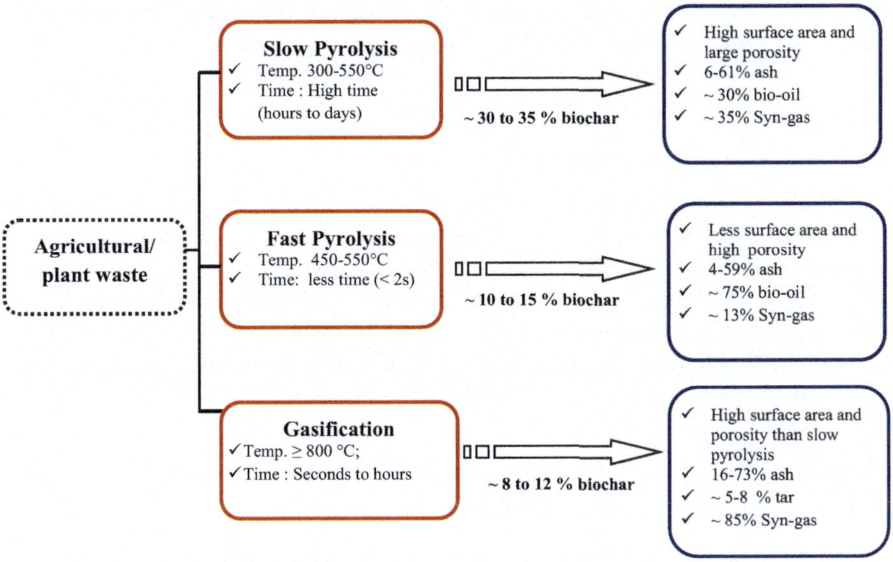

Fig. 2.4 Types of pyrolysis and product distribution

2.2.2 *Slow Pyrolysis*

Production of biochar through pyrolysis is a technology that has been examined for biochar production. Historically, slow pyrolysis of woody biomass through traditional kilns have been the most widespread application for biochar or charcoal production (Van Zwieten et al. 2010). Compared to the fast pyrolysis, rate of heating in the slow pyrolysis uses slower (ranging from 0.0 °C s^{-1} to up to 10 °C s^{-1}).

2.2.3 *Fast Pyrolysis*

This technology uses more stable processes designed to produce a higher fraction of liquid products (Brown et al. 2011). Fast pyrolysis use for conversion of biomass in a few seconds, using high heating rates (> 200 °C s^{-1}) and short residence times (< 5 s).

2.2.4 *Gasification*

This is a technology in which, the major aim is to turn the whole of the organic fraction of the biomass into gaseous fraction through thermo-chemical conversion. Gasification is technology is generally used to avoid the production of char and bio-oil. The process of converting biomass into gas is a highly endothermic reaction, and therefore it requires heat to facilitate this step in the process.

2.3 Factor Affecting Biochar Yield

The composition of the feedstock and pyrolysis conditions are one of the major factors for the biochar yield (Wu et al. 2012) and its physical and chemical properties (Yadav et al. 2018a). Pyrolysis of the agricultural waste mainly performed to convert either to attain energy or a high percentage of the solid product. Conversion of agricultural waste into value-added products subjective to type of agricultural waste, pyrolysis temperature, heating rate, and temperature hold time.

2.3.1 *Effect of Feedstock Material*

The composition of the feedstock with dissimilar proportions of moisture content, cellulose, hemicellulose, and lignin may result in a significant alteration in yield and other properties. Cellulose has more potential for decomposition whereas lignin has lower because of its thermal stability (Yang et al. 2007). The volatiles mainly resulted from lignin decomposition are formed of aromatic compounds as compared to aliphatic compounds from cellulose and hemicellulose (Collard and Blin 2014). The higher the moisture content in the biomass lowers the yield of the product.

2.3.2 *Pyrolysis Temperature*

Pyrolysis of lingo-cellulosic biomass was at a range of 200–700 °C (Zhou et al. 2014). The biochar yield decrease has been reported by many authors with an increase in pyrolysis temperature for different feedstock materials (Leng and Huang 2018). A heavy loss in the biomass weight was found between 300–400 °C, which could be because of total decomposition of cellulose and hemicellulose components, while the initial decomposition of lignin, however, major weight loss might decrease after 500 °C because of the thermal stability of lignin components.

2.3.3 *Temperature Holds Time*

Processes of pyrolysis are commonly categorized on the basis of temperature hold time of residue. Increase in the holding time of the biomass during pyrolysis the CO_2 yield decreased effectively. Crop residue yield was reported to be decreased with an increased temperature hold time (Kumar et al. 2013).

2.3.4 *Rate of Heating*

The high rate of heating leads to fast carbonization of feedstock, hence, resulted in broad cracking and generation of micropores and so, producing higher surface area of the biochar (Mukherjee and Lal 2013). Many studies were performed at different heating rates ranging from 10 °C min^{-1} (Lee et al. 2013) up to 45 °C min^{-1} (Cao et al. 2014).

2.4 Thermodynamics of Pyrolysis

Amongst the different sources of feedstock, which have been used for pyrolysis, cellulosic and lignocellulosic feedstock make up the major portions. During pyrolysis, there were different structural changes in the biomass resulting in enhanced aromaticity which is controlled by the feedstock material and the pyrolysis conditions such as time, temperature and particle size (Enders et al. 2012). Pyrolysis covers a range of thermal decomposition of biomass such as condensation polymerization, dehydration, dehydrogenation, decarboxylation and deoxygenation reactions (Fig. 2.5). The cellulosic components of feedstock started decomposing at about 200–400 °C, while lignin starts to decompose to 300–700 °C (Cao et al. 2014).

2.5 Characteristics of Biochar

The composition of biochar is highly heterogeneous, contains both stable and labile components (Ameloot et al. 2013). Carbon, mineral matter (ash) and surface area are generally regarded as major constituents of biochar to characterization. Table 2.1 summarized the important parameters based on some previous studies on biochar production and their characteristics. The data of the Table 2.1 shows that characteristics of the biochar varied with pyrolysis temperature and feedstock used. Various biomasses as crop residues and wood biomass are used to produce biochar via pyrolysis processes. The pyrolysis temperature used by the researchers varied from 200 °C to 900 °C with heating rates ranging from 2.5 to 20 °C min^{-1}.

2.6 Role of Biochar to Increase the Soil Fertility

The biochar application in agriculture is attracting significant attention globally as a way to increase characteristics of the soil (Fig. 2.6). Soil amendment with biochar has been considered as an suitable option for several purposes such as improving soil nutrient availability (Prasad et al. 2017), biological activity i.e. soil enzyme activity (Awad et al. 2018) and microbial activity (Pressler et al. 2017). Application

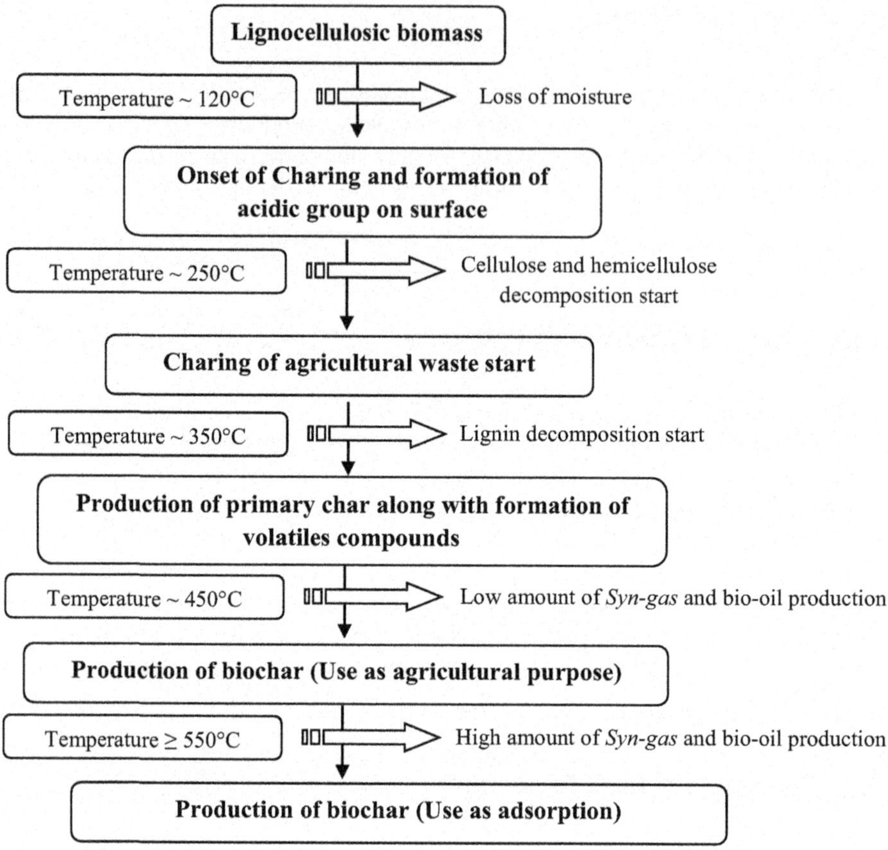

Fig. 2.5 A detailed mechanistic exploration of pyrolysis process and product yield from lignocellulose biomass source

of biochar is a sustainable selection to provide enduring effects in soil fertility enhancements especially in sandy soils as sustainable agriculture is facing great limitation because of lower water holding capacity, and higher leaching of all the nutrients of soil (Uzoma et al. 2011). Table 2.2 summarize affect of biochar on different soil study.

As biochar has ability to hold the nutrients and to enhance the soil water holding capacity, hence, biochar application in soil could be utilized to overcome most of the limitations with pyrolysis temperature. Recent studies showed that biochar addition to soil increase soil pH, improve the cation exchange capacity (CEC), water holding capacity, alter soil bulk density, and increased the exchangeable basic cations soils (Basso et al. 2013) in soil. Many studies examined the increase in the characteristics was mainly because of the higher surface area, porosity, and surface functional group on the surface of the biochar. These properties of biochar would be

Table 2.1 Characteristics of biochars produced from different feedstocks

Feedstock	Pyrolysis temperature (°C)	Yield (%)	Fixed matter (%)	Ash (%)	pH	C (%)	H (%)	O (%)	N (%)	Surface area ($m^2 g^{-1}$)	References
Rice straw	300 °C	49.5	36.9	–	9.19	72.1	5.03	21.3	1.55	–	Wu et al. (2012)
Rice straw	500 °C	39.0	44.9	–	10.48	82.8	3.25	12.1	1.77	–	Wu et al. (2012)
Rice straw	700 °C	34.7	56.8	–	10.77	90.6	1.80	6.2	1.41	–	Wu et al. (2012)
Soybean stover	300 °C	37.0	38.8	10.4	7.3	68.81	4.29	24.99	1.88	5.6	Ahmad et al. (2012)
Soybean stover	300 °C	21.6	67.7	17.2	11.3	81.98	1.27	15.45	1.30	420.3	Ahmad et al. (2012)
Swine solid	350 °C	62.3	17.7	32.5	8.4	51.51	4.91	11.10	3.54	0.9	Cantrell et al. (2012)
Swine solid	700 °C	36.4	33.8	55.9	9.5	44.06	0.74	4.03	2.61	4.1	Cantrell et al. (2012)
Lemongrass	450 °C	50	46	22	–	53	6	31	9	–	Yadav et al. (2013)
Lemongrass	650 °C	40	49	22	–	61	6	27	6	–	Yadav et al. (2013)
Lemongrass	850 °C	24	51	27	–	80	3	10	7	–	Yadav et al. (2013)
Rose	450 °C	31	63	4	–	68	6	5	20	–	Yadav et al. (2013)
Rose	650 °C	29	63	5	–	78	5	1	17	–	Yadav et al. (2013)
Rose	850 °C	26	64	8	–	81	6	1	13	–	Yadav et al. (2013)
Eucalyptous	450 °C	45	50	12	–	76	8	4	12	–	Yadav et al. (2013)
Eucalyptous	650 °C	35	63	14	–	79	4	3	14	–	Yadav et al. (2013)
Eucalyptous	850 °C	25	64	15	–	84	2	4	10	–	Yadav et al. (2013)
Hickory wood	300 °C	43.7	–	–	7.1	69.13	4.85	24.36	0.39	> 0.1	Sun et al. (2014)
Hickory wood	450 °C	28.50	–	–	7.9	83.62	3.24	11.46	0.17	12.9	Sun et al. (2014)
Hickory wood	600 °C	22.70	–	–	8.4	81.81	2.17	14.03	0.73	401	Sun et al. (2014)
Bagasse	300 °C	33.40	–	–	7.3	69.50	4.20	24.36	0.90	5.2	Sun et al. (2014)
Bagasse	450 °C	28.00	–	–	7.5	78.60	3.52	15.46	0.92	13.6	Sun et al. (2014)
Bagasse	600 °C	26.50	–	–	7.5	76.45	2.93	18.33	0.79	88.3	Sun et al. (2014)
Bamboo	300 °C	40.50	–	–	7.9	66.20	4.70	27.72	0.40	1.3	Sun et al. (2014)
Bamboo	450 °C	26.30	–	–	8.5	76.89	3.55	18.11	0.23	10.2	Sun et al. (2014)

(continued)

Table 2.1 (continued)

Feedstock	Pyrolysis temperature (°C)	Yield (%)	Fixed matter (%)	Ash (%)	pH	C (%)	H (%)	O (%)	N (%)	Surface area ($m^2 g^{-1}$)	References
Bamboo	600 °C	24.00	–	–	9.2	80.89	2.45	14.87	0.15	375.5	Sun et al. (2014)
Maize biochar	300 °C	43.63	–	16.3	9.84	48.9	3.54	23.85	1.25	1.0	Wang et al. (2015)
Maize biochar	450 °C	32.61	–	22.2	10.47	53.81	3.11	18.58	1.22	4.0	Wang et al. (2015)
Maize biochar	600 °C	29.54	–	27.1	11.37	62.9	1.98	16.17	1.28	70.0	Wang et al. (2015)
Undaria pinnatifida	400 °C	67.70	–	41.8	8.38	31.92	1.74	13.56	2.35	70.29	Jung et al. (2016)
Undaria pinnatifida	600 °C	47.75	–	48.7	10.36	36.64	0.99	11.76	2.31	61.80	Jung et al. (2016)
Undaria pinnatifida	800 °C	39.29	–	50.4	11.09	40.51	0.70	10.02	2.00	44.49	Jung et al. (2016)
Rapeseed stem	200 °C	–	13.30	3.0	–	47.54	5.92	39.50	0.76	1.0	Zhao et al. (2018)
Rapeseed stem	450 °C	–	68.19	8.2	–	73.46	3.03	12.19	1.09	37.3	Zhao et al. (2018)
Rapeseed stem	700 °C	–	75.18	14.1	–	80.23	1.27	5.34	0.63	45.1	Zhao et al. (2018)
Buckwheat	350 °C	46.3	46.3	4.0	9.23	70.1	4.44	24.4	0.92	11.40	Zama et al. (2017)
Buckwheat	550 °C	34.2	34.2	5.8	10.0	82.8	2.75	13.4	0.90	17.02	Zama et al. (2017)
Buckwheat	650 °C	28.5	28.5	33.1	9.14	83.9	1.81	13.3	0.89	17.80	Zama et al. (2017)
Mulberry wood	350 °C	37.5	37.5	7.5	10.2	67.9	4.53	25.2	2.16	16.56	Zama et al. (2017)
Mulberry wood	550 °C	26.2	26.2	9.8	10.6	77.0	2.41	18.8	1.68	58.03	Zama et al. (2017)
Mulberry wood	650 °C	22.8	22.8	9.7	10.6	80.1	1.63	16.6	1.58	24.46	Zama et al. (2017)
Peanut shells	350 °C	45.7	45.7	7.0	10.4	64.3	4.32	29.4	1.69	14.03	Zama et al. (2017)
Peanut shells	550 °C	32.5	32.5	7.1	10.6	73.7	2.41	21.7	1.58	18.58	Zama et al. (2017)
Peanut shells	650 °C	29.4	29.4	24.4	10.6	74.6	1.81	21.7	1.66	28.11	Zama et al. (2017)
Barley straw	300 °C	38.33	–	11.1	7.39	56.05	2.15	29.96	0.73	1.30	Jazini et al. (2018)
Barley straw	400 °C	31.66	–	11.3	8.72	56.99	2.05	28.78	0.87	2.13	Jazini et al. (2018)
Barley straw	500 °C	25.00	–	12.2	10.28	66.46	1.44	19.08	0.80	1.43	Jazini et al. (2018)
Cymbopogon winterianus	450 °C	33.26	–	–	8.2	58.39	3.94	32.79	1.2	–	Yadav et al. (2018a)
Cymbopogon winterianus	850 °C	24.05	–	–	9.3	67.74	1.89	29.93	1.0	–	Yadav et al. (2018a)

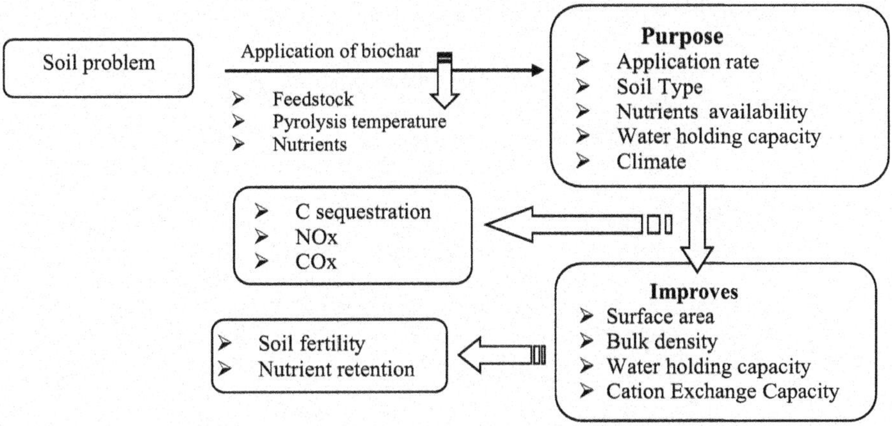

Fig. 2.6 Characteristics of biochar responsible for different applications

expected due to influence of ion adsorption in soil mainly in the rhizospheric soil. Rhizospheric soil is a key point of relation between the soil and growing plants.

Possible mechanisms that are involved in biochar-root interactions in soil include:

1. Biochar changes the soil physicochemical characteristics of soil (DeLuca et al. 2015).
2. Biochar change microbial biodiversity (Edenborn et al. 2018).
3. Biochar decrease residual effects of nutrients in the soil by sorption on its surface (DeLuca et al. 2015).
4. Biochar change flux of secondary metabolite that affects root growth (Spokas et al. 2012)

2.7 Mechanisms of Nutrients Immobilization by Biochar in Soil

The mechanisms of nutrients immobilization by biochars (Fig. 2.7) include (1) Physical trapping of nutrients within pores of biochars, (2) direct electrostatic interactions between cationic nutrients and negatively charged carbon surfaces, (3) ionic exchange between nutrients ions and ionisable protons at the surface of acidic carbon, (4) specific binding of nutrients by surface ligands (functional groups) abundant on biochar surfaces, (5) reaction with mineral impurities (ash) and basic nitrogen groups (e.g., pyridine) of carbonaceous materials, (6) forming hydroxides, carbonates and/or various phosphate-involved precipices and (7) redox reactions with biochar along with sportive reactions (Li et al. 2018).

Table 2.2 Summarized response of biochar application on soil biophysical and chemical properties

Biochar source	Soil type	Effect on soil properties/soil quality changes	References
Pinewood	Fort Collins Loam	Increase in soil characteristics (Physico-chemical) and nematode abundance was observed	Chan et al. (2008)
Peanut hulls, pecan shells, poultry litter	Loamy sand	Biochars produced at higher pyrolysis temperature increased soil pH, while biochar made from poultry litter increased available phosphorous and Na.	Novak et al. (2009)
Wood and peanut shell – Chicken manure – wheat chaff	Sandy soils	Increase in phosphorus availability from 163 to 208%,	Warnock et al. (2010)
Different feedstock	Different soil	Increase in soil pH, CEC, available nutrients (K, Ca and Mg and P); decrease in Al saturation of soils	Schulz and Glaser (2012)
Wood and manure-derived biochars	Different soil types	Increase the soil's saturated hydraulic conductivity and plant's water accessibility, as well as boost the soil's total N concentration and CEC, improving soil field capacity, and reduce NH_4-N leaching.	Stavi (2013) Stavi (2012)
Picea abies, *Pinus sylvestris*	Sandy clay loam texture	Biochar addition increased carbon content however did not significantly affect the soil chemical composition	Tammeorg et al. (2014)
Acacia whole tree green waste	Planosol	Increase in porosity either direct pore contribution, creation of accommodation pores or improved aggregate stability	Hardie et al. (2014)
Palm fronds, pinewood, coconut shells	Arlington sandy loam	Increase in soil pH, WHC, and specific surface area of the soils	Hale et al. (2015)
Eucalyptus, Maize stover	Clay-loam Oxisol, silt loam	Increase in total N derived from the atmosphere up to 78%; higher total soil N recovery with biochar addition	Güereña et al. (2015)
Different biochar sources	Different soil types	Increased crop yield, improved microbial habitat and soil microbial biomass, rhizobia nodulation, plant K tissue concentration, soil pH, soil P, soil K, total soil N, and total soil C compared with control conditions.	Thies et al. (2015)
Cymbopogon flexuosus	Makum coal fields overburden	Overburden alone had negative effect on enzymatic properties. However, amendments of overburden with biochar did not have any negative effect on soil enzymatic activity.	(Jain et al. (2016)
Wheat straw	Fimi-Orthic Anthrosols	Increase in soil pH, organic carbon, total nitrogen, and reduction in yield scaled N2O emissions	Li et al. (2016)

(continued)

Table 2.2 (continued)

Biochar source	Soil type	Effect on soil properties/soil quality changes	References
Salix spp.	Dark reddish-brown Ferralso	Increase maize biomass yield up to 29% with biochar application to soil	Agegnehu et al. (2016)
Green garden waste	Agricultural fields	Biochar degrading bacteria had a positive impact on enhancing rhizospheric effect.	Hussain et al. (2018)
Cymbopogon winterianus	Sandy loam	The amendment biochar of leads amelioration of soil fertility and plant growth, howver biochar prepared at lower pyrolysis (450 °C) more suitable for crop cultivation.	Yadav et al. (2018a)
Cymbopogon winterianus	Sandy loam	Combine biochar application with inorganic fertilizers provide hihj yield of plant biomass along with low nutrients losses/harvest ratio.	Yadav et al. (2019b)
Cymbopogon winterianus	Sandy loam	Mineralization of biochar enhanced in aged biochar soil mixture and its positive effect on soil microorganisms	Yadav et al. (2019a)
Mentha arvensis	Sandy loam	Amendemnt of biochar in metal contaminated soil improves the plant plant production, photosynthetic attributes and reduced the antioxidant enzyme activity.	Nigam et al. (2019a)
Mentha arvensis	Sandy loam	Biochar stabilized the metal (Pb) in the less mobile fraction in soil by complexation and precipitation. The alteration in the metal fractionation path of Pb provide suitable habitat to the microbes.	Nigam et al. (2019b)

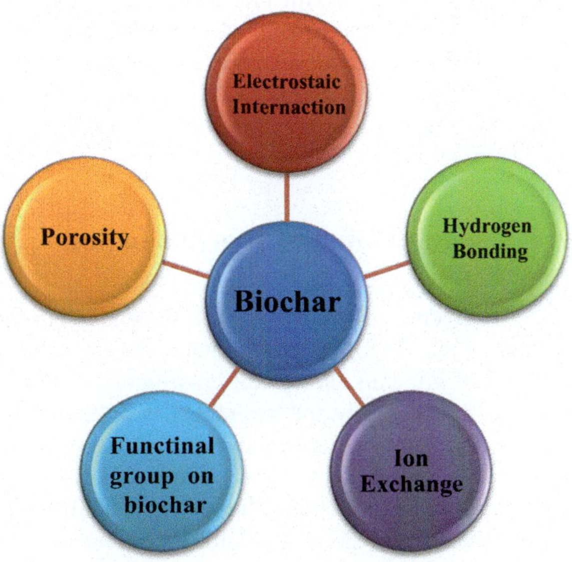

Fig. 2.7 Different factors for nutrient immobilization in soil biochar mixture

2.8 Effect of Pyrolysis Temperature on Soil Physico-Chemical Characteristics

A massive variability in characteristics of the biochar was observed based on the parent feedstock and the conditions opted for biochar production especially the pyrolysis temperatures. Biochar contains an enormous amount of micro or macro-nutrients and carbon content that depends on type of pyrolysis temperature and feedstock (Enders et al. 2012). Therefore, pyrolysis temperature might be main factor to access the role of biochar effect on soil performance and plant growth. The previous study showed once biochar application to soil may possibly perform as source of plant mineralizable nutrients (Gaskin et al. 2008) and consequently biochar application influenced soil mineral accessibility by both directly and indirectly.

Different feedstock like nutshells, grasses, forestry products, and animal manures were utilized for biochar production (Qiu et al. 2014). It is well known that minerals content or chemical characteristics such as pH, cation exchange capacity (CEC), electrical conductivity (EC) or plant-available nutrients of biochars are varied with pyrolysis temperature (Lee et al. 2017). To produce designer biochar which is modified to answer a specific soil problem, feedstocks and pyrolysis should be examined in detail of the existing and original source (feedstock) for production of biochar. Production of biochar at lower (250 to 450 °C) pyrolysis temperature have high yield of biochar and contains high functional groups (C=O and C-H). These functional groups can act as minerals substitute sites after biochar oxidation in soil biochar mixture. Similarly, biochar has large amount of carbon in compact polyaromatic structure are obtain by pyrolyzed the organic feedstocks at pyrolysis higher temperatures (450 °C to 700 °C), potentially limiting biochar effectiveness in maintaining soil minerals (Novak et al. 2009). The dynamics of the nutrients mineralization and soil enzymatic activities for the production of biochar at altered pyrolysis temperature are still inadequately understood although recent advances on biochar study. Only a few, studies have examined the biochar ability to influence the plant yield, nutrient retention and soil enzymatic activity for biochar formed at altered pyrolysis temperature (Ameloot et al. 2013; Lin et al. 2012).

2.9 Sustainability of Biochar in Soil

The effectiveness of biochar in soil depends upon processes occurs on its surface (Huang et al., 2018), including interactions with microbes, mineralization, organic matter and minerals in the soil environment and abiotic and biotic factors (Yang et al. 2018). Generally, the fresh biochar act to improve these properties due to their high micro-porous structures and specific area (Khorram et al., 2017), which results better nutrient availability for the plants and microbes in soil (Singh et al. 2017a, b, c, 2018; Tiwari et al. 2018; Zheng et al. 2018). In addition, biochar contains numbers of fractions of stable and labile compounds that decomposed with time in soil

(Kelly et al. 2017) (Abbruzzini et al. 2017). Decomposition or mineralization of biochar in soil mainly depends upon the two types of factors: (i) Environmental factor such as microbial interactions, humidity, temperature and physical breakdown (Rechberger et al. 2017), (ii) Soil type and native microbial interaction with biochar (Fang et al. 2018). So far, only few reports are available on the aging processes in biochar amended soil (Khorram et al. 2017; Zhang et al. 2016). Little is known about the mineralization and microbial activity aging of biochar amended soil. Oxidation process took on the biochar surface in soil during the aging of biochar particles (DeLuca et al. 2015). These oxidation processes in soil on biochar surface lead to formation of oxygenated functional groups (e.g. carboxylic, phenolic and lactonic) often with acidic characteristics (Suliman et al. 2015), which may further enhance chemical and physical weathering of biochar particle (Naisse et al. 2015). These changes in the soil during aging of biochar may later mineralization process and soil biology as well (Pressler et al. 2017). The bio-chemical alteration during aging of the biochar in the soil environment is still unclear (de la Rosa et al. 2018). However, these alterations may vary with the biochar production type i.e. feedstock (Fernández-Ugalde et al. 2017), pyrolysis process and temperature (Kim et al. 2017).

2.10 Biochar and Soil Biological Activity in Soil

Aromatic compounds, formed during the pyrolysis of plant biomass, are the major components responsible for building up a stable carbon pool in the soil after biochar addition. Yet, the biochar is not inert biologically, when added into the soil and follows the mineralization pattern. The consumption of these labile compounds of biochar supposed changes in the physicochemical soil habitat (Lehmann et al. 2011), soil biological parameters, such as microbial and enzyme activity (Ameloot et al. 2013). Enzymes are biological molecules that are fabricated and consumed by all organisms. Soil enzymes are essential to the survival of organisms as they catalyze the essential metabolic processes in soil by both specifying the molecules to be involved and speeding up the rate of these reactions. Each soil contain specific enzymes that regulate its processes, the activity of these enzymes in soil environments varies greatly with different organic matter content and their composition. There is presently great attention in the use of extracellular enzyme activities as biological indicators of soil quality because they respond rapidly to changes in land management and simple to measure (Schloter et al. 2003). Some of the soil enzymes (Table 2.3) that are principally important in soils include those related to carbon and nutrient mineralization, such as phosphatase (Acidic and Alkaline), β-Glucosidase, dehydrogenase, and Urease.

Nowadays, much of the research has been focused on the uses and application rate of biochar that are economically feasible at present. Further research is needed to evaluate the application of biochar into agricultural systems, as well as, techniques for enriching biochar with fertilizer.

Table 2.3 Soil enzymes as an indicator of soil quality

Soil enzyme	Reaction catalyzed	Enzyme reaction	Indicator of microbial activity
Phosphatase	Phosphate ester + $H_2O \rightarrow$ ROH phosphate	Release of PO_4^-	P-cycling
β-Glucosidase	Glucosidase + $H_2O \rightarrow$ ROH + glucose	Cellobiose hydrolysis	C-cycling
Dehydrogenase	$XH_2 + A \rightarrow X + AH_2$	Electron transport system	C-cycling
Urease	Urea $\rightarrow 2NH_3 + CO_2$	Urea hydrolysis	N-cycling

2.11 Conclusion

Burning of crop residue in the agricultural field is not new to India; resultant society faces global environmental challenges. Further, progressive and intensive use of agricultural land leads to depletion of soil nutrients and organic carbon content. These two major environmental challenges if diagnosed and managed scientifically can provide solution to clean and use renewable solid materials.

Carbonization of woody wastes to produce products like biochar has been suggested to avoid negative impacts of direct biomass burning. As biochar is resistant to the biological decay, it can be preserved in the terrestrial systems for longer time than the other plant residues or compost, hence; the beneficial effects of biochar are extended. Therefore, carbonization of waste biomass through pyrolysis to produce biochar could be a successful technique to avoid the negative impacts on environment and at the same time biochar application to soil could improve soil fertility.

References

Abbruzzini TF, Moreira MZ, de Camargo PB, Conz RF, Cerri CEP (2017) Increasing rates of biochar application to soil induce stronger negative priming effect on soil organic carbon decomposition. Agric Res 6:389–398

Agegnehu G, Bass AM, Nelson PN, Bird MI (2016) Benefits of biochar, compost and biochar–compost for soil quality, maize yield and greenhouse gas emissions in a tropical agricultural soil. Sci Total Environ 543:295–306

Ahmad M, Lee SS, Dou X, Mohan D, Sung J-K, Yang JE, Ok YS (2012) Effects of pyrolysis temperature on soybean Stover-and peanut shell-derived biochar properties and TCE adsorption in water. Bioresour Technol 118:536–544

Ameloot N, Graber ER, Verheijen FG, De Neve S (2013) Interactions between biochar stability and soil organisms: review and research needs. European J Soil Sci 64:379–390

Awad YM, Lee SS, Kim KH, Ok YS, Kuzyakov Y (2018) Carbon and nitrogen mineralization and enzyme activities in soil aggregate-size classes: effects of biochar, oyster shells, and polymers. Chemosphere 198:40–48

Basso AS, Miguez FE, Laird DA, Horton R, Westgate M (2013) Assessing potential of biochar for increasing water-holding capacity of sandy soils. GCB Bio 5:132–143

Brown TR, Wright MM, Brown RC (2011) Estimating profitability of two biochar production scenarios: slow pyrolysis vs fast pyrolysis. Biofuels Bioprod Biorefin 5:54–68

Cantrell KB, Hunt PG, Uchimiya M, Novak JM, Ro KS (2012) Impact of pyrolysis temperature and manure source on physicochemical characteristics of biochar. Bioresour Technol 107:419–428

Cao X, Zhong L, Peng X, Sun S, Li S, Liu S, Sun R (2014) Comparative study of the pyrolysis of lignocellulose and its major components: characterization and overall distribution of their biochars and volatiles. Bioresour Technol 155:21–27

Chan KY, Van Zwieten L, Meszaros I, Downie A, Joseph S (2008) Agronomic values of greenwaste biochar as a soil amendment. Soil Res 45:629–634

Collard FX, Blin J (2014) A review on pyrolysis of biomass constituents: mechanisms and composition of the products obtained from the conversion of cellulose, hemicelluloses and lignin. Renew Sustain Energy Rev 38:594–608

de la Rosa JM, Rosado M, Paneque M, Miller AZ, Knicker H (2018) Effects of aging under field conditions on biochar structure and composition: implications for biochar stability in soils. Sci Total Environ 613:969–976

DeLuca TH, Gundale MJ, MacKenzie MD, Jones DL (2015) Biochar effects on soil nutrient transformations. Biochar Environ Manage Sci Technol Implement 2:421–454

Edenborn SL, Johnson LM, Edenborn HM, Albarran-Jack MR, Demetrion LD (2018) Amendment of a hardwood biochar with compost tea: effects on plant growth, insect damage and the functional diversity of soil microbial communities. Biol Agric Hortic 34:88–106

Enders A, Hanley K, Whitman T, Joseph S, Lehmann J (2012) Characterization of biochars to evaluate recalcitrance and agronomic performance. Bioresour Technol 114:644–653

Fang Y, Singh BP, Luo Y, Boersma M, Van Zwieten L (2018) Biochar carbon dynamics in physically separated fractions and microbial use efficiency in contrasting soils under temperate pastures. Soil Biol Biochem 116:399–409

Fernández-Ugalde O, Gartzia-Bengoetxea N, Arostegi J, Moragues L, Arias-González A (2017) Storage and stability of biochar-derived carbon and total organic carbon in relation to minerals in an acid forest soil of the Spanish Atlantic area. Sci Total Environ 587:204–213

Gaskin J, Steiner C, Harris K, Das K, Bibens B (2008) Effect of low-temperature pyrolysis conditions on biochar for agricultural use. Trans ASABE 51:2061–2069

Güereña DT, Lehmann J, Thies JE, Enders A, Karanja N, Neufeldt H (2015) Partitioning the contributions of biochar properties to enhanced biological nitrogen fixation in common bean (Phaseolus vulgaris). Biol Fertil Soils 51:479–491

Hale L, Luth M, Crowley D (2015) Biochar characteristics relate to its utility as an alternative soil inoculum carrier to peat and vermiculite. Soil Biol Biochem 81:228–235

Hardie M, Clothier B, Bound S, Oliver G, Close D (2014) Does biochar influence soil physical properties and soil water availability? Plant Soil 376:347–361

Huang H, Yao W, Li R, Ali A, Du J, Guo D, Xiao R, Guo Z, Zhang Z, Awasthi MK (2018) Effect of pyrolysis temperature on chemical form, behavior and environmental risk of Zn, Pb and Cd in biochar produced from phytoremediation residue. Bioresour Technol 249:487–493

Hussain F, Hussain I, Khan AHA, Muhammad YS, Iqbal M, Soja G, Reichenauer TG, Yousaf S (2018) Combined application of biochar, compost, and bacterial consortia with Italian ryegrass enhanced phytoremediation of petroleum hydrocarbon contaminated soil. Environ Exp Bot 153:80–88

Jain S, Mishra D, Khare P, Yadav V, Deshmukh Y, Meena A (2016) Impact of biochar amendment on enzymatic resilience properties of mine spoils. Sci Total Environ 544:410–421

Jazini R, Soleimani M, Mirghaffari N (2018) Characterization of barley straw biochar produced in various temperatures and its effect on lead and cadmium removal from aqueous solutions. Water Environ J 32:125–133

Jha P, Biswas A, Lakaria B, Rao AS (2010) Biochar in agriculture–prospects and related implications. Curr Sci 99(9):1218–1225

Jung KW, Kim K, Jeong T-U, Ahn K-H (2016) Influence of pyrolysis temperature on characteristics and phosphate adsorption capability of biochar derived from waste-marine macroalgae (Undaria pinnatifida roots). Bioresour Technol 200:1024–1028

Kelly CN, Benjamin J, CALDERóN FC, Mikha MM, Rutherford DW, Rostad CE (2017) Incorporation of biochar carbon into stable soil aggregates: the role of clay mineralogy and other soil characteristics. Pedosphere 27:694–704

Khorram MS, Lin D, Zhang Q, Zheng Y, Fang H, Yu Y (2017) Effects of aging process on adsorption–desorption and bioavailability of fomesafen in an agricultural soil amended with rice hull biochar. J Environ Sci 56:180–191

Kim H, Kim J, Kim M, Hyun S, Moon DH (2017) Sorption of sulfathiazole in the soil treated with giant Miscanthus-derived biochar: effect of biochar pyrolysis temperature, soil pH, and aging period. Environ Sci Pollut Res 25(26):1–9

Kour D, Rana KL, Yadav N, Yadav AN, Rastegari AA, Singh C, Negi P, Singh K, Saxena AK (2019) Technologies for biofuel production: current development, challenges, and future prospects. In: Rastegari AA et al (eds) Prospects of renewable bioprocessing in future energy systems, Biofuel and Biorefinery Technologies 10. Springer, Cham, pp 1–50

Kumar S, Masto RE, Ram LC, Sarkar P, George J, Selvi VA (2013) Biochar preparation from Parthenium hysterophorus and its potential use in soil application. Ecol Eng 55:67–72

Lee Y, Park J, Ryu C, Gang KS, Yang W, Park Y-K, Jung J, Hyun S (2013) Comparison of biochar properties from biomass residues produced by slow pyrolysis at 500 C. Bioresour Technol 148:196–201

Lee J, Yang X, Cho S-H, Kim J-K, Lee SS, Tsang DC, Ok YS, Kwon EE (2017) Pyrolysis process of agricultural waste using CO 2 for waste management, energy recovery, and biochar fabrication. Appl Energy 185:214–222

Lehmann J, Rondon M (2006) Bio-char soil management on highly weathered soils in the humid tropics. Biol Approaches Sustain Soil Syst 113:e530

Lehmann J, Gaunt J, Rondon M (2006) Bio-char sequestration in terrestrial ecosystems–a review. Mitig Adapt Strateg Glob Chang 11:403–427

Lehmann J, Rillig MC, Thies J, Masiello CA, Hockaday WC, Crowley D (2011) Biochar effects on soil biota–a review. Soil Biol Biochem 43:1812–1836

Leng L, Huang H (2018) An overview of the effect of pyrolysis process parameters on biochar stability. Bioresour Technol 270:627–642

Li G, Zhu W, Zhang C, Zhang S, Liu L, Zhu L, Zhao W (2016) Effect of a magnetic field on the adsorptive removal of methylene blue onto wheat straw biochar. Bioresour Technol 206:16–22

Li L, Wang S, Li X, Li T, He X, Tao Y (2018) Effects of Pseudomonas chenduensis and biochar on cadmium availability and microbial community in the paddy soil. Sci Total Environ 640:1034–1043

Lin Y, Munroe P, Joseph S, Henderson R, Ziolkowski A (2012) Water extractable organic carbon in untreated and chemical treated biochars. Chemosphere 87:151–157

Mukherjee A, Lal R (2013) Biochar impacts on soil physical properties and greenhouse gas emissions. Agronomy 3:313–339

Naisse C, Girardin C, Lefevre R, Pozzi A, Maas R, Stark A, Rumpel C (2015) Effect of physical weathering on the carbon sequestration potential of biochars and hydrochars in soil. GCB Bioenergy 7:488–496

Nigam N, Khare P, Yadav V, Mishra D, Jain S, Karak T, Panja S, Tandon S (2019a) Biochar-mediated sequestration of Pb and Cd leads to enhanced productivity in Mentha arvensis. Ecotoxicol Environ Saf 172:411–422

Nigam N, Yadav V, Mishra D, Karak T, Khare P (2019b) Biochar amendment alters the relation between the Pb distribution and biological activities in soil. Int J Environ Sci Technol:1–12

Novak JM, Busscher WJ, Laird DL, Ahmedna M, Watts DW, Niandou MA (2009) Impact of biochar amendment on fertility of a southeastern coastal plain soil. Soil Sci 174:105–112

Prasad M, Tzortzakis N, McDaniel N (2017) Chemical characterization of biochar and assessment of the nutrient dynamics by means of preliminary plant growth tests. J Environ Manag

Pressler Y, Foster EJ, Moore JC, Cotrufo MF (2017) Coupled biochar amendment and limited irrigation strategies do not affect a degraded soil food web in a maize agroecosystem, compared to the native grassland. GCB Bioenergy 9:1344–1355

Qiu M, Sun K, Jin J, Gao B, Yan Y, Han L, Wu F, Xing B (2014) Properties of the plant-and manure-derived biochars and their sorption of dibutyl phthalate and phenanthrene. Sci Rep (1):4

Quispe I, Navia R, Kahhat R (2017) Energy potential from rice husk through direct combustion and fast pyrolysis: a review. Waste Manag 59:200–210

Rechberger MV, Kloss S, Rennhofer H, Tintner J, Watzinger A, Soja G, Lichtenegger H, Zehetner F (2017) Changes in biochar physical and chemical properties: accelerated biochar aging in an acidic soil. Carbon 115:209–219

Schloter M, Dilly O, Munch J (2003) Indicators for evaluating soil quality. Agric Ecosys Environ 98:255–262

Schulz H, Glaser B (2012) Effects of biochar compared to organic and inorganic fertilizers on soil quality and plant growth in a greenhouse experiment. J Plant Nutri Soil Sci 175:410–422

Shackle S, Sohi S, Ibarrola R, Hammond J, Mašek O, Brownsort P, Haszeldine S (2012) Biochar as a tool for climate change mitigation and soil management. Encyclopedia of Sustainabilty Science and Technology, Springer, New York

Singh C, Tiwari S, Boudh S, Singh JS (2017a) Biochar application in management of paddy crop production and methane mitigation. In: Singh JS, Seneviratne G (eds) Agro-environmental sustainability: managing environmental pollution, 2nd edn. Springer, Cham, pp 123–146

Singh C, Tiwari S, Singh JS (2017b) Impact of rice husk biochar on nitrogen mineralization and methanotrophs community dynamics in paddy soil. Int J Pure App Biosci 5(5):428–435

Singh C, Tiwari S, Singh JS (2017c) Application of biochar in soil fertility and environmental management: a review. Br Environ Pharmacol Life Sci 6(12):07–14

Singh C, Tiwari S, Gupta VK, Singh JS (2018) The effect of rice husk biochar on soil nutrient status, microbial biomass and paddy productivity of nutrient poor agriculture soils. Catena 171:485–493

Singh C, Tiwari S, Singh JS (2019) Biochar: a sustainable tool in soil 2 pollutant bioremediation. In: Bharagava RN, Saxena G (eds) Bioremediation of industrial waste for environmental safety. Springer, Singapore, pp 475–494

Spokas KA, Cantrell KB, Novak JM, Archer DW, Ippolito JA, Collins HP, Boateng AA, Lima IM, Lamb MC, McAloon AJ (2012) Biochar: a synthesis of its agronomic impact beyond carbon sequestration. J Environ Qual 41:973–989

Stavi I (2012) The potential use of biochar in reclaiming degraded rangelands. J Environ Plan Manag 55:657–665

Stavi I (2013) Biochar use in forestry and tree-based agro-ecosystems for increasing climate change mitigation and adaptation. Int J Sustainable Dev World Ecol 20:166–181

Suliman W, Abu-Lail N, Fortuna AM, Harsh J, Garcia-Perez M (2015) Effect of pyrolysis temperature and post-pyrolysis oxidation on the physico-chemical properties of biochars. Chem Eng J 373:44–57

Sun Y, Gao B, Yao Y, Fang J, Zhang M, Zhou Y, Chen H, Yang L (2014) Effects of feedstock type, production method, and pyrolysis temperature on biochar and hydrochar properties. Chem Eng J 240:574–578

Tammeorg P, Simojoki A, Mäkelä P, Stoddard FL, Alakukku L, Helenius J (2014) Biochar application to a fertile sandy clay loam in boreal conditions: effects on soil properties and yield formation of wheat, turnip rape and faba bean. Plant Soil 374:89–107

Thies JE, Rillig MC, Graber ER (2015) Biochar effects on the abundance, activity and diversity of the soil biota. Biochar Environ Manage Sci Technol Implement 2:327–389

Tiwari S, Singh C, Singh JS (2018) Land use changes: a key ecological driver regulating methanotrophs abundance in upland soils. Energy Ecol Environ 3(6):355–371

Tiwari S, Singh C, Boudh S, Rai PK, Gupta VK, Singh JS (2019a) Land use change: a key ecological disturbance declines soil microbial biomass in dry tropical uplands. J Environ Manag 242:1–10

Tiwari S, Singh C, Singh JS (2019b) Wetlands: a major natural source responsible for methane emission. In: Upadhyay AK et al (eds) Restoration of wetland ecosystem: a trajectory towards a sustainable environment. Springer, Singapore, pp 59–74

Uzoma K, Inoue M, Andry H, Fujimaki H, Zahoor A, Nishihara E (2011) Effect of cow manure biochar on maize productivity under sandy soil condition. Soil Use Manag 27:205–212

Van Zwieten L, Kimber S, Morris S, Chan K, Downie A, Rust J, Joseph S, Cowie A (2010) Effects of biochar from slow pyrolysis of papermill waste on agronomic performance and soil fertility. Plant Soil 327:235–246

Wang X, Zhou W, Liang G, Song D, Zhang X (2015) Characteristics of maize biochar with different pyrolysis temperatures and its effects on organic carbon, nitrogen and enzymatic activities after addition to fluvo-aquic soil. S Total Environ 538:137–144

Warnock DD, Mummey DL, McBride B, Major J, Lehmann J, Rillig MC (2010) Influences of non-herbaceous biochar on arbuscular mycorrhizal fungal abundances in roots and soils: results from growth-chamber and field experiments. Appl Soil Ecol 46:450–456

Wu W, Yang M, Feng Q, McGrouther K, Wang H, Lu H, Chen Y (2012) Chemical characterization of rice straw-derived biochar for soil amendment. Biomass Bioenergy 47:268–276

Yadav V, Baruah B, Khare P (2013) Comparative study of thermal properties of bio-coal from aromatic spent with low rank sub-bituminous coals. Bioresour Technol 137:376–385

Yadav V, Khare P, Deshmukh Y, Shanker K, Nigam N, Karak T (2018a) Performance of biochar derived from Cymbopogon winterianus waste at two temperatures on soil properties and growth of Bacopa monneri. Commun Soil Sci Plant Anal 49:2741–2764

Yadav V, Khare P, Deshmukh Y, Shanker K, Nigam N, Karak T (2018b) Performance of biochar derived from Cymbopogon winterianus waste at two temperatures on soil properties and growth of Bacopa monneri. Commun Soil Sci Plant Anal 49(22):1–24

Yadav V, Jain S, Mishra P, Khare P, Shukla AK, Karak T, Singh AK (2019a) Amelioration in nutrient mineralization and microbial activities of sandy loam soil by short term field aged biochar. Appl Soil Ecol 138:144–155

Yadav V, Karak T, Singh S, Singh AK, Khare P (2019b) Benefits of biochar over other organic amendments: responses for plant productivity (Pelargonium graveolens L.) and nitrogen and phosphorus losses. Ind Crops Prod 131:96–105

Yang H, Yan R, Chen H, Lee DH, Zheng C (2007) Characteristics of hemicellulose, cellulose and lignin pyrolysis. Fuel 86:1781–1788

Yang Y, Sun K, Han L, Jin J, Sun H, Yang Y, Xing B (2018) Effect of minerals on the stability of biochar. Chemosphere 204:310–317

Zama EF, Zhu Y-G, Reid BJ, Sun G-X (2017) The role of biochar properties in influencing the sorption and desorption of Pb (II), Cd (II) and As (III) in aqueous solution. J Clean Prod 148:127–136

Zhang X, Sarmah AK, Bolan NS, He L, Lin X, Che L, Tang C, Wang H (2016) Effect of aging process on adsorption of diethyl phthalate in soils amended with bamboo biochar. Chemosphere 142:28–34

Zhao B, O'Connor D, Zhang J, Peng T, Shen Z, Tsang DC, Hou D (2018) Effect of pyrolysis temperature, heating rate, and residence time on rapeseed stem derived biochar. J Clean Prod 174:977–987

Zheng H, Wang X, Chen L, Wang Z, Xia Y, Zhang Y, Wang H, Luo X, Xing B (2018) Enhanced growth of halophyte plants in biochar-amended coastal soil: roles of nutrient availability and rhizosphere microbial modulation. Plant Cell Environ 41:517–532

Zhou Y, Gao B, Zimmerman AR, Chen H, Zhang M, Cao X (2014) Biochar-supported zerovalent iron for removal of various contaminants from aqueous solutions. Bioresour Technol 152:538–542

Chapter 3
The Potential Application of Biochars for Dyes with an Emphasis on Azo Dyes: Analysis Through an Experimental Case Study Utilizing Fruit-Derived Biochar for the Abatement of Congo Red as the Model Pollutant

Kumar Vikrant, Kangkan Roy, Mandavi Goswami, Himanshu Tiwari, Balendu Shekher Giri, Ki-Hyun Kim, Yui Fai Tsang, and Ram Sharan Singh

Abstract The unbridled industrialization and unrestrained expansion of modern textile facilities combined with a deficiency of adequate treatment provisions have escalated the discharge of toxic effluents rich in carcinogenic pollutants such as dyes. As a consequence, there is an alarming need for the development of financially suitable and highly efficient treatment options to protect the immaculate ecosystems, natural resources, and human health. In this respect, adsorption-based treatment options have attracted widespread attention as eco-friendly and cost-

---

Authors Kumar Vikrant and Kangkan Roy have been contributed equally to this chapter.

K. Vikrant
Department of Chemical Engineering and Technology, Indian Institute of Technology, Banaras Hindu University, Varanasi, India

Department of Civil and Environmental Engineering, Hanyang University, Seoul, Republic of Korea

K. Roy · H. Tiwari · B. S. Giri (✉) · R. S. Singh
Department of Chemical Engineering and Technology, Centre of Advanced Study, Indian Institute of Technology, Banaras Hindu University, Varanasi, India

M. Goswami
Department of Bioengineering, Integral University, Lucknow, Uttar Pradesh, India

K.-H. Kim (✉)
Department of Civil and Environmental Engineering, Hanyang University, Seoul, Republic of Korea
e-mail: kkim61@hanyang.ac.kr

Y. F. Tsang
Department of Science and Environmental Studies, The Education University of Hong Kong, Tai Po, New Territories, Hong Kong, China

J. S. Singh, C. Singh (eds.), *Biochar Applications in Agriculture and Environment Management*, https://doi.org/10.1007/978-3-030-40997-5_3

effective approach. Biochar has propelled itself to the forefront of the scientific community as a highly economical sorbent with great adsorption capabilities. Notably, biochars provide a win-win strategy by simultaneously utilizing the waste biomass during its production and a great adsorbent for pollutant removal. Although biochars have been applied for the treatment of various dyes, there have been very few reports of its application for Congo red (CR) dye. In this book chapter we analyze the application of biochar for dyes with particular focus on CR. We try to practically understand the mechanism of interaction between biochar and CR molecules (a model anionic azo dye) by elucidating an experimental case study. The case study will provide valuable insights into the importance of the utilization of locally available bio waste for economic biochar production and the mechanism of removal of anionic dyes through biochars. In brief, the adsorptive removal of CR was investigated using Arjun fruit biochar (AFB) derived from the fruit of locally grown *Terminalia arjuna*. The sorptive removal of CR on AFB was investigated under the following operational conditions (pH, 2–12; biochar dosage, 4–14 g/L; temperature, 30–60 °C; and contact time, 30–480 min). The sorption behavior of CR was well described through the Langmuir monolayer model ($R^2 = 0.9985$) and pseudo-second order kinetics ($R^2 \geq 0.9977$) for all tested CR levels (20–100 mg/L). The results of thermodynamic analysis revealed that the sorption of CR onto AFB proceeded favorably and spontaneously.

Keywords Adsorption · *Terminalia arjuna* · Dye · Biochar · Kinetics · Thermodynamics

3.1 Introduction

Recent years have witnessed prolific rise in endeavors focused towards water quality management due to the ever rising issues concerned with water pollution (Mekonnen Mesfin and Hoekstra Arjen 2017; D'Inverno et al. 2018). The regulation and removal of baleful water pollutants such as dyes have been recognized as prime task by environmental bodies such as the United States Environmental Protection Agency (US EPA) (Vikrant et al. 2018). Hence, extensive investigations have been undertaken to analyze the long as well as short term effects of dyes on ecosystems and human health (Chung 2016).

Dyes are chemical compounds with intense color and extremely high water solubility. A wide variety of dyes and pigments are utilized in paper, textile, paint, drugs, tanning, food, and cosmetic industries (Chen et al. 2018; Vikrant et al. 2018). The ubiquitous presence of highly toxic synthetic dye compounds in water bodies poses grave health concerns as these chemicals are known to be highly carcinogenic and mutagenic (Abu Talha et al. 2018). Dyes impart intense color to water, making it aesthetically displeasing. They also cause an imbalance in the aquatic ecosystems by increasing chemical oxygen demand and lowering light penetration (Vikrant et al. 2018). Also, very low removal rate is generally observed for dyes during secondary and primary wastewater treatment owing to their recalcitrant structures resulting in there swift carryover into water bodies (Meerbergen et al. 2018). As a consequence, toxic dye molecules bioacculmulate in the ecosystem and get transmitted to the potable water supply (Leo et al. 2018). Microbial mediated services are also considered as valuable means for water treatments and restoration of degraded ecosystems (Singh et al. 2017a, b; Vimal et al. 2018; Kumar et al. 2018; Singh 2019; Singh and Singh 2019; Singh et al. 2019a,b; Vimal and Singh 2019).

Azo dyes are the most common variety of dyes in used today and is the oldest industrially synthesized compound (Vikrant et al. 2018). Interestingly, azo dyes comprise around 60% of the total dyes being used in the textile industry, thus posing a significant threat (Munagapati and Kim 2016). Azo dyes attract widespread usage in various industries owing to their multiple advantages such as easy usage, wide availability of color range, great photostability, low requirement of energy, and great covalent adherence with textiles (Brüschweiler and Merlot 2017). The chromophoric -N=N- groups in nonionic and anionic dyes undergoes reductive breakage to result in the formation of highly virulent aromatic amines (Jayapal et al. 2018). Congo red (CR) is a major azo dye utilized extensively in the cellulose industries (e.g., pulp, paper, and cotton textiles) (Chen et al. 2018). The recalcitrant and toxic nature of CR makes its removal from water bodies decidedly imperative (Vikrant et al. 2018).

Due to a synthetic origin and complex structure, the highly recalcitrant tendency of azo dyes renders them exceptionally stable towards heat and light. As such, their removal *via* conventional treatment techniques is not simple (Vikrant et al. 2018). A wide array of methodologies (e.g., photocatalysis (Zhao et al. 2018), ozone treatment (Mella et al. 2018), ultrasound treatment, oxidation (Nidheesh et al. 2018), coagulation-flocculation (Mella et al. 2018), application of membranes (Peydayesh et al. 2018), and biological processes (Vikrant et al. 2018) have been applied for the abatement of pigments and dyes from water/wastewater. Nevertheless, all these techniques suffer from environmental and economic shortcomings in terms of excessive operating and capital financial requirement, excessive sludge generation, and complex operational procedures (Munagapati and Kim 2016). Adsorption has been considered as an environmentally benign and economically advantageous alternative for the abatement of pigments/dyes from aqueous solutions (Chen et al. 2018; Vikrant et al. 2018). Adsorption is favored primarily because of the costs involved in the preparation and procurement of adsorbents and their regeneration (Vikrant et al. 2018). In recent years microbial services has been proposed as potential tool as bioremediation and restoration of polluted soil and environment (Singh 2013, 2014, 2015, 2016; Singh Boudh 2016; Kumar et al. 2017; Kumar and Singh 2017; Tiwari and Singh 2017).

3.2 Biochar for the Removal of Dyes

Sorptive removal of dyes has attracted extensive interest from the scientific community owing to its easy operation, environmental benignity, cost-effectiveness, and great selectivity (Pham et al. 2017; Pham et al. 2018; Sophia and Lima 2018). Selection of a suitable adsorbent both in terms of cost and efficiency is paramount towards designing a sorption-based dye treatment system. At present, pristine and carbonaceous adsorbents are extracted from various agricultural/biological wastes such as coconut shell (Bello and Ahmad 2012), rice husk (Singh and Srivastava 2001), rice straws (El-Sonbati et al. 2016), almond shell (Doulati Ardejani et al. 2008), and wood char (Bangash Fazlullah and Manaf 2013) for the abatement of organic wastes from water/wastewater. In recent years, biochar (a solid carbonaceous substance synthesized through the pyrolysis of biomass under conditions of low oxygen) has attracted widespread scientific attention as a novel tool for the abatement of aqueous contaminants owing to its large surface area, stable carbon matrix, and high porosity (Thines et al. 2017; Vikrant et al. 2018). Biochars provide a win-win strategy as they utilize potentially unwanted biomass (e.g., agricultural waste) and provide excellent sorptive media for toxic pollutant removal (Vikrant et al. 2018).

The *Terminalia arjuna* belongs to the family of Combretaceae and is indigenous to the Indian subcontinent (Amalraj and Gopi 2017). Its bark decoction has been traditionally utilized for curing numerous disorders related to the cardiovascular system such as dyslipidemia, angina, hypertension, and congestive heart failure (Dwivedi and Chopra 2014; Amalraj and Gopi 2017). Moreover, the crude drug made from the bark of *T. arjuna* possesses antioxidant, hypolipidemic, anti-ischemic, and anti-atherogenic activities (Dwivedi and Chopra 2014). However, the fruit of *T. arjuna* remains unused in the pharmaceutical industry (Amalraj and Gopi 2017).

In the presented experimental case study, Arjun fruit biochar (AFB) was synthesized from the fruit of *T. arjuna* and used as an adsorbent. The aim of the present investigation was to determine the feasibility of AFB as a sorbent for the abatement of CR from water. The effects of various operating parameters (e.g., pH of the solution, dose of biochar, contact time, initial concentration of CR, and temperature) on the adsorption behavior of CR and treatment performance were investigated. Kinetic, isothermal, and thermodynamic fits were utilized to understand the sorption phenomenon, underlying mechanisms, and equilibrium of the sorption process.

This case study was carried out to explore the possibility for a practical utilization of *T. arjuna* fruits which are commonly produced as waste biomass by the pharmaceutical industry (Amalraj and Gopi 2017). It is a known fact that the study site (i.e., Varanasi, India) has a dense cluster of local carpet industries that produce glut amounts of CR rich effluents (Abu Talha et al. 2018). In light of these environmental conditions, it was proposed that biochar should be produced from indigenous *T. arjuna* fruits and used subsequently in local dye remediation. Thus, based on this study, a practical solution was sought for the indigenous utilization of biowaste while at the same time pursuing a treatment for wastewater. Furthermore, the performance of the biochar was assessed at the original pH of the textile industry effluents. The results of this study were of enough relevance to showcase an economical and practical solution for the abatement of dye by biochar derived from biowaste. Such experimental studies clearly elucidate the practical application of biochar for CR removal.

3.3 Experimental Case Study – Materials and Methods

3.3.1 Preparation of Biochar

Arjun fruit were obtained from indigenous *T. arjuna* trees. A stainless-steel pyrolysis reactor (inside diameter, 75 mm; length, 1.10 m) was used for biochar production, in which the temperature was controlled by an electric heater. The Arjun fruit were cleansed with Milli-Q water and parched naturally for 10 days. The dried fruit were subsequently crushed followed by sieving (size of particles: 72 BSS mesh, i.e., 210 μm) and then 600 g of the fruit was pyrolyzed at 500 °C (heating rate: 5 °C/min) for 3 h. The pyrolyzed sample (i.e., biochar) was cleansed with hot Milli-Q water and then oven dried at 75 °C for 2 h (Oven Universal NSW-143). The obtained biochars were stored in airtight borosilicate glass vials (70 mL) and used for subsequent characterization and adsorption studies.

3.3.2 Characterization of the Biochar

The porosity and specific surface of AFB were analyzed *via* nitrogen adsorption/desorption isotherm at 77 K (the biochar sample was degassed before measurement *via* helium for 3 h at 553 K) through the BET methodology using a micrometer (TriStar II 3020 V1.03, USA). Scanning electron microscopy was utilized to survey the visual characteristics of the AFB surface (SEM, EVO 18 research SEM, Germany). The KBr pellet method was adopted to recognize the functionalities existing on the AFB samples through Fourier transform infrared (FTIR) spectroscopy (Thermo-Fisher FTIR analyzer Nicolet 5700, Japan) in the 400–4000 cm^{-1} range. The method of solid addition method was adopted to ascertain the point of zero charge (pH_{ZPC}) of the biochar sample (Cheng et al. 2015). Essentially, a series of solutions (0.1 g AFB dissolved in 100 mL 0.01 M NaCl solution) were prepared in 150-mL Erlenmeyer flasks. The initial pH (pH_i) of the solutions were regulated in the range of 2–12 by the addition of suitable amounts of 0.1 M NaOH and 0.1 M HCl. The solution pH was measured *via* a digital pH meter (Ion/pH meter metrohm model-691, USA) in two replicates. A rotary shaker was used to agitate the prepared solutions at 200 rpm and 298 K for 24 h. The resulting suspensions were subsequently filtered, and the final solution pH (pH_f) of the filtrates was determined upon the attainment of equilibrium. A plot between pH_f and pH_i was drawn to measure the pH_{ZPC} of AFB.

3.3.3 Dye Solution

CR (analytical grade, 99% pure) was procured from Sigma-Aldrich, India. The stock solution of CR (1000 mg/L) ($C_{32}H_{22}N_6Na_2O_6S_2$, MW: 696.66 g/mol) was made with Milli-Q water, whereas the desired concentrations for batch adsorption experiments were prepared as per requirement *via* successive dilutions. The concentration of dye was determined *via* a spectrophotometer (UV-Vis, HACH DR5000, Canada) at the maximum wavelength (λ_{max}) of 497 nm for the adsorption studies. The detection limit for CR dye was determined to be 0.049 ppm with a relative standard deviation (RSD) of 0.39% (or relative standard error of 0.22%).

3.3.4 Batch Adsorption Experiments

The sorption behavior of CR on AFB was investigated using batch mode experiments to study the effects of operating conditions (e.g., solution pH, biochar dosage, initial CR concentration, temperature, and contact time), implementing a single condition at a particular time. The values of these operating conditions were as follows: pH, 2–1; AFB dosage, 4–14 g/L; initial CR concentration, 20–100 mg/L; contact time, 30–480 min; and temperature, 30–60 °C. CR solutions (100 mL) with a predetermined quantity of AFB were added to Erlenmeyer flasks (125 mL). Standard experimental conditions of AFB dosage (w = 12 g/L), CR concentration (C_o = 50 mg/L), shaking speed (100 rpm), contact time (t = 3 h), temperature (30 ± 2 °C), and pH (7.0 ± 0.1) were utilized in all the batch experiments unless mentioned otherwise.

A constant temperature shaking system was used to equilibrate the AFB and CR suspensions. After the experiment was completed, the resultant solution samples were decanted and centrifuged for 30 min at 1832 × *g*. The supernatant collected from centrifuged samples were used to analyze the concentration of residual CR *via* a UV-Vis spectrophotometer (HACH DR5000, Canada) at the maximum wavelength (λ_{max}) of 497 nm. The efficiency of removal (R %) and sorption capacity (q) were determined accordingly:

$$R(\%) = \frac{(C_o - C_e)x100}{C_o} \tag{3.1}$$

$$q = \frac{(C_o - C_e) x V}{m} \tag{3.2}$$

where C_o (mg/L) is the initial concentration of CR, C_e (mg/L) is the equilibrium concentration of CR, V (L) is solution volume, and m (g) is the biochar dosage.

3.4 Experimental Case Study – Results and Discussion

3.4.1 Biochar Characterization and Dye Interaction

In Fig. 3.1a and b, the surface morphology of AFB was examined before and after adsorption, respectively using SEM micrograph images. Figure 3.1a shows that the surface morphology of the fresh AFB was relatively porous and irregular with a massive surface area for a surface interaction with CR. The SEM micrograph image of dye-loaded AFB shown in Fig. 3.1b confirmed that the surface became saturated with the dye (i.e., most pores covered with CR molecules) after adsorption. According to the characterization results of the biochar, the BET specific surface, average pore width, and net pore volume of Arjun fruit were 770.68 m^2/g, 2.89 nm, and 0.4 cm^3/g, respectively.

The FTIR spectra of AFB in Fig. 3.2a and b show various characteristic peaks. These peaks represent different functionalities present on the AFB surface. The wide band at 3387 cm^{-1} resembled the hydrogen bonded OH groups of alcohol and H-bonded N-H group (Amir et al. 2010). However, the band underwent a shift to 3385 cm^{-1} after adsorption of CR onto the AFB surface owing to the interaction of the O-H bond of AFB with the –N⁻ bond of CR. The band perceived at 2337 cm^{-1} belonged to the stretching vibrations of C=O bond of the CO_2 molecule, which was shifted to 2348 cm^{-1} with a simultaneous increase in the intensity of the peak, indi-

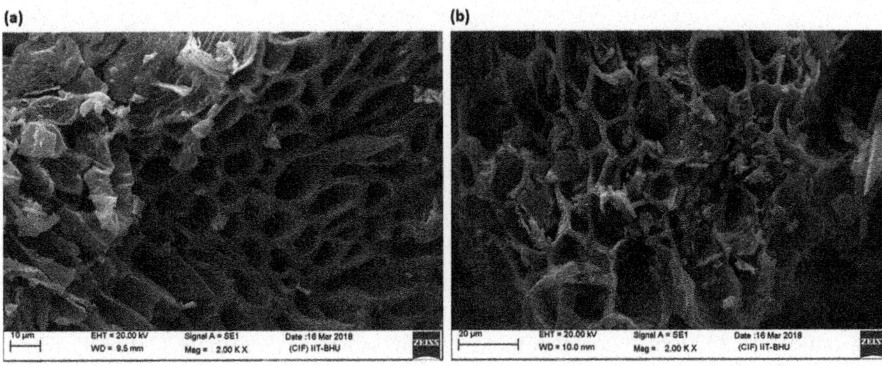

Fig. 3.1 Scanning electron microscopy images of Arjun fruit biochar. (**a**) Before and (**b**) after adsorption of Congo red

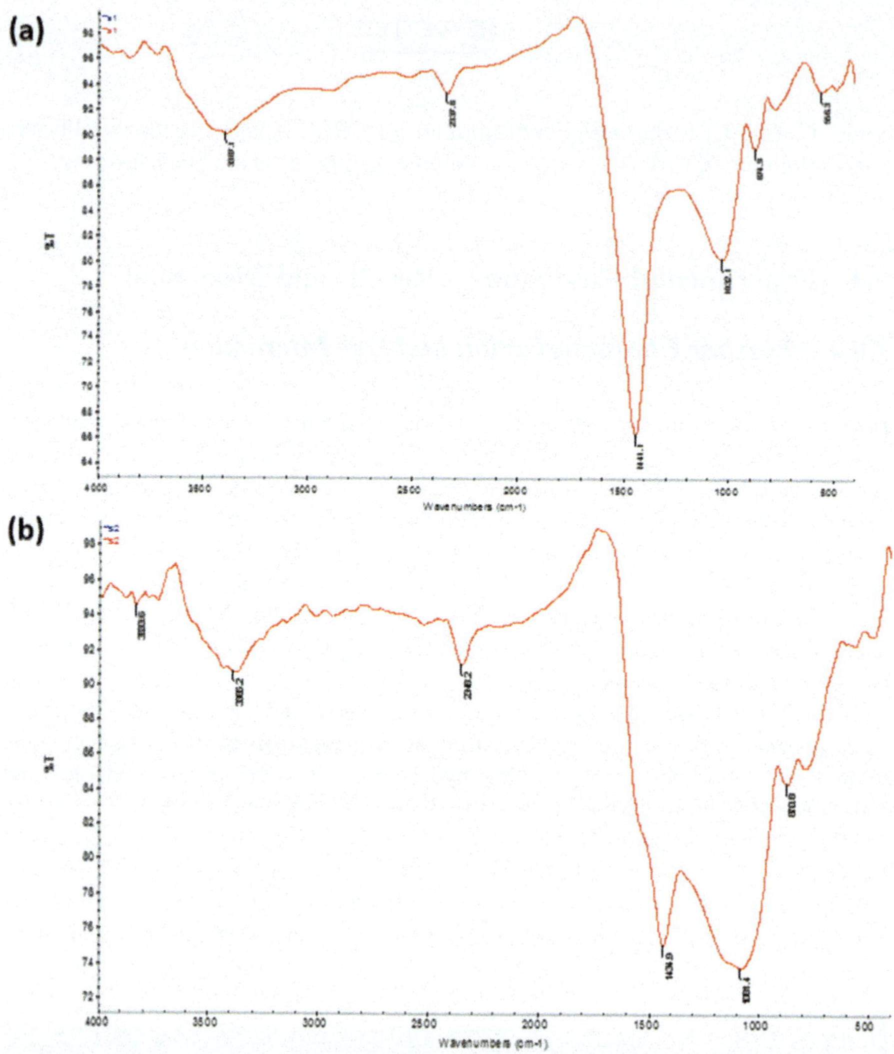

Fig. 3.2 Fourier-transform infrared spectroscopy spectra of Arjun fruit biochar before and after dye adsorption. (**a**) Before adsorption. (**b**) After adsorption

cating a rise in the CO_2 concentration after the sorption process (Pachecka et al. 2017). Another prominent band was perceived at 1441 cm^{-1} and belonged to the O-H of phenol, COO^- (Amir et al. 2010). This band also resulted in a shift to 1439 cm^{-1} because of the adsorption of the CR dye. The shift indicates the participation of the COO^- of AFB in the adsorption mechanism. Furthermore, the stretching band at 1332 cm^{-1} corresponded to the vibrations of the CO bond, which surprisingly shifted to 1081 cm^{-1} with a wider peak. This was due to an increase in the concentration of CO bond vibrations after the adsorption of CR. The absorption peaks at 874 cm^{-1} showcased the existence of a $[CO_3]$ functionality in the AFB, which decreased in concentration after adsorption as the peak reduced to 870 cm^{-1} (Fleet and Liu 2007). These interactions amongst the functionalities present in AFB and CR dye molecule could form weak van der Waals bonds or hydrogen bonds to improve the adsorption capacity and performance of AFB.

3.4.2 *Batch Adsorption Experiments*

3.4.2.1 Effects of Solution pH

The solution pH boosts the sorption of dye molecules on the biochar in multiple ways such as fluctuation in the biochar surface charge and dye molecule ionization. These two phenomena govern the mass transfer between the biochar surface and dye molecules. As a result, information on the optimum pH is imperative in precisely describing the adsorption process (Yu et al. 2018; Zazycki et al. 2018). The solution pH was regulated in the range of 2–12 through the addition of 0.1 M NaOH and 0.1 M HCl. The maximal efficiency of dye removal (96.3%) and sorption capacity of the sorbent (4.81 mg/g) were obtained at the solution pH of 2 ± 0.1 (Fig. 3.3).

The variation in the pH of aqueous solution was also observed after the sorption of CR on AFB (Fig. 3.4). As the initial CR solution pH (pH_i) fluctuated from 2 to 8, the final pH of CR solution at equilibrium (pH_f) was larger than the corresponding pH_i values. The point of zero charge (pH_{ZPC}) of AFB can be utilized to evaluate the impact of solution pH on CR sorption. As shown in Fig. 3.4, the pH_{ZPC} of AFB was 8.1. The surface of AFB is positively charged at pH values lower than pH_{ZPC}. The findings of high efficiency of dye removal and biochar sorption capacity may be ascribed to the prospective electrostatic attraction amongst CR anions and the posi-

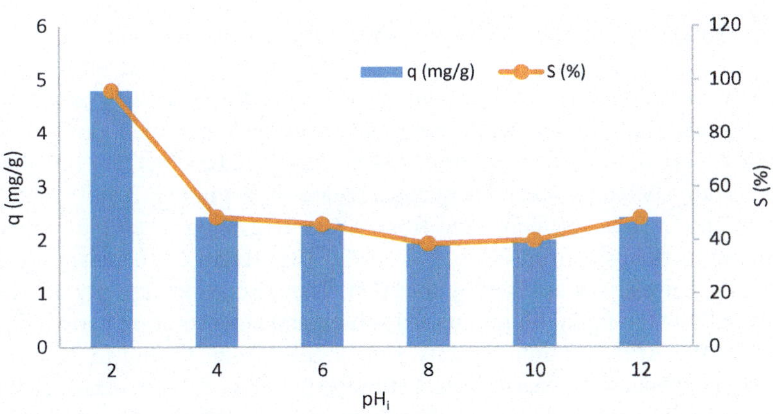

Fig. 3.3 Effect of solution pH on the adsorption of Congo red on Arjun fruit biochar (T = 30 ± 2 °C, C_o = 50 mg/L, w = 10 g/L, t = 3 h, and agitation speed = 100 rpm)

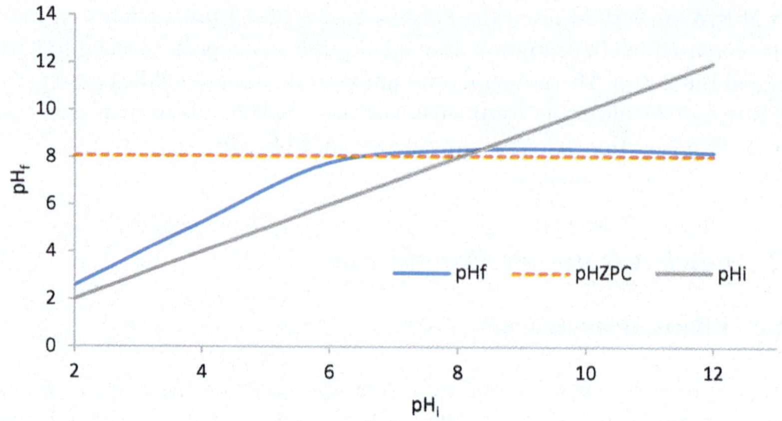

Fig. 3.4 Point of zero charge of Arjun fruit biochar (T = 30 ± 2 °C, C_o = 50 mg/L, w = 10 g/L, t = 3 h, and agitation speed = 100 rpm)

tively charged AFB surface. The adsorption of CR by orange peel powder was observed to be maximum at a pH of 3 (Munagapati and Kim 2016). The adsorbent surface acquired a positive charge at the low pH where the elevated adsorption of CR anions takes place. The pH of industrial effluents rich in CR is generally reported in the range from 7 to 11 (Gharbani et al. 2008). Moreover, as pH adjustments require a significant amount of chemicals, it may render the treatment process highly uneconomical (Jiang et al. 2018; Molinos-Senante and Guzmán 2018). Considering all the above-mentioned factors, the subsequent experiments were operated at a pH of 7 ± 0.1. As this pH value is lower than pH_{ZPC}, a positively charged biochar surface is made to provide a practical and economical solution for the abatement of dyes from water/wastewater.

3.4.2.2 Effects of Contact Time and Initial Dye Concentration

The initial concentration of a dye solution and contact time are key parameters for designing adsorption-based wastewater treatment systems (Oladoja et al. 2017). The effects of initial dye concentration (20–100 mg/L) and contact time (30–480 min) on the sorption behavior and sorbent capacity are shown in Fig. 3.5.

The sorption capacity increassed from 0.8 to 1.72 mg/g when the primary CR concentration changed from 20 to 100 mg/L. Two stages of the adsorption were observed when CR was uptaken by the AFB. The adsorption capacity of AFB elevated rapidly in the initial 30 min with a subsequent slow sorption period, throughout which the AFB sorption capacity slowly rises from 30 min to 120 min. No substantial enhancement was observed after the increase. This two-stage adsorption (rapid adsorption followed by slow adsorption) was also observed for the adsorption of brilliant green dye on biochar produced from hydrolyzed rice straw (Saif Ur

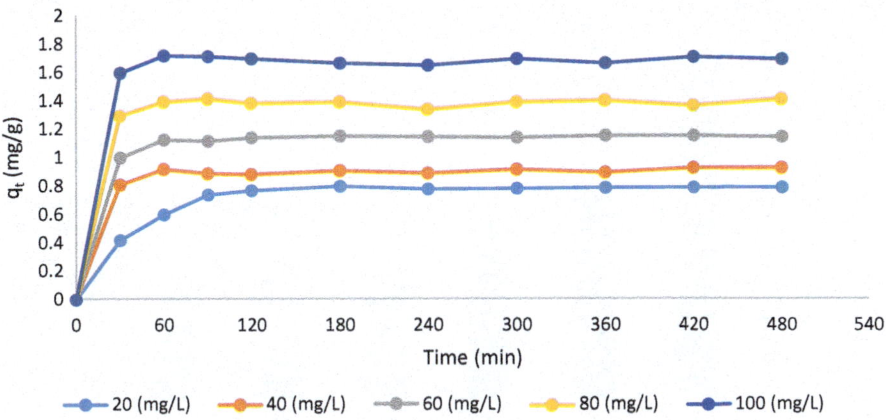

Fig. 3.5 Effect of contact time on the adsorption of Congo red on Arjun fruit biochar (T = 30 ± 2 °C, pH 7.0 ± 0.1, w = 12 g/L, agitation speed = 100 rpm)

Rehman et al. 2016). At low initial concentrations, few CR molecules cover the external active sites of the AFB surface through boundary layer adsorption. However, most external active sites remain uncovered owing to the low availability of CR molecules in the solution. Conversely, at high initial concentrations, a significant amount of available external active sites decreased with time, and the CR molecules reached the internal sites of the AFB through diffusion. Therefore, a slow adsorption process was seen in the second stage until equilibrium was achieved (Oladipo and Ifebajo 2018; Yu et al. 2018).

3.4.2.3 Effects of Biochar Dosage

The effects of the AFB dosage on the adsorption performance are shown in Fig. 3.6. The AFB dosage varied from 4 to 14 g/L. An opposite trend was observed for adsorption efficiency and capacity with an increase in the AFB dosage. The removal efficiency rose from 32.5% to 40%, while the sorption capacity was observed to decrease from 4.07 to 1.29 mg/g. The adsorption capacity of rice straw biochar for brilliant green reduced from 277 mg/g to 20 mg/g when the biochar dosage increased from 0.05 g/L to 1.25 g/L (Saif Ur Rehman et al. 2016). The removal efficiency of kenaf fiber biochar for methylene blue removal significantly increased from 48% to 76% when the biochar dosage was varied from 0.2 to 0.5 g/L (Mahmoud et al. 2012). Equations 3.1 and 3.2 can be used effectively to describe the observed opposite trends. The net amount of CR molecules was fixed (C_0 = 50 mg/L) against an increasing AFB dosage (4–14 g/L). As a consequence, the number of CR molecules adsorbed per unit mass of AFB was reduced resulting in a low adsorption capacity; hence, a large number of active sites were left uncovered at high AFB dosage (Sadaf and Bhatti 2014). An elevation in the AFB dosage gradually increased the CR

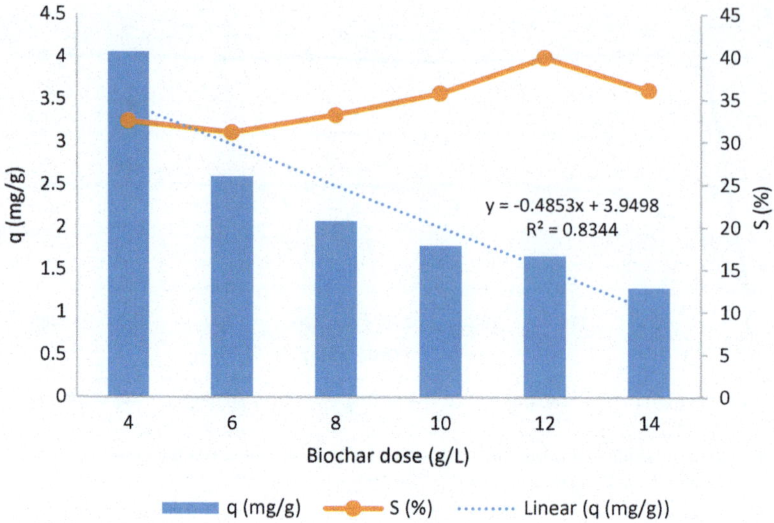

Fig. 3.6 Effect of biochar dosage on the adsorption of Congo red on Arjun fruit biochar (T = 30 ± 2 °C, pH 7.0 ± 0.1, C_o = 50 mg/L, t = 3 h, agitation speed = 100 rpm)

uptake (C_o-C_e) and consequently elevated the removal efficiency (Eq. 3.1). However, q_e was lower than R (%) because the AFB dosage (w) was present in the denominator of Eq. 3.2, resulting in an inverse relation between q and w (Zazycki et al. 2018). Subsequent experiments were conducted using 12 g/L of AFB to achieve the highest CR removal.

3.4.2.4 Effects of Temperature

Temperature plays a detrimental role in adsorption because the adsorption capacity at equilibrium is highly dependent on temperature fluctuations (Vikrant et al. 2018). The AFB adsorption capacity increased from 1.66 mg/g to 2.16 mg/g when the temperature rose from 30 °C to 60 °C, suggesting the endothermic character of the sorption process (Fig. 3.7).

The elevated sorption capacity could be attributed to the temperature-mediated increase in the amount of active sites on the AFB surface (Santos et al. 2017). Also, this phenomenon could be attributed to the increased mobility of CR molecules coupled with their heightened affinity toward the binding sites that are available on the AFB surface (Zhang et al. 2017). A similar endothermic adsorption of CR on cationic modified orange peel powder was reported previously (Munagapati and Kim 2016); a relatable endothermic character was also perceived for the sorptive removal of methylene blue by acid treated kenaf fiber char (Mahmoud et al. 2012).

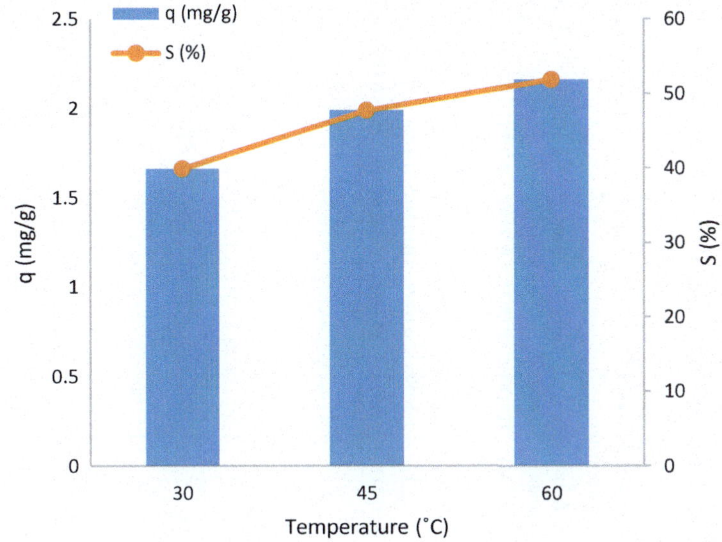

Fig. 3.7 Effect of temperature on the adsorption of Congo red on Arjun fruit biochar (pH 7.0 ± 0.1, C_o = 50 mg/L, t = 3 h, w = 12 g/L, agitation speed = 100 rpm)

3.4.3 Adsorption Kinetics

Adsorption kinetic investigations supplies into adsorption mechanisms. Commonly used models, namely Lagergren pseudo-first order (Eq. 3.3), pseudo-second order (Eq. 3.4), and intra-particle diffusion (Eq. 3.5) models, were used to analyze the prospective adsorption mechanisms (Vikrant et al. 2018).

$$\log\left(q_e - q_t\right) = \log q_e - \frac{k_1 t}{2.303} \tag{3.3}$$

$$\frac{t}{q_t} = \frac{1}{\left(k_2\right)\left(q_e\right)^2} + \frac{t}{q_e} \tag{3.4}$$

$$q_t = K_{pi} t^{\frac{1}{2}} + C_i \tag{3.5}$$

where q_t is the sorption capacity (mg/g) at any instant t, k_1 (min^{-1}) is the first order sorption rate constant, k_2 ($g.(mg.min)^{-1}$) is the second order sorption rate constant, $h = k_2 q_e^2$ depicts the initial sorption rate ($mg.(g.min)^{-1}$), C_i (mg/g) represents the thickness of boundary layer, and K_{pi} ($mg.(g.min)^{-1/2}$) denotes the intra-particle diffusion rate constant. Table 3.1A summarizes the kinetic parameters for CR adsorption on AFB. The pseudo-first order kinetics did not fit well for the obtained batch

Table 3.1 Adsorption parameters of Congo red dye on Arjun fruit biochar. (A) Kinetics, (B) isotherm, and (C) thermodynamics

[A] Kinetics														
			Pseudo-first order parameter					Pseudo-second order parameter			Intra-particle diffusion parameter			
Order	Dye Conc. (mg/L)	q_{exp} (mg/g)	k_1 (h^{-1})	q_{cal} (mg/g)	R^2	q_{DEV} (mg/g)		k_2 (g/mgh)	q_{cal} (mg/g)	R^2	q_{DEV} (mg/g)	K_{pi}	C_i	R^2
1	20	0.77	0.00852	0.2652	0.8045	0.5048		0.1041	0.8146	0.9977	0.0446	0.0301	0.2743	0.6841
2	40	0.884	0.00806	0.1418	0.5668	0.7422		0.2912	0.9289	0.9994	0.0449	0.0274	0.465	0.4697
3	60	1.135	0.00783	0.0833	0.3449	1.0517		0.4567	1.1487	0.9999	0.0137	0.0344	0.5782	0.4731
4	80	1.39	0.00898	0.2113	0.5154	1.1787		0.527	1.4029	0.9993	0.0129	0.0399	0.7479	0.4209
5	100	1.698	0.00484	0.1211	0.2443	1.5769		0.7115	1.6955	0.9997	0.0025	0.0475	0.9197	0.4068
[B] Isotherm														
Order	Langmuir					Freundlich								
	R^2	q_m (mg/g)	K_a (L/mg)	R_L		K_f (mg/g)	n	R^2						
1	0.998	2.534	0.411	0.023–0.108		16.683	0.413	0.257						
[C] Thermodynamics														
Order	Temperature (°C)	- ΔG (kJ Mol^{-1})	ΔH (kJ Mol^{-1})	ΔS (kJ Mol^{-1})										
1	30	3.161	5.886	0.009										
2	45	2.951												
3	60	2.893												

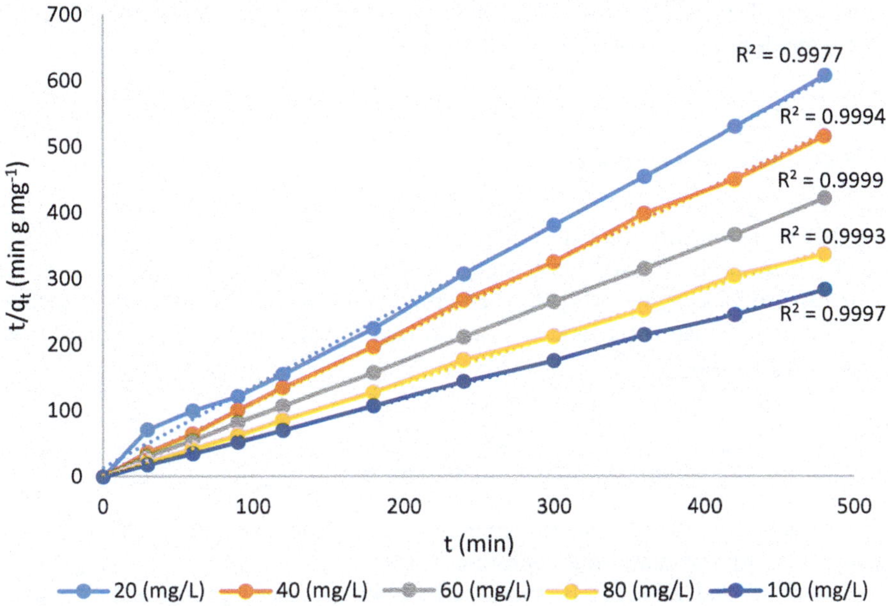

Fig. 3.8 Pseudo-second order kinetic model of Congo red on Arjun fruit biochar (T = 30 ± 2 °C, pH 7.0 ± 0.1, C_o = 20–100 mg/L, t = 3 h, w = 12 g/L, agitation speed = 100 rpm)

adsorption data because the R^2 values ranged between 0.2443 and 0.8045 along with a large absolute deviation of the experimental adsorption capacity (q_{DEV}) (0.5048–1.5769 mg/g), whereas the pseudo-second order kinetics fits the adsorption data exceptionally well ($R^2 > 0.9977$) (Fig. 3.8). The q_{DEV} values were obtained by using Eq. 3.6.

$$q_{DEV} = \left| q_{e,\exp} - q_{e,cal} \right| \tag{3.6}$$

Where $q_{e,exp}$ represents the experimental sorption capacity (mg/g) and $q_{e,cal}$ denotes the calculated sorption capacity (mg/g). High correlation coefficients (R^2) indicate that the pseudo-second order model shows a better fit than other models to explain the sorption of CR on AFB. Along similar lines, the pseudo-second order model fitted the sorption of CR onto cationic modified orange peel powder (Munagapati and Kim 2016), ZnO-modified SiO_2 nanospheres (Zhang et al. 2018), and chestnut husk-like $NiCo_2O_4$ hollow microspheres (Chen et al. 2018) with high accuracy.

For a better understanding of sorption kinetics, the intra-particle diffusion model was also fitted to the kinetics data (Table 3.1A). The R^2 values for the intra-particle diffusion kinetic model were in the range of 0.4068–0.6841, indicating its non-applicability. The C_i values (0.2743–0.9197) indicate the contribution of pore diffusion to a limited extent toward adsorption of CR on AFB. However, these values may not be the only rate-determining stage involved in the sorption activity (e.g., rapid film adsorption may also be involved) (Mane et al. 2007; Mane and Babu

2011). Similar observations were also reported for the abatement of brilliant green dye by hydrolyzed rice straw (Saif Ur Rehman et al. 2016).

3.4.4 Adsorption Isotherms

Investigations on sorption equilibrium are necessary to learn how to optimize and scale up adsorption-based wastewater treatment systems (Molinos-Senante and Guzmán 2018). Langmuir (Eq. 3.7) and Freundlich (see supplementary information and Table 3.1B) isotherm models were utilized to analyze the data obtained from batch experiments.

$$\frac{C_e}{q_e} = \frac{C_e}{q_m} + \frac{1}{K_a q_m} \tag{3.7}$$

Where q_m represents the maximum monolayer sorption capacity and K_a is the Langmuir sorption equilibrium constant (L/mg).

Figure 3.9 and Table 3.1(B) show the Langmuir isotherm profile and associated parameters, respectively. The very high R^2 value (0.9985) indicates that Langmuir model should be more fitting than various other models to analyze the sorption of CR at equilibrium condition. A separation factor (R_L) value in the range of 0 to 1 also indicates the feasibility of adsorption of CR onto AFB (Eq. 3.8) (Saif Ur Rehman et al. 2016).

$$R_L = \frac{1}{\left(1 + K_a C_o\right)} \tag{3.8}$$

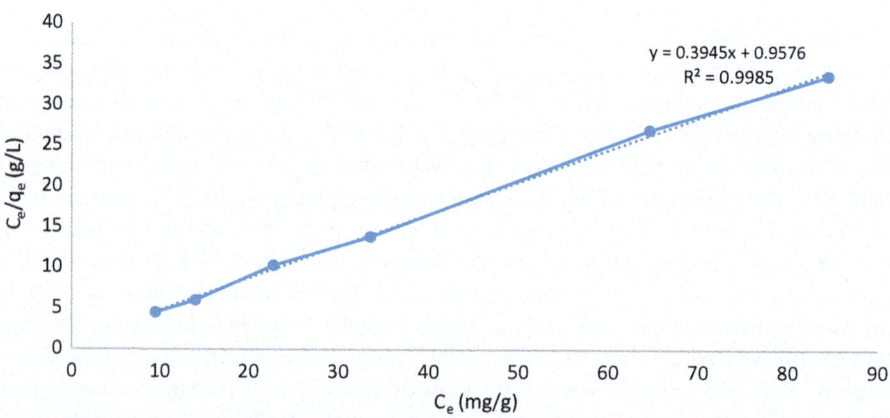

Fig. 3.9 Langmuir isotherm model fitted to the data of Congo red adsorption on Arjun fruit bio-char (T = 30 ± 2 °C, pH 7.0 ± 0.1, C_o = 20–100 mg/L, t = 3 h, w = 12 g/L, agitation speed = 100 rpm)

Table 3.2 Comparison between different biomass-based adsorbents for the removal of Congo red dye

Order	Adsorbent	Max. Adsorption efficiency (%)	Max. Adsorption capacity (mg/g)	pH	References
1	Jute stick powder	–	35.7	7	Panda et al. (2009)
2	Sugarcane bagasse	89.9	38.2	5	Zhang et al. (2011)
3	Rubber seeds	98.4	9.82	6	Zulfikar et al. (2015)
4	Modified orange peel powder	–	163	3	Munagapati and Kim (2016)
5	Arjun fruit biochar	96.25	4.81	2	This study

The calculated R_L values were observed to be in the range of 0.023–0.108 for all tested concentrations, thus confirming that CR could be successfully removed by adsorption on AFB. The Langmuir isotherm was also a good fit for CR adsorption onto cationic modified orange peel powder (Munagapati and Kim 2016) and chestnut husk-like $NiCo_2O_4$ hollow microspheres (Chen et al. 2018). These findings suggest that a monolayer adsorption was prevalent. The formation of a monolayer can be attested by calculating the coverage factor (θ), which is given by Eq. 3.9.

$$\theta = \frac{\left[q_m N\sigma 10^{-20}\right]}{S_{BET}} \tag{3.9}$$

The value of θ varies between 0 (no coverage) to 1 (full coverage). The θ value was 0.68, which confirms the formation of a monolayer. The partial coverage of the AFB surface occurred because CR is a large sized molecule, which can obstruct its movement onto the micropores (Al-Degs et al. 2008). A comparison between the performance of processes from different studies on the adsorptive removal of CR is summarized in Table 3.2.

3.5 Adsorption Thermodynamics

The following equations were utilized to analyze the thermodynamics of CR sorption onto AFB (Munagapati and Kim 2016):

$$\Delta G = -RTlnK \tag{3.10}$$

$$lnK = \frac{\Delta S}{R} - \frac{\Delta H}{RT} \tag{3.11}$$

where R is the ideal gas constant (8.314 J mol^{-1} K^{-1}), T (K) is the absolute temperature, K = q_e/C_e, ΔG is the Gibbs free energy, ΔS is the entropy change, and ΔH is the enthalpy change. The values of ΔS and ΔH were calculated from the intercept and slope of lnK vs. 1/T (Saif Ur Rehman et al. 2016). ΔG values were calculated using Eq. 3.10. The feasibility and spontaneity of CR adsorption on AFB were affirmed by the negative values of ΔG at all tested temperatures (30–60 °C) (Table 3.1C) (Mane et al. 2007). The ΔG value continuously decreased for a rise in the solution temperature, thereby indicating the endothermic tendency of the sorption process as elucidated in Sect. 3.4.2.4. The values of ΔG were observed to be in the −20–0 kJ mol^{-1} range, suggesting that CR sorption on AFB is a physically driven process (Zhang et al. 2018).

The positive value of ΔH (i.e., +5.886 kJ mol^{-1}) confirms the endothermic character of the adsorption process. ΔH values can be used to understand the physical adsorption of analytes on adsorbents such as 5 kJ mol^{-1} (hydrophobic bonding), 4–10 kJ mol^{-1} (van der Waals forces), 2–40 kJ mol^{-1} (hydrogen bonding), 2–29 kJ mol^{-1} (dipole bonding), and 40 kJ mol^{-1} (coordination exchange) (Saif Ur Rehman et al. 2016). Based on this classification, the ΔH value of CR adsorption on AFB affirms its physical nature and suggests there is a complex combination of van der Waals, hydrophobic, dipole, and hydrogen bonding forces. The positive ΔS value (+0.009 kJ mol^{-1}) indicates that a temperature increase should have promoted CR adsorption by dislocating the water molecules present on the biochar surface (Auta and Hameed 2012). The adsorption of CR on cationic modified orange peel powder was also spontaneous and endothermic (Munagapati and Kim 2016). However, the thermodynamic analysis of CR adsorption onto ball-milled sugarcane bagasse implies that the process is spontaneous and exothermic (Zhang et al. 2011). Interestingly, the sorption of CR onto rubber seeds was observed to be endothermic and non-spontaneous (Zulfikar et al. 2015). It can be perceived that the sorption thermodynamic results are indispensable toward the detailed understanding of sorption behavior as well as investigating the spontaneity of the process.

3.5.1 Desorption and Regeneration Experiments

Desorption studies showcase the regenerative ability of adsorbents and the potential of recovering dye molecules. The selection of proper eluents is important for a successful desorption operation. The choice of eluents is typically based upon the type of adsorption mechanism and the nature of the adsorbent. For the desorption experiment, various eluents were prepared such as 0.1 M NaOH, 0.1 M CH_3COOH, deionized water, and 0.1 M HCl. Figure 3.10a presents the results of desorption investigations. The superior eluent, when compared to three other eluents, was noted to be 0.1 M NaOH, which displayed a CR recovery of 86%.

The regeneration of biochar is favorable from a financial perspective and for the recovery of dye molecules. Based on the results of Fig. 3.10a, 0.1 M NaOH was utilized as the eluent for regeneration cycle experiments (Fig. 3.10b). A progressive reduction in CR sorption capacity was observed for AFB with a rise in the number of regeneration cycles. After the subsequent five cycles, the CR desorption capacity of AFB decreased from 86% to 79%. The decrease in the sorption capacity was

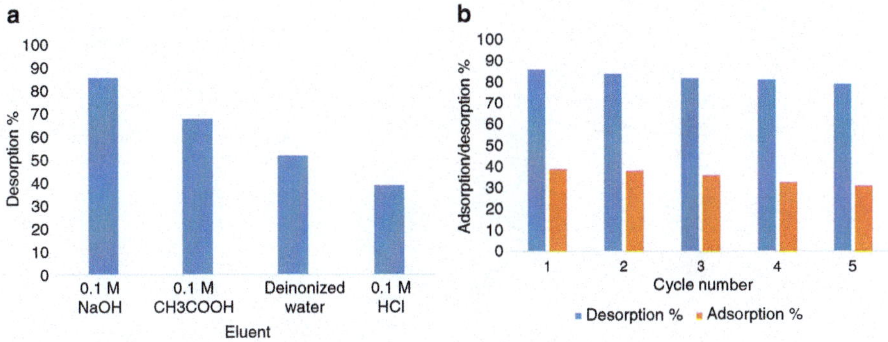

Fig. 3.10 Desorption studies of Congo red. (**a**) Desorption; (**b**) Number of adsorption-desorption cycles

found to be less than 10% which might be a consequence of biochar loss during the regeneration process. As a result, AFB can be used for at least five cycles of CR adsorption effectively.

3.6 Adsorption Mechanisms

The adsorption of CR onto the surface of AFB can be mainly explained by three forces, namely electrostatic, Van der Waals, and hydrogen bonding. Observed two probable mechanisms for adsorption of CR onto coir pith carbon, including (1) electrostatic adsorption owing to the interactions amongst the positive charge bearing protonated functionalities of carbon and the negatively charged deprotonated acidic dye, and (2) the physical adsorption or chemical reaction amongst the adsorbate and adsorbent. Analogous adsorption processes might also occur during the adsorption of CR onto AFB.

Figure 3.11 shows the adsorption mechanism of CR onto AFB in solutions of different pH (acidic, neutral, and alkaline). In the solution with a pH below pH_{ZPC}, the surface of the AFB was positively charged. This highly favors the electrostatic interactions amongst the AFB surface (bearing positive charge) and the anionic CR molecules with SO_3^- as a functional group. The FTIR analysis of the AFB discloses the presence of acidic functionalities (e.g., hydroxyls and carboxyls), which should primarily furnish the electrostatic interactions amongst the SO_3^- group of CR and the functional groups of AFB at acidic conditions. When the solution pH changed from acidic to alkaline, the number of positively charged sites of AFB decreased. The shift in solution pH resulted in an increase of negatively charged surface sites. This phenomenon resulted in the repulsion of the AFB surface and anionic dye. However, Fig. 3.3 shows outstanding removal efficiency and maximum adsorption capacity at a high pH. Hence, the adsorption can be physically controlled when the solution is alkaline. This physisorption process was also supported by the ΔG val-

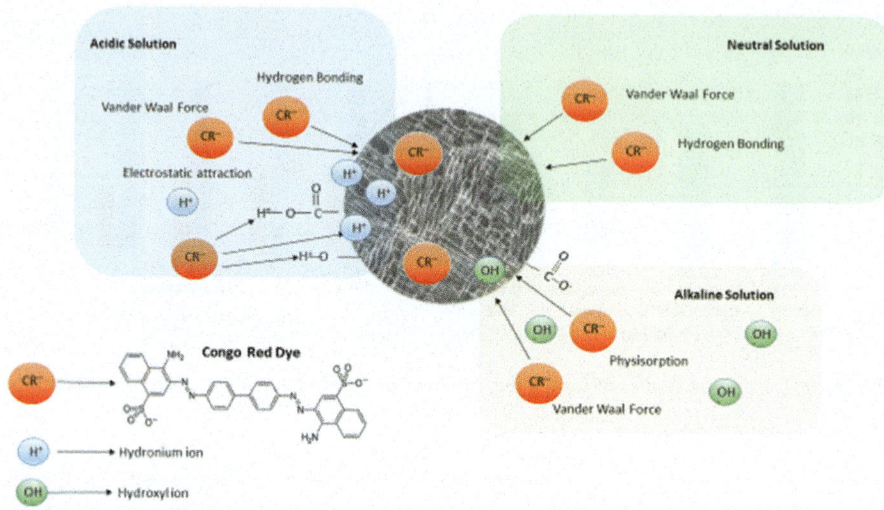

Fig. 3.11 Proposed adsorption mechanism for Congo red dye on the Arjun fruit biochar surface at different pH conditions

ues obtained in Sect. 3.6. The FTIR data also demonstrate a change in the band from 3387 cm^{-1} to 3385 cm^{-1} after the sorption of CR onto AFB due to the interaction of the O-H bond of AFB with the $-N^{-}$ bond of CR. Another band at 1441 cm^{-1} underwent a shift to 1439 cm^{-1}, showing the participation of the COO^{-} group of AFB. The drastic change in the peak from 1032 cm^{-1} to 1081 cm^{-1} clearly explains the adsorption of CR onto the AFB surface. The shift in the peaks of FTIR data is likely to reflect the formation of new bonds for weak van der Waals forces and hydrogen bonding between the functional groups of the AFB and CR dye molecule. Also, the transfer of the CR molecules from the bulk of aqueous solution to the AFB surface might contain these four steps: (1) transport of CR molecule from the bulk aqueous solution to the AFB surface, (2) transfer of CR molecules to the AFB surface through diffusion across the boundary layer, (3) adsorption at the surface sites of the sorbent, and (4) intra-particular diffusion through the surface of the sorbent (Sen et al. 2011).

3.7 Conclusions

Till date, varied physicochemical treatment options have been designed for the removal of dyes. However, the performance of these conventional systems is limited due to high cost, large energy and chemical requirements, generation of sludge and toxic byproducts. This chapter acknowledges adsorption as an apt option for the removal of dyes, due to their advantageous properties such as environmental benignity, economical operation, easy and safe operation, and no sludge generation. The experimental case study showcased the applicability of a biochar derived from the fruit of *T. arjuna* as an efficacious sorbent for the abatement dyes from water/wastewater. The maximum removal efficiency and adsorption capacity of CR by AFB were 96.25% and 4.81 mg/g, respectively. The kinetic, equilibrium, and thermodynamic analyses of the adsorption process revealed that CR sorption on AFB is a spontaneous and feasible method that is physical in nature. The reusability of biochar was reliable with minimal loss of adsorption capacity when tested over five regeneration cycles. The experimental observations of the present investigation showcased that the fruit of *T. arjuna*, which remains largely unused by the pharmaceutical industry, can be used in treatment facilities for the efficient removal of pigments and dyes from water/wastewater. Such innovative usage of indigenously produced waste biomass holds a great potential for sustainable waste management as well as a cost-effective pollution control process.

Acknowledgements This study was supported by a grant from the National Research Foundation of Korea (NRF) funded by the Ministry of Science, ICT, & Future Planning (No. 2016R1E1A1A01940995). The authors also acknowledge Project Varanasi and the Department of Chemical Engineering and Technology, IIT (BHU) Varanasi for providing the necessary facilities for conducting this research.

References

Abu Talha M, Goswami M, Giri BS, Sharma A, Rai BN, Singh RS (2018) Bioremediation of Congo red dye in immobilized batch and continuous packed bed bioreactor by Brevibacillus parabrevis using coconut shell bio-char. Bioresour Technol 252:37–43

Al-Degs YS, El-Barghouthi MI, El-Sheikh AH, Walker GM (2008) Effect of solution pH, ionic strength, and temperature on adsorption behavior of reactive dyes on activated carbon. Dyes Pigments 77:16–23

Amalraj A, Gopi S (2017) Medicinal properties of Terminalia arjuna (Roxb.) Wight & Arn.: a review. J Tradit Complement Med 7:65–78

Amir S, Jouraiphy A, Meddich A, El Gharous M, Winterton P, Hafidi M (2010) Structural study of humic acids during composting of activated sludge-green waste: elemental analysis, FTIR and 13C NMR. J Hazard Mater 177:524–529

Auta M, Hameed BH (2012) Modified mesoporous clay adsorbent for adsorption isotherm and kinetics of methylene blue. Chem Eng J 198-199:219–227

Bangash Fazlullah K, Manaf A (2013) Dyes removal from aqueous solution using wood activated charcoal of Bombax Cieba tree. J Chin Chem Soc 52:489–494

Bello OS, Ahmad MA (2012) Coconut (Cocos nucifera) Shell based activated carbon for the removal of malachite green dye from aqueous solutions. Separation Sci Technol 47:903–912

Brüschweiler BJ, Merlot C (2017) Azo dyes in clothing textiles can be cleaved into a series of mutagenic aromatic amines which are not regulated yet. Regul Toxicol Pharmacol 88:214–226

Chen H, Zheng Y, Cheng B, Yu J, Jiang C (2018) Chestnut husk-like nickel cobaltite hollow microspheres for the adsorption of Congo red. J Alloys Comp 735:1041–1051

Cheng Z, Zhang L, Guo X, Jiang X, Li T (2015) Adsorption behavior of direct red 80 and congo red onto activated carbon/surfactant: process optimization, kinetics and equilibrium. Spectrochim Acta A Mol Biomol Spectrosc 137:1126–1143

Chung KT (2016) Azo dyes and human health: a review. J Environ Sci Health C Environ Carcinog Ecotoxicol Rev 34:233–261

D'Inverno G, Carosi L, Romano G, Guerrini A (2018) Water pollution in wastewater treatment plants: an efficiency analysis with undesirable output. Eur J Oper Res 269:24–34

Doulati Ardejani F, Badii K, Limaee NY, Shafaei SZ, Mirhabibi AR (2008) Adsorption of Direct Red 80 dye from aqueous solution onto almond shells: effect of pH, initial concentration and shell type. J Hazard Mater 151:730–737

Dwivedi S, Chopra D (2014) Revisiting Terminalia arjuna – An Ancient Cardiovascular Drug. J Tradit Complment Med 4:224–231

El-Sonbati AZ, El-Deen IM, El-Bindary MA (2016) Adsorption of hazardous Azorhodanine dye from an aqueous solution using Rice Straw Fly Ash. J Dispers Sci Technol 37:715–722

Fleet ME, Liu X (2007) Coupled substitution of type A and B carbonate in sodium-bearing apatite. Biomaterials 28:916–926

Gharbani P, Tabatabaii SM, Mehrizad A (2008) Removal of Congo red from textile wastewater by ozonation. Int J Environ Sci Technol 5:495–500

Jayapal M, Jagadeesan H, Shanmugam M, Danisha JP, Murugesan S (2018) Sequential anaerobic-aerobic treatment using plant microbe integrated system for degradation of azo dyes and their aromatic amines by-products. J Hazard Mater 354:231–243

Jiang Y, Dinar A, Hellegers P (2018) Economics of social trade-off: balancing wastewater treatment cost and ecosystem damage. J Environ Manag 211:42–52

Kumar A, Singh JS (2017) Cyanoremediation: a green-clean tool for decontamination of synthetic pesticides from agro-and aquatic ecosystems. In: Singh JS, Seneviratne G (eds) Agro-environmental sustainability: Vol (2) managing environmental pollution. Springer, Cham, pp 59–83

Kumar A, Kaushal S, Saraf S, Singh JS (2017) Cyanobacterial biotechnology: an opportunity for sustainable industrial production. Clim Change Environ Sustain 5(1):97–110

Kumar A, Kaushal S, Saraf S, Singh JS (2018) Microbial bio-fuels: a solution to carbon emissions and energy crisis. Front Biosci (Landmark) 23:1789–1802

Leo L, Loong C, Ho XL, Raman MFB, Suan MYT, Loke WM (2018) Occurrence of azo food dyes and their effects on cellular inflammatory responses. Nutrition 46:36–40

Mahmoud DK, Salleh MAM, Karim WAWA, Idris A, Abidin ZZ (2012) Batch adsorption of basic dye using acid treated kenaf fibre char: equilibrium, kinetic and thermodynamic studies. Chem Eng J 181-182:449–457

Mane VS, Babu PVV (2011) Studies on the adsorption of Brilliant Green dye from aqueous solution onto low-cost NaOH treated saw dust. Desalination 273:321–329

Mane VS, Mall ID, Srivastava VC (2007) Use of bagasse fly ash as an adsorbent for the removal of brilliant green dye from aqueous solution. Dyes Pigments 73:269–278

Meerbergen K, Willems KA, Dewil R, Van Impe J, Appels L, Lievens B (2018) Isolation and screening of bacterial isolates from wastewater treatment plants to decolorize azo dyes. J Biosci Bioeng 125:448–456

Mekonnen Mesfin M, Hoekstra Arjen Y (2017) Global anthropogenic phosphorus loads to freshwater and associated Grey water footprints and water pollution levels: a high-resolution global study. Water Resour Res 54:345–358

Mella B, Barcellos BSC, da Silva Costa DE, Gutterres M (2018) Treatment of leather dyeing wastewater with associated process of coagulation-flocculation/adsorption/ozonation. Ozone Sci Eng 40:133–140

Molinos-Senante M, Guzmán C (2018) Benchmarking energy efficiency in drinking water treatment plants: quantification of potential savings. J Clean Prod 176:417–425

Munagapati VS, Kim D-S (2016) Adsorption of anionic azo dye Congo Red from aqueous solution by cationic modified Orange Peel powder. J Mol Liquids 220:540–548

Nidheesh PV, Zhou M, Oturan MA (2018) An overview on the removal of synthetic dyes from water by electrochemical advanced oxidation processes. Chemosphere 197:210–227

Oladipo AA, Ifebajo AO (2018) Highly efficient magnetic chicken bone biochar for removal of tetracycline and fluorescent dye from wastewater: two-stage adsorber analysis. J Environ Manag 209:9–16

Oladoja NA, Unuabonah EI, Amuda OS, Kolawole OM (2017) Operational principles and material requirements for coagulation/flocculation and adsorption-based water treatment operations. In: Oladoja NA, Unuabonah EI, Amuda OS, Kolawole OM (eds) Polysaccharides as a green and sustainable resources for water and wastewater treatment. Springer, Cham, pp 1–11

Pachecka M, Sturm JM, Lee CJ, Bijkerk F (2017) Adsorption and dissociation of CO2 on Ru(0001). J Phys ChemC 121:6729–6735

Panda GC, Das SK, Guha AK (2009) Jute stick powder as a potential biomass for the removal of congo red and rhodamine B from their aqueous solution. J Hazard Mater 164:374–379

Peydayesh M, Mohammadi T, Bakhtiari O (2018) Effective treatment of dye wastewater via positively charged TETA-MWCNT/PES hybrid nanofiltration membranes. Sep Purif Technol 194:488–502

Pham TD, Do TT, Ha VL, Doan THY, Nguyen TAH, Mai TD, Kobayashi M, Adachi Y (2017) Adsorptive removal of ammonium ion from aqueous solution using surfactant-modified alumina. Environ Chem 14:327–337

Pham T, Bui T, Nguyen V, Bui T, Tran T, Phan Q, Pham T, Hoang T (2018) Adsorption of polyelectrolyte onto Nanosilica Synthesized from Rice husk: characteristics, mechanisms, and application for antibiotic removal. Polymers 10:220

Sadaf S, Bhatti HN (2014) Batch and fixed bed column studies for the removal of Indosol Yellow BG dye by peanut husk. J Taiwan Inst Chem Eng 45:541–553

Saif Ur Rehman M, Kim I, Rashid N, Adeel Umer M, Sajid M, Han J-I (2016) Adsorption of brilliant green dye on biochar prepared from Lignocellulosic bioethanol plant waste. Clean (Weinh) 44:55–62

Santos RMMD, Gonçalves RGL, Constantino VRL, Santilli CV, Borges PD, Tronto J, Pinto FG (2017) Adsorption of acid yellow 42 dye on calcined layered double hydroxide: effect of time, concentration, pH and temperature. Appl Clay Sci 140:132–139

Sen TK, Afroze S, Ang HM (2011) Equilibrium, kinetics and mechanism of removal of methylene blue from aqueous solution by adsorption onto pine cone biomass of Pinus radiata. Water Air Soil Pollut 218:499–515

Singh JS (2013) Anticipated effects of climate change on methanotrophic methane oxidation. Clim Change Environ Sustain 1(1):20–24

Singh JS (2014) Cyanobacteria: a vital bio-agent in eco-restoration of degraded lands and sustainable agriculture. Clim Change Environ Sustain 2:133–137

Singh JS (2015) Biodiversity: current perspective. Clim Change Environ Sustain 3(1):71–72

Singh JS (2016) Microbes play major roles in ecosystem services. Clim Change Environ Sustain 3:163–167

Singh JS (ed) (2019) New and future developments in microbial biotechnology and bioengineering: microbes in soil, crop and environmental sustainability. Elsevier, Amsterdam

Singh JS, Boudh S (2016) Climate change resilient crops to sustain Indian agriculture. Clim Change Environ Sustain 5:97–110

Singh JS, Singh DP (eds) (2019) New and future developments in microbial biotechnology and bioengineering: microbial biotechnology in agro-environmental sustainability. Elsevier, Amsterdam

Singh DK, Srivastava B (2001) Basic dyes removal from wastewater by adsorption on rice husk carbon. Ind J Chem Technol 8:133–139

Singh C, Tiwari S, Boudh S, Singh JS (2017a) Biochar application in management of paddy crop production and methane mitigation. In: Singh JS, Seneviratne G (eds) Agro-environmental sustainability: Vol (2) managing environmental pollution. Springer, Cham, pp 123–145

Singh JS, Koushal S, Kumar A, Vimal SR, Gupta VK (2017b) Book review: microbial inoculants in sustainable agricultural productivity- Vol. II: Functional application. Front Microbiol 7:2015

Singh JS, Kumar A, Singh M (2019a) Cyanobacteria: a sustainable and commercial bioresource in production of bio-fertilizer and bio-fuel from waste waters. Environ Sustain Indic 3-4:100008

Singh MK, Rai PK, Rai A, Singh S, Singh JS (2019b) Poly-β-hydroxybutyrate production by the cyanobacterium Scytonema geitleri Bharadwaja under varying environmental conditions. Biomol Ther 9(198):1–10

Sophia AC, Lima EC (2018) Removal of emerging contaminants from the environment by adsorption. Ecotoxicol Environ Saf 150:1–17

Thines KR, Abdullah EC, Mubarak NM, Ruthiraan M (2017) Synthesis of magnetic biochar from agricultural waste biomass to enhancing route for waste water and polymer application: a review. Renew Sustain Energy Rev 67:257–276

Tiwari AP, Singh JS (2017) Plant growth promoting rhizospheric *Pseudomonas aeruginosa* strain inhibits seed germination in *Triticum aestivum* (L) and *Zea mays* (L). Microbiol Res 8(7233):73–79

Vikrant K, Giri BS, Raza N, Roy K, Kim KH, Rai BN, Singh RS (2018) Recent advancements in bioremediation of dye: current status and challenges. Bioresour Technol 253:355–367

Vimal SR, Singh JS (2019) Salt tolerant PGPR and FYM application in saline soil Paddy agriculture sustainability. Clim Change Environ Sustain 7(1):23–33

Vimal SR, Gupta J, Singh JS (2018) Effect of salt tolerant Bacillus sp. and *Pseudomonas* sp. on wheat (*Triticum aestivum* L.) growth under soil salinity: a comparative study. Microbiol Res 9(1):1–14

Yu J, Zhang X, Wang D, Li P (2018) Adsorption of methyl orange dye onto biochar adsorbent prepared from chicken manure. Water Sci Technol 77(5-6):1303–1312

Zazycki MA, Godinho M, Perondi D, Foletto EL, Collazzo GC, Dotto GL (2018) New biochar from pecan nutshells as an alternative adsorbent for removing reactive red 141 from aqueous solutions. J Clean Prod 171:57–65

Zhang Z, Moghaddam L, O'Hara IM, Doherty WOS (2011) Congo red adsorption by ball-milled sugarcane bagasse. Chem Eng J 178:122–128

Zhang B, Dong Z, Sun D, Wu T, Li Y (2017) Enhanced adsorption capacity of dyes by surfactant-modified layered double hydroxides from aqueous solution. J Indust Eng Chem 49:208–218

Zhang J, Yan X, Hu M, Hu X, Zhou M (2018) Adsorption of Congo red from aqueous solution using ZnO-modified SiO2 nanospheres with rough surfaces. J Mol Liquids 249:772–778

Zhao G, Li C, Wu X, Yu J, Jiang X, Hu W, Jiao F (2018) Reduced graphene oxide modified NiFe-calcinated layered double hydroxides for enhanced photocatalytic removal of methylene blue. Appl Surf Sci 434:251–259

Zulfikar MA, Setiyanto H, Rusnadi Solakhudin L (2015) Rubber seeds (Hevea brasiliensis): an adsorbent for adsorption of Congo red from aqueous solution. Desalin Water Treat 56:2976–2987

Chapter 4
Potential of Biochar for the Remediation of Heavy Metal Contaminated Soil

Amita Shakya and Tripti Agarwal

Abstract Soil contamination with heavy metals has become a serious concern. Due to their non degradable nature, speciation and bioavailability to the living organisms, heavy metals are prone to enter the food chain. Heavy metals are also known as trace elements because they are found in trace concentration into the environment (<10 ppm or ppb). Accumulation of heavy metals and metalloids such as As, Pb, Cd, Hg, Se, Cd in soil is an issue of growing concern due to their lethal, carcinogenic, mutagenic effects, and multiple organ dysfunction in the living organisms. However some heavy metals like Cr, Zn, Bo, Cu, Co, Fe, Mn and Ni are necessary in trace amount for different biochemical functions of animal and plant system. Rapid urbanization and increasing industrialization leads to the accumulation of heavy metals into the soil. Contaminated soil can causes many physiochemical and biochemical changes into plants resulting in reduced growth, affecting the yield which eventually leads to food insecurity. Many strategies like isolation, immobilization, extraction, phytoremediation, soil washing, etc. are used for remediation of heavy metal contaminated soil.

Biochar has been applied to soil as a novel carbon rich material for heavy metal remediation. Biochar is a solid product with porous structure, obtained when biomass is thermo-chemically treated in a closed container in oxygen-limited environment. Physicochemical properties of biochar made it a potential candidate for long term carbon storage. Low production cost, high cation exchange capacity, pH, surface functional groups and porous structures are some of the intrinsic properties of biochar making it a choice as an adsorbent for heavy metal remediation from soil. Biochar has capacity to make complex with heavy metals present in soil, which reduce their bioavailability.

Keywords Biochar · Pyrolysis · Remediation · Contaminated soil · Heavy metal

---

A. Shakya · T. Agarwal (✉)
Department of Agriculture and Environmental Sciences, National Institute of Food Technology Entrepreneurship and Management, Kundli, Sonipat, Haryana, India

J. S. Singh, C. Singh (eds.), *Biochar Applications in Agriculture and Environment Management*, https://doi.org/10.1007/978-3-030-40997-5_4

4.1 Introduction

Soil pollution is among the prime most issues of concern worldwide for the present century. Soil pollution is referred as the presence of any chemical, compound or substance in soil at undesirable (excess) concentration which directly or indirectly affects the normal physicochemical activities of the non targeted organisms. While, soil contamination is known as the occurrence of any chemical, compound or substance higher than its natural concentration into soil which may or may not cause harm (Rodríguez-Eugenio et al. 2018). However soil contamination ultimately leads to the soil pollution if not managed on time, so both the terms can be used synonymously. After World War II notable advancement in agricultural practices took place to meet the demand of food supply for increasing population, which is still insufficient to meet the actual demand (Kavitha et al. 2018). Use of inadequate amount of pesticides and fertilizers for achieving high yield, rapid industrialization and urbanization and poor management of house hold and industrial effluents created the scenario of environmental pollution with many organic and inorganic pollutants. Soil contamination also affects the soil micro-biota leading to the compromised soil-biodiversity and soil system (Kumar and Singh 2017; Singh et al. 2019; Singh 2019). Inorganic pollutants are of major concerns due to their non degradable and highly toxic nature. Heavy metals and metalloids are the major inorganic soil pollutants. Soil contamination with heavy metals has become a global concern due to degradation in soil quality and fertility. It has raised the issues of food security by reducing the yields and quality of the crops as well as by making consumption of contaminated crop as a concern of human and animal toxicity. Unlike organic contaminants, metal(loid)s do not undergo microbial or chemical degradation, and their total concentration persist for a long time after their introduction in soils (Bolan et al. 2014). Due to their mobility, bioavailability and persistent nature these heavy metals can easily introduced and accumulates into food chain. Continuous anthropogenic practices are the major sources for soil contamination with heavy metals. Heavy metal uptake by crops is crucial from socio-economic point of view, since it directly affects human health. Soil pollution, cannot be directly visible or accessible which makes it a concealed risk to living world. Various physical, chemical and microbial methods are used to remediate metal polluted soils. Factors such as cost, availability, effectiveness, remediation potential are some of the parameters which decide the feasibility and applicability of a remediation technique. Use of adsorptive materials like biochar and activated carbon (AC) is considered as cost effective and more efficient method for on-site remediation of contaminated soil (Beesley et al. 2011). Biochar is charred organic material deliberately produced from thermochemical conversion (generally <700 °C) of raw biomass in the oxygen limited environment for intentional soil application to improve soil properties (Lehmann 2009). Though, biochar is produced from residual or waste biomass and can be used without any activation process, it own the properties similar to AC which made it a cheaper alternative than AC (<1000$/ton biochar vs. 2500$/ton AC) Sizmur et al. (2016). Since Biochar has lesser density than AC, hence it holds higher number of particles per unit mass, relative to AC at the same dose and particle size, this could favor the kinetics of contaminant mass transfer from the soil to the biochar particles Sizmur et al. (2016). The potential of biochar to adsorb organic as well as inorganic contaminants makes it an appealing cost-effective alternative for soil remediation.

4.2 Soil Pollution with Heavy Metals and Sources

4.2.1 Heavy Metals and Their Toxic Effects

Though no clear definition for heavy metals is available but metals with relatively higher specific density compared to water (>5 g/cm^3) are considered as heavy metals (Järup 2003). Their ability to induce toxicity at lower level of exposure made them pollutants of immediate concerns. Naturally, heavy metals are found over the earth's crust in trace amount so also considered as trace metals (Tchounwou et al. 2012). In past few decades the amount of heavy metals is found to be increasing dramatically besides their natural occurrence. Parallely, public health concerns due to the toxicity of these metals increased worldwide. Heavy metals can be classified as essential and non essential according to their role. Metals like chromium (Cr), cobalt (Co), copper (Cu), Molybdenum (Mo), Nickel (Ni), Zinc (Zn), Iron (Fe), Manganese (Mn), Magnesium (Mg) are considered as essential trace elements for living organism which participate in their physiological and biochemical functions (WHO 1973). Heavy metals like lead (Pb), mercury (Hg), arsenic (As), Lithium (Li), Titanium (Ti), antimony (Sb), cadmium (Cd), are the non essential metals and among the most toxic metals with highly detrimental effects on human and animal organs and plant system (Nagajyoti et al. 2010). Exposure to these metals beyond their permissible limit have life threatening effects on biological world. Soil and water contamination due to industrial runoff, municipal sludge; direct exposure during mining, tanning, coal and petroleum burning, nuclear power stations, electronics, textiles, dye, paper and wood industry; use of metals in domestic as well as small and large scale agricultural practices; bioaccumulation of metals in food chain are the prime manifestations of heavy metals to humans. Heavy metals can bind with many cellular organelles like cell membrane, mitochondria, lysosome, endoplasmic reticulum creating imbalance to their functionality also they interact with DNA, nuclear and cellular proteins which may lead to cell cycle modulation, carcinogenesis, mutagenesis or apoptosis (Tchounwou et al. 2012). Heavy metals affect the number, diversity, and activities of soil microorganisms (Chibuike and Obiora 2014). Hg, Pb, As, Cd and Cr are classified as top priority toxic metal pollutants of significant concern. Neurological disorders, cardiovascular and renal dysfunction, acute immune disorders, dermatologic problems, reproductive problems are some of the human disorders associated with Hg, As, Be, Cd contamination (Mamtani et al. 2011). Metal contamination has shown various detrimental effects on plant growth too. Heavy metals which are present as soluble components in the soil solution (pore water) or those that are easily solubilized by root exudates are available for plant uptake (Chibuike and Obiora 2014). Due to hyper-accumulation of metals such as Zn, Hg, As, Cd, Cr, Co, Cu, Mn, Ni reduction in biomass, root and shoot growth, seed production, plant nutrient is reported in various fruit, vegetable and crop plants (Chibuike and Obiora 2014; Nagajyoti et al. 2010).

4.2.2 Sources of Heavy Metal Contamination to Soil

The soil parent material itself contains most of the heavy metals in trace amount which is not bioavailable rather, anthropogenically added heavy metals have high bioavailability (Bolan et al. 2014; Wuana and Okieimen 2011). Some intentional or unintentional reasons that made heavy metals as soil contaminant can be: rapid anthropogenic generation cycle compared to natural one, transmittance from mines to other locations, discarded products with high concentration of metals, changes in chemical forms of heavy metals in the different environmental conditions which make them more bio-available and creates high risk of direct exposure (Wuana Okieimen 2011). Soil pollution can be point source or diffuse pollution. When contaminants are released into the soil of a particular area from specific source or serials of events, which could be identified easily is known as point-source pollution and anthropogenic activities represent the main sources (Rodríguez-Eugenio et al. 2018). While, diffuse soil pollution is referred as the accumulation of contaminants over very wide areas in soil, and has more than one source which could not be easily identified. Infusion, emission or transformation of contaminant in other media prior to soil application is the major cause of diffusion pollution in soil (Rodríguez-Eugenio et al. 2018). The chemical form and metal speciation are the key factors on which the fate and transport of heavy metals in soil depends. Once metal entered into the soil, it got distributed into divergent forms having different bioavailability, mobility, and toxicity. There are different identified sources adding heavy metal contamination to soil including agricultural practices, industrial and domestic effluents, natural and atmospheric sources. Most of the mismanaged anthropogenic activities are responsible for rapid contamination of soil with various toxic metals. Release of industrial waste, waste water and sludge deposited due to industrial applications of metals in electroplating, milling, and etching, tannery, textile and dye, paint and color, metal casting and smelting, print and chemical, wood preservation and processing, glass processing, photography, pharmaceutical printed circuit board (PCB) manufacturing industries, generate adequate amount of Cd, Zn, Cr, Ni, V, Pt etc. (Barakat 2011; Nagajyoti et al. 2010). Phosphate and nitrate fertilizers also contain variable amount of Cr, Cd, Ni, Pb, Hg and Zn in which Cd is of main concern due to its accumulation in plant leaves (Rodríguez-Eugenio et al. 2018). Pesticides, fungicides extensively used in agriculture, horticulture and animal husbandries are the mixture of different compounds containing metals like Cu, Hg, Fe, Pb, Zn. Fungicidal spray Bordeaux mixture (mixture of copper sulphate and copper oxychloride) is the best example (Wuana and Okieimen 2011). All these practices contribute to elevate the background concentration of heavy metals in soil. Mining operations like tailing, refinement, ore transportation, metal finishing etc. are the direct sink of metals to the onsite soil. While, runoff from erosion of mine waste, direct disposal of mine waste to land, corrosion and leaching of heavy metals, dusts produced during the transport of crude ores contaminate the off site locations too (Nagajyoti et al. 2010). Use of cattle, poultry and pig manures for agricultural practices can add high amount of Cu, Cd, Zn, Mn, Cr, Pb in soil. Since, compounds containing various metals are used in diet of animals in pig and poultry industry and repeated use of manure from such animals for land application, can cause considerable buildup of such metals in the soil in the long run (Wuana and Okieimen 2011). A study conducted in China reported that the application of pig and chicken manure led to a certainty extend pollution of copper and zinc in soil, which becomes severity in the release of heavy metals in plantation (Jiang et al. 2011). Municipal sewage sludge is also contain plenty of heavy metals including Pb, Cr, Cd, As, Zn, Ni, Fe which mainly accumulate due to their household usage in the form of steel, sewage pipes, paints, building materials, pigments etc. The direct application of untreated sludge can accumulate heavy metals into soil and eventually enter into the food chain. Metals like Ag, Cr, Hg, Ti, As, Cd, Mo, Co, Se, Ni, Zn, B, Cu, Pb and Mn are identified which enter the food chain when sewage sludge was applied to soil (Rodríguez-Eugenio et al. 2018). Natural sources like weathering of rocks may also add metals to soil as they are geological parent material of soil specially ingenious and sedimentary rocks. The amount of As in igneous rocks varies upto 100 mg/kg, while in sedimentary rocks it varies from trace amount in limestone and sandstone and upto 15,000 mg/kg in some Mn ores (Nriagu 1994). Besides it volcanic activities can also add Hg into soil.

4.2.3 Approaches for Heavy Metal Remediation from Soil

Several techniques are used for heavy metal remediation from soil. The overall intention of any such approach is to create a final solution which is protective to environment and human health (Wuana and Okieimen 2011). Also, the selection of appropriate remediation technique has a crucial importance, which can be influenced by chemical and physical nature of the heavy metal as well as soil, type and concentration of heavy metals. Soil remediation from heavy metals and metalloids can be done by source control and contaminant remediation, either by the mean of in situ (onsite*)* or ex situ treatment technologies. In situ treatment involves treatment of contaminated soil at its original place; unmoved, unexcavated; remaining at the site or in the subsurface, while for Ex situ process, the contaminated soil is moved, excavated, or removed from the site or subsurface (Wuana and Okieimen 2011). In situ stabilization of the contaminant by the mean of sorbent amendment has been gained substantial attention due to its cost effectiveness, high remediation efficiency and on-site treatment strategy.

Immobilization and extraction are the two main processes applied either in situ or ex situ (on site or off site) for heavy metal contaminated soils. Metal immobilization is referred to stabilizing or fixing metal ions to reduce its leaching into the soil matrix and transform it to less soluble, toxic or bioavailable form in the soil reducing the associated risks to the environment (Tajudin et al. 2016). While, the extraction processes involve separation of the metals from the soil's composition. Extraction reduces the concentration and volume of metals as well as the entire contaminated medium. However, immobilization approaches are more acceptable due to many advantages over extraction in terms of cost and handing. Use of activated carbon or biochar as an adsorbent (immobilization agent) has shown promising and effective results for heavy metal remediation from soil. With low energy demand and global warming potential impact than activated carbon (AC) biochar become prime choice than activated carbon (Hjartardottir 2017). Table 4.1 summarized the techniques used for the remediation of heavy metals from soil.

Table 4.1 Techniques used for heavy metal remediation from soil

Remediation technology	Process	Agent
Solidification (stabilization or fixation)	Physical encapsulation of the metals. Convert metals to a less soluble form to reduce its leachability by the addition of cementing agent.	Cementing agent like Calcium Aluminate cement (CAC), pozzolanic cement (PC), Ordinary Portland cement (OPC), mixture of fly ash and cement etc
Immobilization (stabilization or fixation)	Conversion of contaminant into soil in more chemically stable form so it could not be accessible to plants Redistribution of heavy metals from solution phase to solid phase by the mean of adsorption, precipitation, and complexation	Binding agents (clay, lime, bentolite, polyethylene, paraffins, bitumen); phosphate compounds; liming material; organic compost, Biochar
Vitrification (extraction)	Extremely high temperature treatment (≈2000 °C) to the contaminated area Formation of vitreous material in the form of an oxide solid Volatilization or destroying volatile metal species like Hg, Sb due to high temperature application	For in situ treatment: Heating devices like plasma torches and electric arc furnaces For ex situ treatment: Heating electrodes
Soil washing (extraction)	Physical extraction: removal of highly contaminated soil particles Chemical extraction: aqueous solutions are used for removal of the contaminant	Water, aqueous chemicals (can be performed in situ as well as ex situ.)
Soil capping and isolation	Physical method which involves placing of a cover over the contaminated soil or barricading of contaminated site	Vertical engineered barriers (VEB), caps, liners
Phytoremediation (stabilization or fixation)	Use of metal accumulating plants and associated microbiota for metal remediation In situ remediation involves phytoextratction, phytoaccumulation, phytostabilization, phytofilterartion	Plants and microbes
Bio-methylation (stabilization or fixation)	Microorganisms present in soil produce methylated derivatives of metals which are less toxic in nature and often volatile	Microbes

Summarized with Tajudin et al. (2016) and Wuana and Okieimen (2011)

4.2.4 *Potential of Biochar for Heavy Metal Remediation from Soil*

Biochar is a solid, carbon rich product produced through the thermal treatment of biomass under oxygen limited conditions. Thermal treatments may include pyrolysis, hydrothermal carbonization, torrification or gasification. However, with pyrolysis the highest yield of biochar can be obtained and for other treatments it is a byproduct. Primary component of biochar is carbon (70–80%) with hydrogen, nitrogen, oxygen and mineral content including potassium, magnesium, major nutrients of plant growth (Kookana et al. 2011). Biochar has high surface area, higher porosity, variable charge, and functional groups that can increase soil water-holding capacity, pH, cation exchange capacity (CEC), base saturation, crop resistance to disease, surface sorption capacity and reduce the mobility of plant available pollutants when added to soil (Anawar et al. 2015; Safaei Khorram et al. 2018). Only ionic form of metal, dissolved into pore water of soil is bioavailable to the soil organisms and plants. When biochar is added to soil the transfer of metal ions from contaminated soil to biochar will takes place. Abundant surface functional groups, highly porous structure, high surface area attracts metal ions to associate with stronger sorption sites of biochars and dissociate for weaker attraction with soil molecules. When applied to the contaminated soil, biochar acts as a barrier which hinders the metal ion linkage or association to soil molecules by making strong affinity towards them (Fig. 4.1). Metal ions accumulate into plant via transfer from pore water of soil to plant roots Sizmur et al. (2016). Biochar adsorbs metal ions present in the pore water of soil and made them non-bioavailable to plant roots. To access the effectiveness of biochar amendment in soil end point measurements such as bioavailable metal concentration, total metal concentration before and after amendment, uptake by plants or animals are used (Lehmann and Joseph 2015).

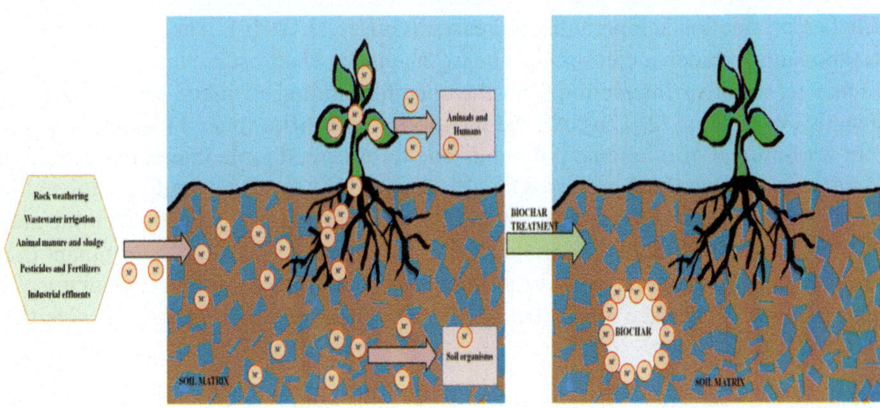

Fig. 4.1 Schematic diagram representing metal remediation with biochar by breaking source (soil) and receptor (plant roots) pathway (pore water). [Adapted and modified from Sizmur et al. (2016)]

4.3 Factors Affecting Heavy Metal Retention in Biochar

4.3.1 Characteristics of Biochar Affecting Metal Remediation

4.3.1.1 Feedstock and Pyrolysis Conditions

The parent material and preparation temperature are the key factors governing the properties of biochars (Kavitha et al. 2018). Almost any form of organic material, such as crop residues, forestry byproducts, urban yard wastes, industrial byproducts, animal manure, and sewage sludge can be converted to biochar by pyrolysis under a wide range of conditions (Xu et al. 2017). Generally, biochars produced from seaweeds, manures and crop residues are richer in nutrients, have higher pH and less stable carbon than biochars produced with lignocellulose rich woody biomass (Gul et al. 2015). Also, biochars prepared from livestock manure and poultry litter have significantly different properties and compositions compared to plant derived biochars in which ash content is the most important (Uchimiya et al. 2010).

Biochar can be produced from various thermochemical treatments like, pyrolysis (slow and fast), gasification and hydrothermal carbonization at 250–1000 °C temperature range with varying heating rates and resident time. Slow pyrolysis yields highest fraction of solid char (30–35%) as compared to fast pyrolysis (12%) and gasification (10%) (Mohan et al. 2014). For slow pyrolysis, biomass is heated from ambient to about 500 °C in absence of air, while residence time vary from minutes to hours. Biochar produced at lower pyrolysis temperature (<500 °C) possess attributes which are considered favorable for heavy metals stabilization leading to metal immobilization into soil (Ahmad et al. 2013). It results biochar with high organic carbon content, specific porous structure and higher O-containing numerous functional groups, which interacts with heavy metals in several ways (Oliveira et al. 2017). Also, the nutrients such as P, K, Ca, surface area, pH, C:N and C:O ratio of biochar increases, while dissolved organic matter concentration decreases when biochar production temperature increases (Gul et al. 2015). The organic carbon components of biochar can stabilize heavy metals by electrostatic interactions, ionic exchange, sorptive interaction, and the specific binding of metal ions by surface ligands (Wang et al. 2017b). All the mentioned properties make biochar a suitable tool for removal of inorganic pollutants due to increased ionic interactions through O-containing functional groups (Oliveira et al. 2017). A recent study conducted in paddy soil polluted with Cu, Cd, Zn and Pb collected nearby from Pb-Zn mine area in China advocated that appropriate choice of feedstock was more determinant factor with respect to the mobility of pollutants than altering pyrolysis temperature or modification of surface properties for immobilization of metals from soil (Lu et al. 2018).

4.3.1.2 Surface Area and Pore Distribution

Generally, surface area of biochar increases with the increase of pyrolysis temperature. Large surface area, highly porous structure and high pore volume is considered favorable for biochar's adsorption capacity. Release of volatiles and deformation of raw material components during pyrolysis increase the surface area of biochar and generate honey comb like porous structure. The pore size of biochar differs depending on the material used for biochar production and usually ranges from nano (<0.9 nm), micro (<2 nm), meso (2–50 nm) to macropores (>50 nm) (Shaaban et al. 2018). High pyrolysis temperature biochars (>500 °C) are more stable with high surface area (>400 m^2/g) compared to low temperature biochars (Keiluweit et al. 2010). The BET method given by Stephen Brunauer, Paul Hugh Emmett, and Edward Teller is the most common and well-known method used for specific surface areas evaluation of biochar. Direct sorption of metal ions into pores of biochar is well known mechanism. Small particle size with large surface area posses more active sites to bind with metal ions. Highly porous structure allows biochar for physical sorption of metals. Scanning Electron Microscopy (SEM) and Energy Dispersive Spectroscopy (EDS) are most commonly used to techniques to analyze the metal mobilization and binding to biochar surface.

4.3.1.3 Inorganic Composition and pH

Most biochars are alkaline and with the rise in pyrolysis temperature the ash content increases so the alkaline nature of biochar increases, resulting in higher pH of biochar. The ash content is non organic fraction of biochar consisting of elements such as Mg, Ca, O, N, S, K etc. A large number of alkaline salts, alkali metals (Na, K, Ca, and Mg) and $CaCO_3$ are associated with higher pH of the biochars (Cao and Harris 2010). Generally, biochar pH is found to be >7. Reportedly, the raise in soil pH due to biochar addition reduces the mobility and bioavailability of Cd, Zn, and Pb, and the phenomena is known as "liming effect (Wu et al. 2017). A 3 year long pot experiments were conducted to determine the long-term impact of holm oak-derived biochar, compost, and peat on Cd and Zn immobilization in the soil from the Campine region (Egene et al. 2018). Authors reported decrease in metal concentration in case of biochar amended soil while unexpected increase was reported for peat and compost treated soils. They reported metal immobilization due to consistently higher pH in the biochar treatments leading to precipitation of metal oxides (Egene et al. 2018). Application of biochar increases the soil pH leading to the formation of metal hydroxide precipitation and adsorption of heavy metals. Biochar with alkaline pH (pH 7–9) when added to acidic soil will increase the soil pH leading to the decrease in the mobility of cationic metals in soils due to reduced competition between H^+ ions and metal ions for cation exchange sites either directly on the surface of biochar or as a general liming effect on the soil matrix (Beesley et al. 2011). The type of raw material and pyrolysis temperature are the most important factors which influence the ash content and pH of biochar. Biochar originated from animal manure and sludge, grass, grain husk and straw contain high amount of ash content than that of woody biochar (Wang et al. 2018b).

4.3.1.4 Cation Exchange Capacity (CEC)

CEC of biochar can be defined as the capacity to exchange or adsorb positively charges ions and ranged from negligible to 200 cmol(+)/kg depending upon type of feedstock and pyrolysis condition (Palansooriya et al. 2019). Low temperature biochars generally have high CEC as they are rich in oxygenated functional groups, indicative of high complex formation intensity with metal cations. Biochar amendment increases the CEC of soil also. Formation of carboxylic groups in biochar during pyrolysis and oxidation of aromatic carbon in soil could be the reason of increase of CEC of soil (Palansooriya et al. 2019). Uchimiya et al. (2011a) suggested CEC as the primary mechanism by which biochar enhance the Cu retention from sandy loam soil (Uchimiya et al. 2011b). Reduction in bio-available Pb, Cd and Zn was also reported in mine trailing with orchard prune derived biochar due to increased pH and CEC (Fellet et al. 2011). Mineral or ash content present in the biochar increase the pH of biochar which favors the adsorption of cationic metal ions like Pb^{2+}, Zn^{2+}, Cd^{2+}, Ni^{2+} etc. by increasing the cation exchange.

4.3.1.5 Surface Functional Group

During thermal treatment of biomass, reformation and rearrangement of chemical bonds take place resulting in the formation of several functional groups (e.g. hydroxyl -OH, amino-NH_2, ketone -OR, ester -(C=O)OR, methyl –CH_3, nitro -NO_2, aldehyde -(C=O)H, carboxyl -(C=O)OH) on biochar surface (Verheijen et al. 2010; Zama et al. 2018). With their nature as electron donors or acceptors, the biochar properties can be of acidic or basic or hydrophobic or hydrophilic. Biochar produced at lower pyrolysis temperature (<400 °C) contain more oxygen containing surface functional groups (-OH, -CO, -COOH) which forms complex with metal cations. X-ray photoelectron spectroscopy (XPS) and Fourier Transform Infrared (FTIR) Spectroscopy are the most widely used technique to identify the surface moieties of biochar and soil. Complex formation with oxygenated functional groups is evidenced in many reports in case of Zn, Pb, Cr, Ni, Cd (Ahmad et al. 2013; Kumar et al. 2018; Qian et al. 2019; Uchimiya et al. 2011a). The negative charge of biochar surface increases the soil pH leading to increase in adsorption of metal ions from soil due to deprotonation of soil surface. Biochar surface oxygenated functional groups may impact on the oxidation of redox sensitive metals (Beesley et al. 2015). The formation of metal-ligand complex in soil could cause mineral

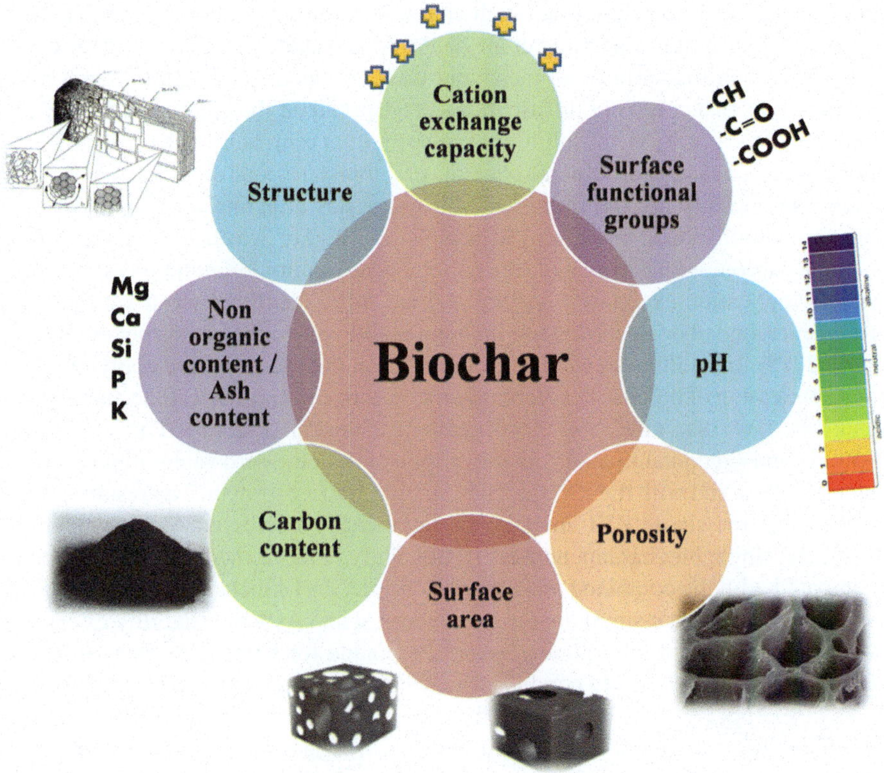

Fig. 4.2 Characteristic physico-chemical properties of biochar

precipitation on the soil particle surface which changes soil porosity and modifies soil physical structure. Study conducted by Xu et al. (2018) shows that dairy manure and sawdust biochars application in soil enhances the immobilization capacity of biochars for Cd due to the enhanced surface complexation with O-containing groups (Xu et al. 2018) (Fig. 4.2).

4.4 Other Factors

Besides characteristic properties of biochars the heavy metal retention onto biochar is also affected by speciation of heavy metals into soil, physical and chemical properties of heavy metal ions and the amount of biochar addition (Wang et al. 2018b). Soil properties, metal speciation and characteristics as well as environmental conditions are other factors that affect the biochar-metal interactions in soil matrix. Properties of metal ion are critical character in sorption process with biochar. Metal ions having small ionic radii and high charges are strongly attracted or repulsed by charged biochar surfaces (Kong et al. 2014). Also, small ions and molecules can be absorbed by diffusion into the meso and micro pores of the biochar. Large molecules or ions that could not diffuse may get adsorbed on the biochar surface, which may block the pores. Metal solubility is also an important factor which allows binding it with biochar. Soil mineral content, pH, moisture, soil organic matter, cation exchange capacity are the important soil properties affecting metal-biochar complex formation. Soils from tropical regions are highly weathered, acidic, low in organic carbon, with dominated by kaolinite and iron or aluminum oxyhydroxides, posses low cation exchange capacity and are more readily phyto-toxic than soils from temperate regions due to their inherent inability to retain heavy metals (Beesley et al. 2015). In addition to soil type, modification in biochar before applying in soil and application of biochar in combination to fertilizers, clay, limestone and compost and their ratio also affect the metal uptake by biochar (Fig.4.3). Trials conducted with As and Cd contaminated soil with application of the zero valent iron (ZVI) and biochar alone and with the ZVI-biochar mixture for rice growth. The results demonstrated that the single ZVI amendment significantly decreased As bioavailability, while the single biochar amendment significantly reduced the bioavailability of Cd compared with the combined amendments. While, ZVI-biochar mixture may have a synergistic effect that simultaneously reduces Cd and As bioavailability by increasing the formation of amorphous Fe and Fe plaque for Cd and As immobilization (Qiao et al. 2018). A study conducted in sandy loam soil collected from surroundings of a copper smelter to examine the effect of bamboo and rice straw biochar on mobility and redistribution of Cd, Cu, Pb and Zn. The results demonstrated that the rice straw biochar is more effective than bamboo biochar and efficiently immobilize

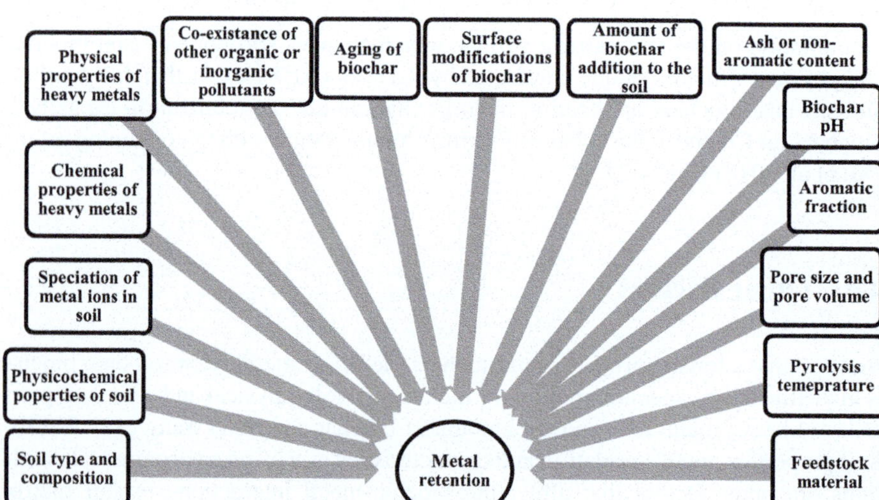

Fig. 4.3 Factors influencing the biochar-metal retention in soil

the heavy metals, and reduces their mobility and bioavailability in contaminated soils (Lu et al. 2017). Experiments conducted with the composting of a mixture of chicken manure and sawdust (3:2, v:v) with 5% dry weight addition of biochar produced from waste of fungal cultures or rice straw demonstrated that both the total and the bioavailable concentration of Zn, Cu and As (exchangeable and reducible fractions) was reduced compared to material composted without any addition of biochar (Cui et al. 2016).

Biochar is a panacea of many characteristic properties which contributes for adsorption of pollutants from soil. With the time, changes in the characteristic as well as sorption properties of biochars may occur which is referred as 'ageing of biochars'. Oxidation of biochar surfaces, mineralization and other biochemical interactions between biochar and soil's natural organic matter (SOM) and non-organic matter (NOM) could be the reason of biochar aging. Aging reduce the adsorption capacity of biochar as SOM and NOM block the pores of biochars (Uchimiya et al. 2010). Aging of pollutants can takes place over the period of time, which could make it more recalcitrant and less bio-available. With increase of time, soil-metal association may also get stronger resulting in non-extractable fraction of pollutant (Hjartardottir 2017). This metal-soil binding could be an obstacle in remediation process.

4.5 Heavy Metal Removal Mechanism of Biochar in Soil

Unlike organic pollutant contamination, the heavy metal contamination of soils and its management are a challenging issue because they are hard to mineralize into other forms, and their persistence causes numerous adverse effects on the soil ecosystem. With porous structure, larger surface area, ample surface functional groups and good CEC biochar shows effective and promising results for heavy metal management in soil. The remediation mechanism of biochar is different for different heavy metal pollutants. For the same heavy metal ion, the adsorption mechanism is different when the biochar is different (Wang et al. 2018b). Biochar act on the bioavailable fraction of soil heavy metals and can be able to reduce also their leachability (Paz-Ferreiro et al. 2014). The mobility and bioavailability of heavy metals and metalloids are controlled by a myriad of reactions either synergistically or alone, which are discussed below (Fig. 4.4).

4.5.1 Complexation

Formation of multi-atom structures (i.e., complexes) with specific metal-ligand interactions over outer or inner sphere is the complexation (Inyang et al. 2016). Biochar is rich in oxygen containing acidic functional groups (phenolic, carbonyl, lactonic, carboxylic, phenolic and hydroxyl) as well as basic functional groups

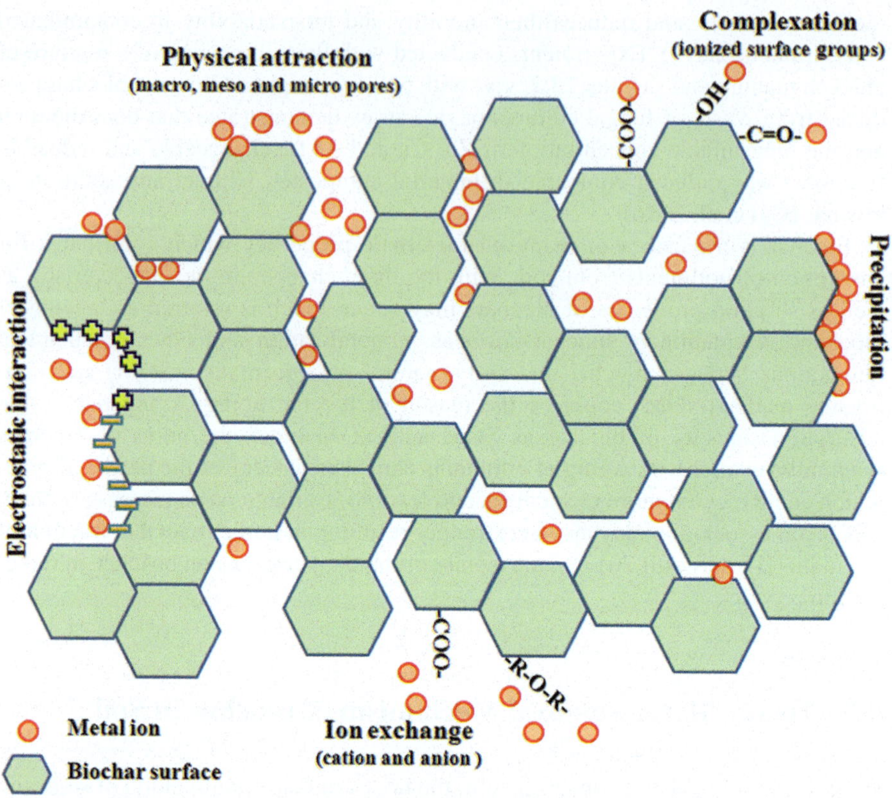

Fig. 4.4 Schematic diagram of metal-biochar interaction in soil

(ketone, pyrone), which play significantly important role in binding (complexation) of heavy metals and metalloids onto the biochar surface as well as inner pores. Also, freshly prepared biochar applied into the soil, exposed to the atmosphere, the oxygenation of the biochar surface occur, which leads to the formation of oxygen containing functional groups (Beesley et al. 2015). Choppala et al. reported that for Cr contaminated soil the functional groups donates electrons for metal binding and high pH biochars may prevent dissociation and oxidation of phenolic and hydroxyl groups, which prevent the proton supply for Cr(VI) reduction (Choppala et al. 2012). Biochar produced through low temperature pyrolysis contains more polar functional groups rather higher temperature pyrolysis biochar. It result in effective metal immobilization due to the formation of inner and outer sphere complexes with oxygenated (acid) functional groups (Beesley et al. 2015; Wang et al. 2018a). Complexation is considered as most common mechanism for adsorption with biochar for metal with soluble nature (Zn, Ni, Pb etc.).

4.5.2 Physical Sorption

Biochar is the carbon material with a well distributed pore networks including micropores (<2 nm), mesopores (2–50 nm), and macropores (>50 nm) (Lehmann and Joseph 2015). Physical or surface adsorption involves the removal of heavy metals by diffusion of metal ions into the pores of sorbent. This mechanism is not implicated with any chemical bond formation. For both animal and plant biochars, increasing temperatures of carbonization (>300 °C) will favor high surface areas and pore volumes in biochars (Inyang et al. 2016). Larger pores allow the metal ions directly immobilize into the pores by diffusion process and retained within the pore. Generally, with increase in pyrolysis temperature, the surface area and porosity of the biochar also increases, which increases adsorption capacity of biochar (Inyang et al. 2016; Zhang et al. 2015). However, other mechanisms are more prominent for heavy metals adsorption from soil.

4.5.3 Electrostatic Attraction

Heavy metal immobilization via electrostatic attractions takes place between metal ions and biochar's surface charge (Inyang et al. 2016). Biochar surface is negatively charged when produced at higher pyrolysis temperatures. The positively charged metal ions interact with negatively charged biochar surface through strong electrostatic interaction in soil, leading to the reduction of metal bioavailability. This mechanism highly influenced by pH of soil, as it changes the metal speciation in soil matrix. Also, biochar application increases the soil pH, leading to reduction in metal mobility by attaching with metal ions.

4.5.4 Ion Exchange

Exchange of ionic species to biochar surface from soil matrix is another way to immobilize metal with biochar. Surface functional groups and size of metallic species are the most important factors related to the efficiency of the ion exchange process in retaining heavy metal contaminants on biochar (Inyang et al. 2016). Ionic exchange of metal species on biochar occurs by a selective replacement of cations on biochar surfaces with target metal species (Inyang et al. 2016; Wang et al. 2018b). Since, CEC of biochar is highly influenced by biochar's surface functional groups. Low temperature biochars posses high amount of these groups which bind with metal ions onto biochar surface. Metal ions may also form complex with minerals like Na, K, Mg, P, Ca etc. present into the inorganic constitute of the biochar matrix.

4.5.5 *Precipitation*

Precipitation is considered as the most common accountable mechanisms for heavy metal immobilization by biochar (Inyang et al. 2016). During the sorption process formation of solid(s), either in solution or on a surface is known as precipitation. Metals like Cr, Cu, Zn, Pb with intermediate ionization potentials (2.5–9.5) and soluble nature are often precipitate on biochar surfaces as insoluble phosphate and carbonate salts (Kong et al. 2014). Biochar addition (with pyrolysis temperature > 300 °C) remarkably increase the pH of soil and the heavy metal ions present into the soil may react with the oxide, phosphate and carbonate (OH, PO_4, CO_3, HCO_3) fractions of the biochar leading to the reduction of heavy metal mobilization into the soil by precipitation (Ahmad et al. 2013; Tang et al. 2013). Generally, high temperature biochar has high pH as well as high mineral content (eg. Ca, Mg, Si, Fe) which is significantly higher in case of animal waste derived biochar like, poultry litter, bone char than the biochars derived from plant materials (Inyang et al. 2016; Shakya and Agarwal 2017). The minerals are present in biochars either in free form or embedded the carbon matrix which can directly react with the heavy metal ions to form metallic precipitate. A study observed a new precipitate on Pb-loaded sludge derived biochar as $5PbO{\cdot}P_2O_5{\cdot}SiO_2$ (lead phosphate silicate) at initial pH 5 and confirmed by XRD and SEM-EDX (Lu et al. 2012).

4.6 Risk Associated with Biochar Application in Soil

4.6.1 *Risk to Environment*

Being cost effective and economic, majority of researches supports use of biochar technology in soil for different prospects. However, limitations are also there. The selection of feedstock and production process, highly affect the organic as well as inorganic composition of the biochar. Use of sewage sludge, municipal waste or contaminated raw material may leads to accumulation of toxic metals in biochar. This increases the threat of phytotoxicity as well as soil contamination if applied to soil. Formation of polyaromatic hydrocarbons (PAH), volatile organic compounds (VOC), dioxins, carbonaceous nanoparticles may occur in biochar during the pyrolysis process (Dutta et al. 2017). Slow pyrolysis and longer residence time result in lower PAH yields than the fast pyrolysis and shorter residence time (Wang et al. 2017a). Pyrolysis temperature has a critical impact in determining the quantity and type of compounds released from biochar (Singh et al. 2014). Generally, low molecular weight PAHs are formed at low temperature (<500 °C) whereas the high molecular weight PAHs commonly appear under high temperature (>500 °C) (Wang et al. 2017a). Biochar aging in soil is also a critical factor in respect to soil biota, nutrient availability and retention. With time the capacity of biochar to immobilize the heavy metals may be modified as the sorption sites may get occupied with native soil organic matter or competing contaminant (Beesley et al. 2011). Recent study reported significant negative impact of biochar aging on exposed earthworm growth and reduction to root biomass (Anyanwu et al. 2018). Biochar prepared at high temperature or with animal origin feedstocks having high ash content, which may affect the plant growth by altering the nutrient bioavailability if applied to soil. Joseph et al. (2018) conducted a study to investigate the mechanisms of nutrient retention in a high temperature wood biochar. Authors compared non-composted biochar (BC) with composted biochar (BCC) and reported that during composting dissolved nutrients are first taken up into biochar pores complex reactions of concentration gradient, surface sorption and retention which block biochar pores and result in deposition of a nutrient-rich organomineral (plaque) layer. They concluded that biochar may react towards nutrients in a soil-fertilizer system, making it a competitor for, rather than provider of, nutrients for plant growth (Joseph et al. 2018). Biochar application to the soil with positive effects also may lead to the growth of invasive plants which act as competitor for crops. With the addition of biochar, improved soil quality and plant growth with significantly elevated weed growth was reported (Safaei Khorram et al. 2018). The adsorption property of the biochar which effectively immobilize the heavy metals can also immobilize various essential micro and macro nutrients and decrease soil fertility leading to reduced bioavailability to plants. Biochar has shown affinity towards pesticides and herbicides, which may reduce the effectiveness of these chemicals which may force to increase their use. Also, such chemicals when bound with biochar may remain in the agroecosystem for a long time which may compromise their applications for generations (Ok et al. 2015). The alkaline nature and inadequate amount of biochar addition to soil may also imbalance the soil properties which may have negative impact on soil micro and macro biota. Continuous use of biochar can accumulate it into the soil matrix and so the bind metals, which may also alter the soil environment in long term. The beneficial effects of biochar addition, also depends upon soil type. So, it is necessary to analyze the soil type and quality before biochar addition. Analysis of feedstock with respect to composition, moisture content and any other contamination is needed to avoid the soil contamination or the availability or mobility of such contaminants to plants. Biochar application for soil remediation is still in laboratory research phase, which need real trails in fields to know the real picture of biochar effects. Certainly, utmost care is required during selection of raw material for biochar preparation; characterize the biochar for its physicochemical properties and judicial amount of biochar before soil application to reduce the associated risks. In addition, with huge differences in its properties a single biochar cannot be considered effective for different metal remediation. The longevity and fate of biochar in soil for long term is still unclear, which is a lack in knowledge for future effects or associated risks of biochar.

4.6.2 *Risk to Human Health*

During production or implication in fields, inhalation of fine particles of biochars can cause serious health issues or respiratory problems. Biochar source material like sludge, sediments, municipal waste, litter and manure may sometimes itself contain metal and other contaminations which may lead to health issues if not handled carefully. Metal accumulation in biochar due to source material or production process may leach into soil or water which further increases their concentration in environment leading to biomagnifications or bio-accumulation into food chain. An urgent need of policies is required to regulate the production, use and application of biochars for various soil applications. Although the heterogeneous nature of biochar in terms of its properties provides many opportunities to explore its usages in various fields for environmental management. However its possible negative imapcts can be a challenge for biochar industry.

4.7 Conclusion and Future Scope

Biochar has emerged as a promising and efficient tool to remediate heavy metals which are absorbed and bind into soil matrix with different strength from various anthropogenic sources. With its strong metal affinity, biochar fixes metal into the soil matrix. Biochar's ability to alter the physicochemical properties of the soil can also help to improve soil fertility. Various factors govern the sorption properties of biochar into soil matrix like biochar raw material, preparation method, nature of targeted metal, properties of soil, environmental conditions, and co-occurrence of other pollutants. The intergraded approach to use biochar with compost, fertilizer and manure can help to achieve the soil remediation as well as soil conditioning to improve yield. Use of engineered, magnetic or designer biochar for metal remediation from soil is also a step ahead approach however most of the research is in infant stage. The properties of biochar significantly depend upon the raw material and production process, need of extensive research to standardize and regulation in this regard is also necessary. Interaction and effect of biochar on biota in heavy metal contaminated soil is still poorly understood. Since, the biochar application into soil is a non reversible approach, so systematic monitoring is necessary requirement in the view of soil, human and environment health as well as food safety. Irreversibility of biochar application from soil matrix and its aging is yet to be studied on larger scale. Though, ongoing research is providing new insights of biochar application for metal mitigation from soil, data is still insufficient to its use for large scale application. This is a need of research to explore the potential of biochar for new and arising species of metal contaminants.

References

Ahmad M, Rajapaksha AU, Lim JE, Zhang M, Bolan N, Mohan D, Vithanage M, Lee SS, Ok YS (2013) Biochar as a sorbent for contaminant management in soil and water: a review. Chemosphere 99:19–33

Anawar HM, Akter F, Solaiman ZM, Strezov V (2015) Biochar: an emerging panacea for remediation of soil contaminants from mining, industry and sewage wastes. Pedosphere 25(5):654–665

Anyanwu IN, Alo MN, Onyekwere AM, Crosse JD, Nworie O, Chamba EB (2018) Influence of biochar aged in acidic soil on ecosystem engineers and two tropical agricultural plants. Ecotoxicol Environ Saf 153:116–126

Barakat MA (2011) New trends in removing heavy metals from industrial wastewater. Arab J Chem 4(4):361–377

Beesley L, Moreno-Jiménez E, Gomez-Eyles JL, Harris E, Robinson B, Sizmur T (2011) A review of biochars' potential role in the remediation, revegetation and restoration of contaminated soils. Environ Pollut 159(12):3269–3282

Beesley L, Moreno-Jimenez E, Fellet G, Carrijo L, Sizmur T (2015) Biochar and heavy metals

Bolan N, Kunhikrishnan A, Thangarajan R, Kumpiene J, Park J, Makino T, Kirkham MB, Scheckel K (2014) Remediation of heavy metal (loid) s contaminated soils–to mobilize or to immobilize. J Hazard Materi 266:141–166

Cao X, Harris W (2010) Properties of dairy-manure-derived biochar pertinent to its potential use in remediation. Bioresour Technol 101(14):5222–5228

Chibuike GU, Obiora SC (2014) Heavy metal polluted soils: effect on plants and bioremediation methods. Appl Environ Soil Sci 2014

Choppala GK, Bolan N, Megharaj M, Chen Z, Naidu R (2012) The influence of biochar and black carbon on reduction and bioavailability of chromate in soils. J Environ Qual 41(4):1175–1184

Cui E, Wu Y, Zuo Y, Chen H (2016) Effect of different biochars on antibiotic resistance genes and bacterial community during chicken manure composting. Bioresour Technol 203:11–17

Dutta T, Kwon E, Bhattacharya SS, Jeon BH, Deep A, Uchimiya M, Kim KH (2017) Polycyclic aromatic hydrocarbons and volatile organic compounds in biochar and biochar-amended soil: a review. GCB Bioenergy 9(6):990–1004

Egene CE, Van Poucke R, Ok YS, Meers E, Tack FMG (2018) Impact of organic amendments (biochar, compost and peat) on Cd and Zn mobility and solubility in contaminated soil of the Campine region after three years. Sci Total Environ 626:195–202

Fellet G, Marchiol L, Delle Vedove G, Peressotti A (2011) Application of biochar on mine tailings: effects and perspectives for land reclamation. Chemosphere 83(9):1262–1267

Gul S, Whalen JK, Thomas BW, Sachdeva V, Deng H (2015) Physico-chemical properties and microbial responses in biochar-amended soils: mechanisms and future directions. Agric Ecosyst Environ 206:46–59

Hjartardóttir S (2017) Remediation of PCB-polluted soil using biochar: the uptake of PCBs in earthworms, plants and passive samplers (Master's thesis)

Inyang MI, Gao B, Yao Y, Xue Y, Zimmerman A, Mosa A, Pullammanappallil P, Ok YS, Cao X (2016) A review of biochar as a low-cost adsorbent for aqueous heavy metal removal. Critical Rev Environ Sci Technol 46(4):406–433

Järup L (2003) Hazards of heavy metal contamination. British med bull 68(1):167–182

Jiang X, Dong R, Zhao R (2011) Meat products and soil pollution caused by livestock and poultry feed additive in Liaoning, China. J Environl Sci 23:S135–S137

Joseph S, Kammann CI, Shepherd JG, Conte P, Schmidt H-P, Hagemann N, Rich AM, Marjo CE, Allen J, Munroe P (2018) Microstructural and associated chemical changes during the composting of a high temperature biochar: mechanisms for nitrate, phosphate and other nutrient retention and release. Sci Total Environ 618:1210–1223

Kavitha B, Reddy PVL, Kim B, Lee SS, Pandey SK, Kim K-H (2018) Benefits and limitations of biochar amendment in agricultural soils: a review. J Environ Manag 227:146–154

Keiluweit M, Nico PS, Johnson MG, Kleber M (2010) Dynamic molecular structure of plant biomass-derived black carbon (biochar). Environ Sci Technol 44(4):1247–1253

Kong L-L, Liu W-T, Zhou Q-X (2014) Biochar: an effective amendment for remediating contaminated soil. In: Reviews of environmental contamination and toxicology, vol 228. Springer, Cham, pp 83–99

Kookana RS, Sarmah AK, Van Zwieten L, Krull E, Singh B (2011) Biochar application to soil: agronomic and environmental benefits and unintended consequences. In: Advances in agronomy, vol 112. Elsevier, pp 103–143

Kumar A, Singh JS (2017) Cyanoremediation: a green-clean tool for decontamination of synthetic pesticides from agro-and aquatic ecosystems. In: Singh JS, Seneviratne G (eds) Agro-environmental sustainability: vol (2) managing environmental pollution. Springer, Cham, pp 59–83

Kumar A, Joseph S, Tsechansky L, Privat K, Schreiter IJ, Schüth C, Graber ER (2018) Biochar aging in contaminated soil promotes Zn immobilization due to changes in biochar surface structural and chemical properties. Sci Total Environ 626:953–961

Lehmann J (2009) In: Joseph S (ed) Biochar for environmental management (pp. 1–5). Earthscan, London

Lehmann J, Joseph S (2015) Biochar for environmental management: science, technology and implementation. Routledge, Oxon

Lu H, Zhang W, Yang Y, Huang X, Wang S, Qiu R (2012) Relative distribution of Pb2+ sorption mechanisms by sludge-derived biochar. Water Res 46(3):854–862

Lu HP, Li ZA, Gascó G, Méndez A, Shen Y, Paz-Ferreiro J (2018) Use of magnetic biochars for the immobilization of heavy metals in a multi-contaminated soil. Sci Total Environ 622–623:892–899

Lu K, Yang X, Gielen G, Bolan N, Ok YS, Niazi NK, Xu S, Yuan G, Chen X, Zhang X, Liu D, Song Z, Liu X, Wang H (2017) Effect of bamboo and rice straw biochars on the mobility and redistribution of heavy metals (Cd, Cu, Pb and Zn) in contaminated soil. J Environ Manag 186:285–292

Mamtani R, Stern P, Dawood I, Cheema S (2011) Metals and disease: a global primary health care perspective. J Toxicol 2011

Mohan D, Sarswat A, Ok YS, Pittman CU (2014) Organic and inorganic contaminants removal from water with biochar, a renewable, low cost and sustainable adsorbent–a critical review. Bioresour Technol 160:191–202

Nagajyoti PC, Lee KD, Sreekanth TVM (2010) Heavy metals, occurrence and toxicity for plants: a review. Environ Chem Lett 8(3):199–216

Nriagu JO (1994) Arsenic in the environment. Part I: cycling and characterization. Wiley, New York

Ok YS, Uchimiya SM, Chang SX, Bolan N (2015) Biochar: production, characterization, and applications. CRC Press, Boca Raton

Oliveira FR, Patel AK, Jaisi DP, Adhikari S, Lu H, Khanal SK (2017) Environmental application of biochar: current status and perspectives. Bioresour Technol 246:110–122

Palansooriya KN, Ok YS, Awad YM, Lee SS, Sung J-K, Koutsospyros A, Moon DH (2019) Impacts of biochar application on upland agriculture: a review. J Environ Manag 234:52–64

Paz-Ferreiro J, Lu H, Fu S, Méndez A, Gascó G (2014) Use of phytoremediation and biochar to remediate heavy metal polluted soils: a review. Solid Earth 5(1):65–75

Qian T-T, Wu P, Qin Q-Y, Huang Y-N, Wang Y-J, Zhou D-M (2019) Screening of wheat straw biochars for the remediation of soils polluted with Zn (II) and Cd (II). J Hazard Materi 362:311–317

Qiao J-T, Liu T-X, Wang X-q, F-b L, Y-h L, Cui J-H, Zeng X-D, Yuan Y-Z, Liu C-P (2018) Simultaneous alleviation of cadmium and arsenic accumulation in rice by applying zero-valent iron and biochar to contaminated paddy soils. Chemosphere 195:260–271

Rodríguez-Eugenio N, McLaughlin M, Pennock D (2018) Soil Pollution: a hidden reality. Food and Agriculture Organization of the United Nations (FAO). Book published in Rome

Safaei Khorram M, Fatemi A, Khan MA, Kiefer R, Jafarnia S (2018) Potential risk of weed outbreak by increasing biochar's application rates in slow-growth legume, lentil (Lens culinaris Medik.). J Sci Food Agric 98(6):2080–2088

Shaaban M, Van Zwieten L, Bashir S, Younas A, Núñez-Delgado A, Chhajro MA, Kubar KA, Ali U, Rana MS, Mehmood MA, Hu R (2018) A concise review of biochar application to agricultural soils to improve soil conditions and fight pollution. J Environ Manag 228:429–440

Shakya A, Agarwal T (2017) Poultry litter biochar: an approach towards poultry litter management–a review. Int J Curr Microbiol App Sci 6(10):2657–2668

Singh JS (ed) (2019) New and future developments in microbial biotechnology and bioengineering: microbes in soil, crop and environmental sustainability. Elsevier, New York

Singh B, Macdonald LM, Kookana RS, van Zwieten L, Butler G, Joseph S, Weatherley A, Kaudal BB, Regan A, Cattle J (2014) Opportunities and constraints for biochar technology in Australian agriculture: looking beyond carbon sequestration. Soil Res 52(8):739–750

Singh C, Tiwari S, Singh JS (2019) Biochar: a sustainable tool in soil pollutant bioremediation. In: Bharagava RN, Saxena G (eds) Bioremediation of industrial waste for environmental safety. Springer, New Delhi, pp 475–494

Sizmur T, Quilliam R, Puga AP, Moreno-Jiménez E, Beesley L, Gomez-Eyles JL (2016) Application of biochar for soil remediation. Agricultural and environmental applications of biochar: advances and barriers (sssaspecpub63) 295–324

Tajudin SAA, Azmi MAM, Nabila ATA (2016) Stabilization/solidification remediation method for contaminated soil: a review. 2016. IOP Publishing, pp 012043

Tang J, Zhu W, Kookana R, Katayama A (2013) Characteristics of biochar and its application in remediation of contaminated soil. J Biosci Bioeng 116(6):653–659

Tchounwou PB, Yedjou CG, Patlolla AK, Sutton DJ (2012) Heavy metal toxicity and the environment. In: Molecular, clinical and environmental toxicology. Springer, Basel, pp 133–164

Uchimiya M, Lima IM, Thomas Klasson K, Chang S, Wartelle LH, Rodgers JE (2010) Immobilization of heavy metal ions (CuII, CdII, NiII, and PbII) by broiler litter-derived biochars in water and soil. J Agric Food Chem 58(9):5538–5544

Uchimiya M, Chang S, Klasson KT (2011a) Screening biochars for heavy metal retention in soil: role of oxygen functional groups. J Hazard Mater 190(1–3):432–441

Uchimiya M, Klasson KT, Wartelle LH, Lima IM (2011b) Influence of soil properties on heavy metal sequestration by biochar amendment: 1. Copper sorption isotherms and the release of cations. Chemosphere 82(10):1431–1437

Verheijen F, Jeffery S, Bastos A, Van der Velde M, Diafas I (2010) Biochar application to soils. A critical scientific review of effects on soil properties, processes, and functions. EUR, 24099, 162

Wang C, Wang Y, Herath HMSK (2017a) Polycyclic aromatic hydrocarbons (PAHs) in biochar – their formation, occurrence and analysis: a review. Org Geochem 114:1–11

Wang T, Sun H, Ren X, Li B, Mao H (2017b) Evaluation of biochars from different stock materials as carriers of bacterial strain for remediation of heavy metal-contaminated. Soil Sci Rep 7(1):12114

Wang H, Yang X, He L, Lu K, Müller K, McGrouther K, Xu S, Zhang X, Li J, Huang H, Yuan G (2018a) Using biochar for remediation of contaminated soils. In Twenty Years of Research and Development on Soil Pollution and Remediation in China (pp. 763–783). Springer, Singapore

Wang S, Xu Y, Norbu N, Wang Z (2018b) Remediation of biochar on heavy metal polluted soils. IOP conference Series: Earth and Environmental Science. IOP Publishing, pp 042113

WHO (1973) Trace elements in human nutrition: report of a WHO expert committee [meeting held in Geneva from 9 to 17 April 1973]

Wu S, He H, Inthapanya X, Yang C, Lu L, Zeng G, Han Z (2017) Role of biochar on composting of organic wastes and remediation of contaminated soils—a review. Environ Sci Pollut Res 24(20):16560–16577

Wuana RA, Okieimen FE (2011) Heavy metals in contaminated soils: a review of sources, chemistry, risks and best available strategies for remediation. Isrn Ecol 2011

Xu X, Zhao Y, Sima J, Zhao L, Mašek O, Cao X (2017) Indispensable role of biochar-inherent mineral constituents in its environmental applications: a review. Bioresour Technol 241:887–899
Xu Z, Xu X, Tsang DCW, Cao X (2018) Contrasting impacts of pre- and post-application aging of biochar on the immobilization of Cd in contaminated soils. Environ Pollut 242:1362–1370
Zama EF, Reid BJ, Arp HPH, Sun G-X, Yuan H-Y, Zhu Y-G (2018) Advances in research on the use of biochar in soil for remediation: a review. J Soil Sediment:1–18
Zhang J, Lü F, Zhang H, Shao L, Chen D, He P (2015) Multiscale visualization of the structural and characteristic changes of sewage sludge biochar oriented towards potential agronomic and environmental implication. Sci Rep 5:9406

Chapter 5
Biochars and Its Implications on Soil Health and Crop Productivity in Semi-Arid Environment

P. Kannan, D. Krishnaveni, and S. Ponmani

Abstract Land degradation and climate change are important associated processes necessitating appropriate management options to solve alarming food security threats in developing nations. Biochar produced from plant matter and applied to the soil has become increasingly recognized to address multiple contemporary concerns, such as agricultural productivity and contaminated ecosystem amelioration, primarily by removing carbon dioxide from the atmosphere and improving soil health. Biochar is an anaerobic pyrolysis product derived from organic material, resistant to easy degradation and stored carbon in the long-term in the terrestrial ecosystem and capable of reducing greenhouse emission from soil to the atmosphere. Further, it has the potential to adsorb and degrade heavy metals accumulated in the industrial and contaminant sites. The different source of biochars and graded levels of application has positive and negative effects on crop yield under different soil types. Most of the results in biochar are a greenhouse and laboratory-based experiment and lack of field experimental evidence in the semiarid environment. In this chapter need for biochar production, characterization, soil health changes, environmental clean-up potential, and crop yield dynamics under changing climate and research on biochar in the near future will be focused on sustainable crop and environmental management.

Keywords Biochar · Waste management · Soil health · Environment

P. Kannan (✉)
Department of Soils and Environment, Agricultural College and Research Institute, Madurai, Tamil Nadu, India

D. Krishnaveni
Alagappa Government Arts and Science College, Alagappa University, Karaikudi, Tamil Nadu, India

S. Ponmani
Mother Terasa College of Agriculture, Pudukkottai, Tamil Nadu, India

J. S. Singh, C. Singh (eds.), *Biochar Applications in Agriculture and Environment Management*, https://doi.org/10.1007/978-3-030-40997-5_5

5.1 Introduction

Global environmental change, including land degradation, loss of biodiversity, changes in hydrology and changes in climate patterns resulting from enhanced anthropogenic emission of greenhouse gases, will have serious consequences for world food security, particularly affecting the more vulnerable socio-economic sectors (Ericksen et al. 2009; Lal 2010). The World Bank suggests that at least a doubling of cereal yields and a 75% increase in meat production by 2030 are required to maintain the current level of nutrition globally (Fresco 2009). To significantly increase food production, when large areas of agricultural lands will be adversely affected by climate change or converted into forestry for C sinks may not be possible unless new technologies and sustainable practices are rapidly adopted (Singh and Singh 2019; Singh et al. 2019a, b; Vimal and Singh 2019). Biochar is a charred by-product of biomass pyrolysis produced from biological wastes, crop residues, animal poultry manure, or any type of organic waste material. Pyrolysis is the chemical breakdown of a substance under extremely high temperatures in the absence of oxygen. Biochar application has been promoted in agricultural practice that creates a win-win situation by improving soil quality and enhancing agricultural sustainability concomitant with mitigating GHG emissions. Recently biochar application got momentum because its capability of carbon sequestration, reducing soil compaction, improves soil physical condition and nutrient uptake from the soil.

The application of biochar to agricultural soils may play a crucial role in global climate change mitigation through the reduction of greenhouse gas production and the sequestering of atmospheric carbon in soils (Gaunt and Cowie 2009; McCarl et al. 2009; McHenry 2009). The agronomic benefits of biochar in soils could assist in the adaptation of agriculture to meet rising demands for food and fibre. Furthermore, improving soil health with biochar applications may increase the resilience of agricultural systems and enable the continuation of farming on marginal lands (Chan et al. 2007; Steiner et al. 2008)

Application of biochar to soil has been shown to have many advantages including enhanced soil health characteristics, reduced metal contamination risks and consequently increased plant growth (Namgay et al. 2010; Reichenauer et al. 2009) as well as reduced greenhouse gas emissions from soil (Singh et al. 2010a, b; Van Zwieten et al. 2009; Yanai et al. 2007). The competing and often conflicting demands of land use primarily stem from growing populations requiring housing and food, coupled with community desires for greater allocation of land to ecological reserves and the increasing production of energy crops to displace greenhouse gas-emitting fossil fuels (Koomen et al. 2005; Simon and Wiegmann 2009). In addition to the challenge of a changing climate, the increasing claim for this scarce land-use resource will force the necessity for greater productivity from less land, meaning farmers will need to undertake activities that result in significant yield increases (Singh et al. 2017a, b, c; Vimal et al. 2018; Kumar et al. 2018; Singh 2019). Land managers from more developed countries historically had greater access to technological innovations and training, thereby improving the productivity of agricultural systems compared with those from developing countries (Singh 2013; Singh 2014; Singh 2015; Singh 2016; Singh Boudh 2016; Kumar et al. 2017; Kumar and Singh 2017; Tiwari and Singh 2017). With the escalating effects of climate change, technological adaptation will become increasingly vital to sustainably augment production systems globally (Bryan et al. 2009).

Recently, biochar has attracted much attention due to its promising role in many environmental management issues. Biochar can be mainly used as a large-scale soil amendment a wide range of agricultural applications including chemical contaminants, soil fertilization as well as carbon sequestration. It improves soil quality by increasing soil pH, moisture-holding capacity, attracting more beneficial fungi and microbes as well as improving cation exchange capacity (Laird 2008). The presence of many functional groups such as carboxylic, alcohol, and hydroxyl on the biochar surface makes an opportunity to form complexes between these groups and heavy metal ions (Woolf 2008). Due to the specific surface area and high sorption capacity in regard to heavy metals ions, the biochar could be used as an effective adsorbent of these contaminants (Matthews 2008). Being a renewable resource and due to its economic and environmental benefits, biochar is a promising resource for environmental technology used for water contaminants treatment. Most studies have reported that biochar showed excellent ability to remove contaminants such as heavy metals, organic pollutants and other pollutants from aqueous solutions. Meanwhile, several biochars exhibit comparable or even better adsorption capacity than commercially activated carbon (Zhang et al. 2013; Yang et al. 2014). Therefore, it is very imperative to emphasize the importance of biochar production, characterization and various applications of biochar in this book chapter.

5.2 Biochar and Its Characteristics

5.2.1 Biochar

Biochar is a carbonaceous material produced from organic waste through thermochemical decomposition under different temperatures in the presence of little/no oxygen by the process of pyrolysis and used for soil conditioning, carbon conservation and greenhouse abatement (Kannan et al. 2016).

5.2.2 Pyrolysis

The chemical breakdown of a substance under high temperatures in the absence of oxygen is called as pyrolysis (Fig. 5.1).

Organic chemical transformation in pyrolysis (Fig. 5.2).

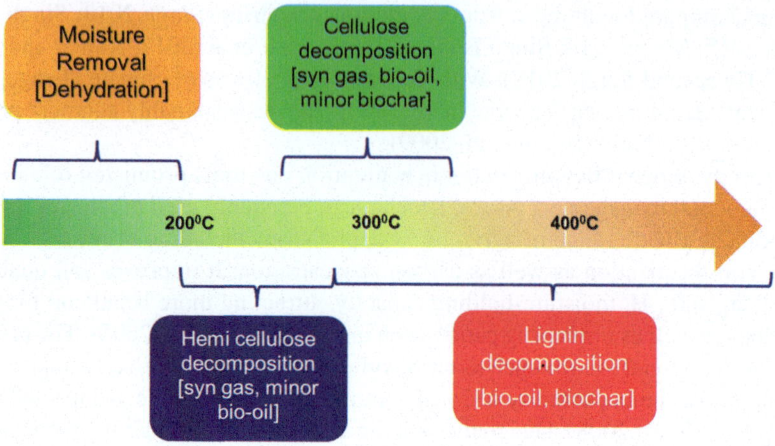

Fig. 5.1 Pyrolysis temperature and conversion nature of proximate components of plant

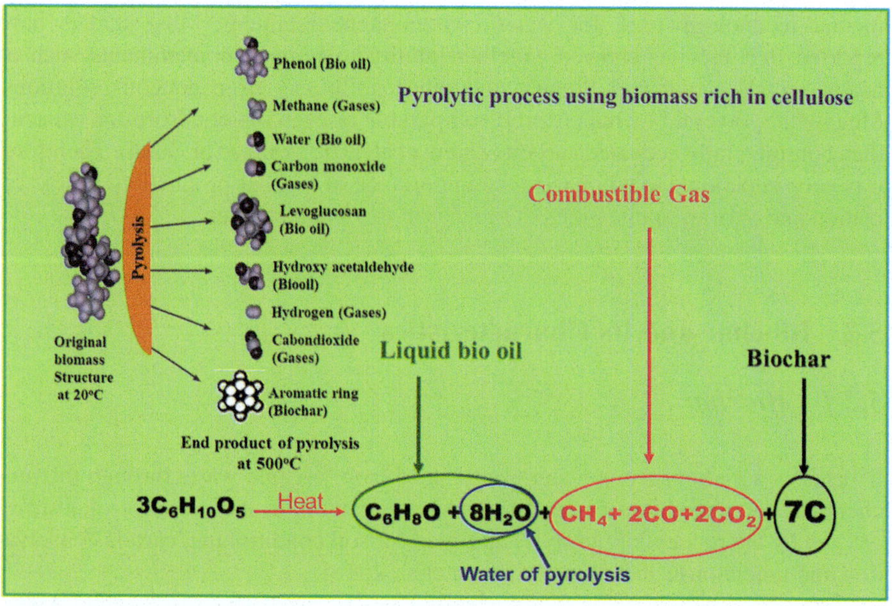

Fig. 5.2 Chemical transformation process of proximate components of the plant into biochar

5.2.3 Biochar Production

Biochar can be produced by thermochemical decomposition of biomass at temperatures of 200–900 °C in the presence of little or no oxygen, which is commonly known as pyrolysis (Demirbas and Gonenc 2002). Pyrolysis is generally divided into fast, intermediate, and slow depending on the residence time and temperature.

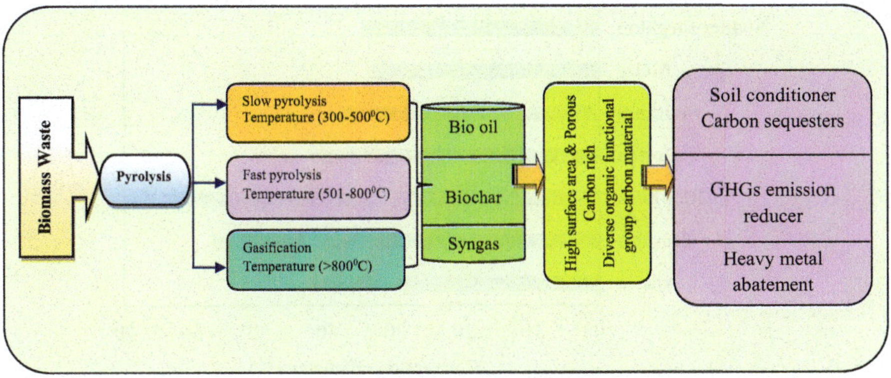

Fig. 5.3 Biomass conversion, properties and application potential

Table 5.1 Effect of pyrolysis method, temperature and residence time on the end product

Process	Temperature (°C)	Residence time	Products Liquid (bio-oil) (%)	Solid (biochar) (%)	Gas (syngas) (%)
Fast pyrolysis	300–1000	Short (<2 s)	75	12	13
Intermediate pyrolysis	500	Moderate (10–20 s)	50	25	25
Slow pyrolysis	100–1000	Long (5–30 min)	30	35	35
Gasification	>800	Moderate (10–20 s)	5	10	85

Fast pyrolysis with a very short residence time (<2 s) is often used to produce bio-oil from biomass yielding about 75% bio-oil (Mohan et al. 2006). Slow and intermediate pyrolysis processes with a residence time of a few minutes to several hours or even days are generally favored for biochar production (25–35%) (Brown 2009). Gasification is different from the general pyrolysis process. For gasification, the biomass is converting into gases rich in carbon monoxide and hydrogen by reacting the biomass at high temperature (>700 °C) in a controlled oxygen environment and/or steam. The resulting gas mixture is known as synthetic gas or syngas (Mohan et al. 2006) (Fig. 5.3 and Table 5.1).

5.2.4 Conversion Efficiency

Conversion of woody biomass to biochar has shown an average recovery of 54% of the initial carbon in the biochar (Lehmann et al. 2003). But our study reported 32–67% recovery from different biomass and these variations mainly attributed to the nature of feedstock used and thickness of the feedstock. The conversion

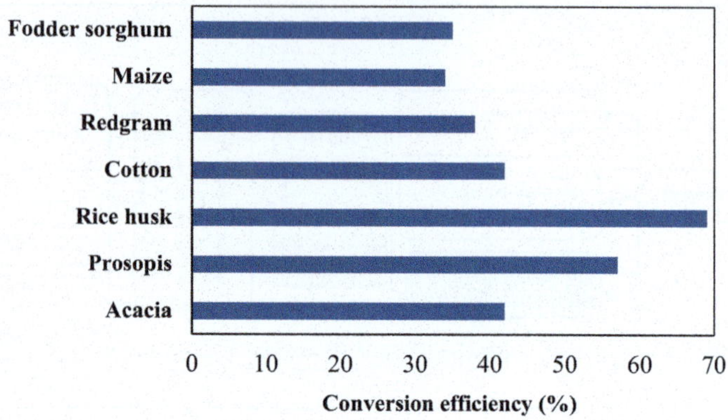

Fig. 5.4 The conversion efficiency of different agricultural waste into biochar

efficiency of different agricultural wastes was given in Fig. 5.4. The crop wastes of 500 kg were converted into its biochar respective under the anaerobic condition at 300–400 °C by slow pyrolysis process using slow pyrolysis unit (Kannan et al. 2016)

Biochar is a fine-grained and porous substance, similar in its appearance to charcoal produced by natural burning. It is produced by the combustion of biomass under oxygen-limited conditions (International Biochar Initiative 2008). As a soil amendment, biochar creates a recalcitrant soil carbon pool that is carbon-negative, serving as a net withdrawal of atmospheric carbon dioxide stored in highly recalcitrant soil carbon stocks (Lehmann and Joseph 2009).

Biochar is the carbon-rich solid product resulting from the heating of biomass in an oxygen-limited environment. Due to its highly aromatic structure, biochar is chemically and biologically more stable compared with the organic matter from which it was made. It has been proposed as a technology to play a useful role in building soil health and mitigating climate change. Properties of biochar vary widely, depending on the biomass source used and the conditions of production of biochar (Lehmann and Joseph 2009).

The pH of the biochar produced from different agricultural feedstocks ranged from 7.9 to 11.2 which are of alkaline range. Among the various feedstocks used, fodder sorghum stalk biochar (11.2) and prosopis biochar (10.8) are high in alkaline nature. Rice husk biochar registered a low level of salinity (0.22 dS/m) whereas the other feedstocks biochar (0.58–2.1 dS/m) are likely to develop a moderate level of salinity. Among the various agricultural feedstocks, redgram and maize stalk biochar registered the highest organic carbon content of 76 g/kg followed by cotton biochar 69 g/kg (Table 5.2).

The nutrient composition of biochars varied with the source of feedstocks. The total N varied from 0.43 to 2.06%. The biochar produced from maize stalk registered the highest total N of 2.06% and a total P of 0.84% and fodder sorghum recorded a maximum total K of 2.7%. The lowest total N of 0.31% was in cotton stalk biochar and the lowest total P of 0.23% and total K of 0.20% were recorded in the rice husk biochar.

Table 5.2 Chemical properties of Biochar (Kannan et al. 2016)

Biochar	pH	EC (dS/m)	OC (g/kg)	Total (C %)	Total (N %)	Total (P %)	Total (K %)
Prosopis	9.4–10.8	0.83–1.25	25–32	62–89	0.70–1.23	0.05–0.26	0.2–0.5
Rice husk	7.9–8.1	0.22–0.52	34–57	57–64	0.63–1.78	0.07–0.23	0.1–0.2
Maize	8.9–10.0	0.65–1.09	21–76	56–71	0.43–2.06	0.08–0.84	0.3–0.8
Cotton	8.8–10.2	0.58–0.85	17–69	54–85	0.31–0.67	0.15–0.39	1.1–1.4
Redgram	8.4–10.3	0.63–1.0	24–76	61–75	0.53–1.65	0.18–0.46	0.8–2.5
Fodder sorghum	9.5–11.2	1.2–2.1	8–15	51–54	0.32–1.02	0.16–0.24	1.1–2.7
Acacia	8.9–9.8	0.55–0.90	45–57	65–72	0.60–1.54	0.22–0.65	1.1–1.7

5.3 Implications of Biochar on Soil Health

The presence of biochar in the soil has a significant effect on the physical nature of the system, affecting the texture, depth, porosity, structure, and consistency, particle-size distribution, surface area, pore size distribution, and packing. Also, biochar affects the physical properties of soil, physical properties that may subsequently have a direct effect on plant growth. Since the availability and penetration depth of water and air into the root zone is determined mainly by the physical compositions of the soil horizons and physical makeup (Lehmann and Rondon 2006; Chan et al. 2008). The presence of biochar will directly influence the response of soil to water, its aggregation, permeability, and swelling-shrinking dynamics, in addition to its capacity to hold cations and its reaction to ambient-temperature variations (Brady and Weil 2008). Biochar also changes the physical nature of soil, causing a net increase in the total soil-specific surface area and improves the soil structure and aeration. Compositional proportions vary according to feedstock materials, pyrolysis conditions, processing temperatures adopted, heating rates and pressures (Fig. 5.5).

As a consequence of improved soil physical properties (structure, surface area, porosity, bulk density, and water holding capacity), plant water availability, nutrient retention capacity, root penetration, and aeration do increase (Chia et al. 2015). Sandy soils amended with biochar have higher water holding capacities than do loamy and clay soils, while increased soil aeration is mainly observed in fine-textured soils (Mukherjee et al. 2014).

The application of biochar increases microbial activity and biomass and changes the microbial community composition and abundance. However (Tiwari et al. 2019a; Tiwari et al. 2019b; Singh et al. 2019; Kour et al. 2019), the beneficial effect of biochar on the soil environment depends on the type of biochar, application rate, soil type and plant response (Lehmann et al. 2011). The bacterial community in soils of cotton that have continuously been cropped for 2 years, 6 years, 11 years

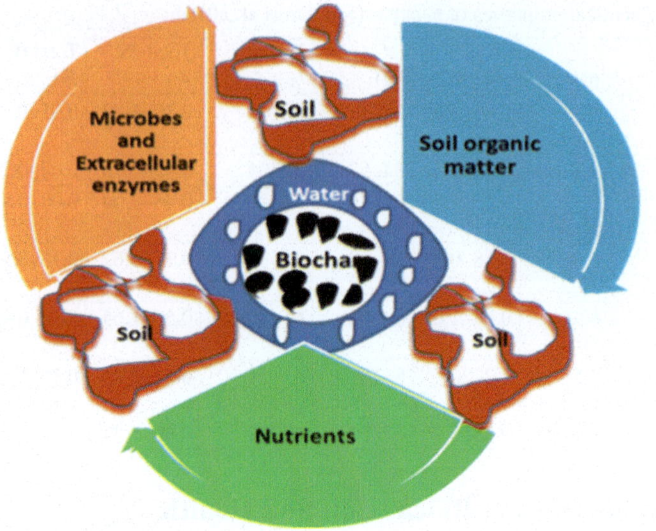

Fig. 5.5 Schematic diagram of biochar and soil interaction. Biochar adsorbs microbes, inorganic nutrients, and soil organic matter

and 14 years and treated with biochar (B0, 0 t/ha; B1, 12.5 t/ha and B2, 20 t/ha) was investigated using next-generation sequencing. The relative abundance of Sphingomonas and Pseudomonas in the biochar-treated soils was significantly higher than that in the soil without biochar treatment.

The results suggest that the biochar application has a significant impact on the soil bacterial community, which may improve the microbial diversity of continuous cropping systems in cotton soils (Han et al. 2017).

Application of biochar could influence the mycorrhizal fungi functioning with the following mechanisms by: (i) modifying physicochemical characteristics of the soil, (ii) indirectly influencing the mycorrhizae by affecting soil microbes in the surroundings, (iii) interfering with plant–fungus signaling and allelochemical detoxification on biochar, and (iv) the provision of refugia from fungal grazers. The porous structure of biochar enhances the habitat of mycorrhizal fungi and other soil microbiota, which improved the soil quality (Warnock et al. 2007).

Biochar may increase the cation and anion exchange capacity of the soil (Singh et al. 2010a, b; Liang et al. 2006) improving the soil properties through raising in pH, increases in total N and P, encouraging greater root development and decreasing available aluminum(Cheng et al. 2006; Chan et al. 2008). Moreover, biochar holds the capability to reduce the effects of drought by raising soil moisture content, therefore decreasing nutrient leaching and soil erosion (Lorenz et al. 2007). Higher water bioavailability and moisture retention are thought to be an important factor for achieving superior yields in biochar-amended soil. Conversely, the biochar surface contain numerous chemically reactive groups, such as OH, COOH, and ketones that provide enormous potential for the adsorption of toxic chemicals, such as aluminum

(Al) and manganese (Mn) in acid soils, and arsenic (As), cadmium (Cd), copper (Cu), nickel (Ni) and lead (Pb) in heavy metal-contaminated soils (Berek et al. 2011; Uchimiya et al. 2010). In landfill sites, biochar particle size increases the soil porosity and promotes the airflow through the landfill cover and increased oxygen diffusion within the landfill cover, which leads to potentially higher levels of microbial degradation (Yaghoubi and Reddy 2011). There is a significant interest in the application of biochar to soils that have gained, significant interest due to the multifold benefits of biochar such as nutrients retention and water-holding capacity and it also promoting plant growth (Quayle 2010; De Gryze et al. 2010). It has been found that biochar can: (a) decrease soil tensile strength, (b)increase improve soil structure and pH, (c) improve fertilizer use efficiency, (d) decrease aluminum toxicity to plant roots and microbiota, and (e) improves soil conditions for earthworm populations. Furthermore, biochar reduces the leaching of soil nutrients, which enhance the availability of nutrients for plants and reduces the bioavailability of heavy metals (Lehmann et al. 2003).

Biochar produced at low-temperatures (300 or 400 °C) is acidic whereas at high temperatures (700 °C) it is alkaline in nature. This is an important finding as the agricultural use of biochar can have a two-fold application. If the soil intended for biochar application is acidic in nature, then the biochars produced at 700 °C or higher temperatures can be used to neutralize the soil and improve soil fertility. Alternatively, biochars formed at lower temperatures might be suitable for alkaline soils to correct for alkalinity problems. It also serves as a valuable soil amendment by supplying plant nutrients with carbon sequestration (Hossain et al. 2011). Biochar is a potential tool for improving the quality, agronomic value of soil, and minimize the harmful effects of heavy metals present in the soil. The multi-fold benefits of biochar application in the soil are listed below for easy understanding of the reader. The application of grass derived biochar and oak-derived biochar in unburned increased the bacterial population (118.7 ± 121.0 and 87.7 ± 4.4 CFUs are per gram of soil, respectively) as compared to control (31.8 ± 1.4 CFUs are per gram of soil) application of biochar (Khodadad et al. 2011).

5.4 Implications of Biochar on Crop Growth and Yield

A combination of higher biochar application rates along with NPK fertilizer increased crop yield on tropical Amazonian soils (Steiner et al. 2007) and semi-arid soils in Australia. Biochar application in low pH soil (<5.2) under steep slope enhanced the carrots and beans yields over the control (Rondon et al. 2004). According to the Lehmann and corkers, increasing yields with increasing biochar applications up to 140 Mg C ha^{-1} on highly weathered soils in the humid tropics (Lehmann et al. 2006). This was not true for all crops however they found that biomass growth of beans increased with biochar applications up to 60 Mg C ha^{-1} but fell to the same value as for control plots when biochar application was increased to

90 Mg C ha^{-1} (Rondon et al. 2004). Crops respond positively to bio-char additions up to 50 Mg C ha^{-1} and may show growth reductions at a very high application rate (Lehmann 2007). Application of redgram stalks biochar @ 5 t ha^{-1} increased the dry matter production 24% and groundnut pod yield 29% in acidic red soil under rain-fed situation (Kannan et al. 2016). Application of biochar at the rate of 25 ha^{-1} in combination with FYM at the rate of 10 ha^{-1} and N at the rate of 30 kg ha^{-1} is recommended for improving mung bean growth and yield (Hussain et al. 2017). Biochar @ 10 t ha^{-1} increased above ground biomass by 23% and grain yield by 10% of durum wheat compared to control (Steiner et al. 2007).

Application of biochar prepared from wheat straw (1.9 ha^{-1}) along with recommended doses of NPK at 180:80:80 kg ha^{-1} significantly increased the yield of maize in Inceptisol of IARI farm and this treatment was superior to either crop residue incorporation or crop residue burning (Purakayastha 2010).

The production of plant biomass through photosynthesis removes CO_2 from the atmosphere, and therefore any increase in plant biomass due to biochar additions in soil systems will contribute to the mitigation of rapidly rising atmospheric CO_2 levels. Specifically, biochar increases plant nutrient availability and enhances the soil environment (CEC, soil pH, aeration), which in turn indirectly contributes to enhanced plant growth (Chan et. 2008; Steiner et al. 2007; Zackrisson et al. 1996). In a field trial (Cowpea and Rice) in Amazon basin Anthrosol soils with high carbon levels and Ferralsols with added wood biochar, significantly increased phosphorus, calcium, manganese and zinc availability, with a 38–45% increase in biomass of the two crops in the Anthrosol (Lehmann et al. 2003). The application of biochar doubled the crop yield in maize under degraded cropping soils and the improvement could be explained by biochar nutrient availability and other soil properties improvement (Kimetu et al. 2008). Biochars generally appear to increase nutrient availability through increased ion retention in soils (Liang et al. 2006) and therefore potentially enhance plant yields. Biochar applications produced from manures may directly contribute high levels of nutrients to soils. Field experimental results reported that the application of 10, 25 and 50 t/ha of poultry manure biochar enhanced the yield of radish (Chan et al. 2008)

Lin et al. (2015) observed yield increases of 11% in soybean grain yield and of 28% in wheat grain yield following maize stalk biochar application to coastal saline soil. Genesio et al. (2015) reported an even greater grape yield increase (66%) in the same field after applying biochar from orchard prunings. On the other hand, Schmidt et al. (2014) reported neither a grape yield nor quality effect after wood BC was applied to Swiss vineyard soils during a 4-year field trial. The application of biochar @ 100 Mg dry weight ha^{-1}, which increases the yield were 1.17 and 0.43 Mg ha^{-1} for maize and soybean, respectively (Katterer et al. 2019). Berihun et al. (2017) reported that the Application of *Lantana* biochar @ 18 t ha^{-1} increased the grain yield of maize (528 kg ha^{-1}) than compared to control (134 kg ha^{-1}) (Table 5.3).

Table 5.3 Effective of different rate of biochar application on crop yield under different soil type

Crops	Soil type	Biochar rate (t/ha)	Fertilizer rate (kg/ha)	Yield / biomass increase over control (%)	Additional information
Wheat	Ferrosol	10	1.25 g nutricote per 250 g soil (nutricote contain 15.2% N. 4.7% P, and 8.9% K)	+250	A similar response was observed for the biomass yield of Soya bean and radish. Calcarosol amended with fertilizer and biochar however gave varied crop responses (Van Zwieten et al. 2010)
Radish	Alfisol	100	N (100)	+266 (biomass)	In the absence of nitrogen fertilizer application of Biochar did not increase the dry matter production of radish even at a higher rate(100 t/ha) (Chan et al. 2008)
Rice	Inceptisol	30	Nil	+294	Sole effect of biochar (Noguera et al. 2010)
	Oxisol	88	Nil	+800	Interaction effect of earthworm and biochar
	Oxisol	88	N (40), P (20), K (20)	−21	Interaction effect of earthworm and biochar
Maize	Oxisol	20	N (156–170), P (30–43), K (83–138)	+28 (1st year) +30 (2nd year) +140 (3rd year)	In the first year after biochar application. No significant effect on crop yield was observed (Major et al. 2010)
Rice	Ferralsol	11	N (30), P (35), K (50)	+29 (stover) +73(grain)	While charcoal addition alone did not affect Crop production, a synergistic effect occurred when both charcoal and inorganic fertilizer were applied (Steiner et al. 2008)
Groundnut	Alfisol	5	N (10), P (10) and K (45)	+29(pod yield)	Biochar addition mainly influence soil moisture retention and increase the soil pH, thereby enhance the nutrient availability in rainfed Alfisol (Kannan et al. 2016)

(continued)

Table 5.3 (continued)

Crops	Soil type	Biochar rate (t/ha)	Fertilizer rate (kg/ha)	Yield / biomass increase over control (%)	Additional information
Maize	HaplicLuvisol	20	Nil	+66 (Cob yield)	Maize grain yield did not significantly increase in the first year after the biochar application, but it increased at plots with the 20 t/ha of biochar over the control by 28, 30 and 140% in three following years (Vitkova et al. 2017).

5.5 Implications of Biochar on Climate Change

5.5.1 Biochar Effect on Carbon Sequestration

Carbon in biochar can persist in soils over a long time. Beyond the carbon sequestered in the internal structure, biochar incorporated in soils also offers numerous other potential climate benefits.

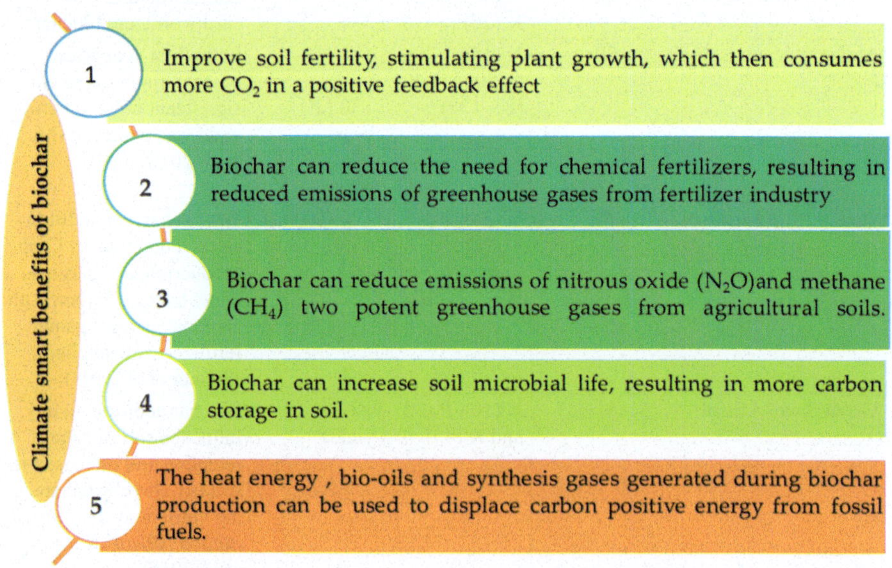

Carbon sequestration is a process in which carbon is captured and stored to prevent it from being emitted into the atmosphere (Duku et al. 2011). It is essential that the carbon is transferred to a passive carbon pool that is stable or inert, in order to decrease C emission to the atmosphere. Hence, biochar provides an easy route from the active carbon pool to the passive pool. Transferring, even a small amount of the carbon that cycles between the atmosphere and plants, to a much slower biochar cycle, would impact greatly on atmospheric CO_2 concentrations because of the annual uptake of CO_2 by plants from the atmosphere through photosynthesis is eight times greater than anthropogenic GHGs emissions. Biochar is biologically and chemically more stable than the original carbon form, due to its molecular structure and its origins.

It is difficult for the sequestered carbon to be released as CO_2, making this a good method for carbon sequestration (Shafie et al. 2012; Lehmann 2009). The diversion of even 1% of the net annual uptake of carbon by plants into biochar would mitigate almost 10% of current anthropogenic carbon emissions (Lehmann and Joseph 2009). It is assumed that 3 billion tonnes of biochar are produced annually. This, in turn, reduces approximately 3 billion tonnes of atmospheric carbon emissions if all of the biomass (60.6 billion tonnes) is regenerated in the form of biochar through pyrolysis (Fernandez-Lopez et al. 2015). An estimated 1 billion tonnes of carbon will be sequestered annually by 2030, which is a rationally conservative approximation of the potential of biochar (Shackley et al. 2009). The thermochemical conversion of biomass into biochar through pyrolysis increases the recalcitrance of the carbon that originated in the biomass. The addition of biochar of similar carbon content to soil leads to steady soil carbon levels, due to its stability in soil (Lehmann et al. 2006). The biochar acts as a carbon sink that remains in the soil for long periods of time, possessing high levels of resistance to chemical and biological degradation, in turn increasing terrestrial carbon stocks. It is estimated that 20% of the total carbon biomass can be captured by conversion into biochar (Lehmann 2007). The common consensus is that soil is a finite C sink at best; the application of biochar provides an opportunity for reducing C emissions and sequestering C for soil remediation (Freibauer et al. 2004; Lal 2004).

5.5.2 Biochar for the Reduction of GHG Emissions

In the carbon cycle, atmospheric CO_2 is fixed by photosynthetic organisms (e.g. plants) and then it is converted into biomass that is mixed with soil when these organisms die. The biomass in soil is mineralized and microbial respiration causes the evolution of CO_2 to the atmosphere. When biochar is applied to soil, its recalcitrant nature causes it to stay in the soil for long periods of time, thereby reducing GHG emissions. Pyrolysis products (bio-oil, syngas) are burned as fuel, releasing CO_2 into the atmosphere that will be utilized by plants, and ultimately converted into biomass again. The carbon cycle of biochar-production has some indirect GHG-emission sources as well. They conducted a life cycle assessment of pyrolysis and concluded that the operation and maintenance of the pyrolysis process contribute to 89% of its GCG emissions, while building works, equipment, and transportation contribute 7.2, 3.33, and 0.23%, respectively(Yang et al. 2016). Global warming potential of several biochars derived from agriculture, poultry litter, sewage sludge, cattle manure, and food waste was compared and reported negative values for almost all of the biochar cycles, averaging −0.9 kg CO_2eq/kg, indicating that more GHG is consumed than emitted (Alhashimi and Aktas 2017).

Apart from CO_2, methane (CH_4) and nitrous oxide (N_2O) emitted from soil have the potential to influence climate. Agriculture is a primary contributor to atmospheric GHGs. CH_4 is generated by soil microorganisms under anaerobic conditions through the methanogenesis. CH_4 is approximately 20 times more powerful than CO_2 in absorbing thermal radiation trapped in the Earth's troposphere, and this augments global warming. Methane emissions were close to zero when biochar was applied at a rate of 2% to the soil. Reduction in CH_4 emissions involves increased soil aeration that may lead to reductions in the frequency and extent of the anaerobic conditions under which methanogenesis occurs (Verheijen et al. 2010).

Another study showed that the biochar amendment significantly reduced total indirect CO_2 while increasing CH_4 emissions from paddy soil (Zhang et al. 2010). The CH_4 emission mainly depends on the physical and chemical properties of the biochar, soil type, soil microorganisms, water and fertilizer management (Van Zwieten et al. 2009). Nitrous oxide (N_2O) is produced by soil microorganisms through nitrification and denitrification. N_2O is 300 times more potent than CO_2 in absorbing thermal tropospheric radiation that enhances global warming. The production of N_2O is greatly affected by the presence of moisture in the soil, as higher moisture content (>70%) promotes the anaerobic conditions that favor denitrification, whereas lower moisture content (<50%) is strongly associated with nitrification. Around 8–23 times more N_2O is generated under conditions of high moisture content (80%) than at lower moisture content (40%) (Bruun et al. 2011). High soil moisture content, with 73–83% water-filled pore space (WFPS), promotes N_2O production, while this was not detected at the lower moisture content (64% WFPS). The addition of 10% biochar to soil (78% WFPS) reduced N_2O emissions by 89% (Yanai et al. 2007).

Biochar-induced reductions in N_2O are affected by the amount of biochar applied; higher application rates (20–60%) reduced N_2O by up to 74% while no reduction was observed at lower application rates (2–10%) (Spokas et al. 2009). Similarly, no reduction in N_2O emissions was found after addition of 4% biochar to soil. They observed that failed to find an immediate decline in N_2O emissions when soils were amended with low levels of biochar (10%) under 85% WFPS conditions, however, they observed a 73% reduction in N_2O over two subsequent rewetting cycles. Clearly, biochar under these circumstances, eventually exhibits improve sorption capacity (Yanai et al. 2007). Biochar soil amendment can affect nitrogen (N) transfer and soil N cycling processes that reduce N_2O emissions. Biochar aids in the biological immobilization of inorganic N that helps to retain N and decrease ammonia volatilization, as biochar contains low N concentrations and high C/N ratios. Biochar efficiently adsorbs NH_3 from the soil and acts as a buffer, thereby potentially decreasing ammonia volatilization from agricultural fields (Oya and Iu 2002). They found a reduced NO_3-N pool in biochar-amended soil plots and assumed that biochar particles enhanced the adsorption and uptake of NH_3. The impact of biochar on soil N_2O fluxes is variable and depends on factors such as soil type, soil water content, additional fertilizer application, biochar feedstock, and pyrolysis temperature. Moreover, biochar is an efficient adsorber of dissolved ammonium, nitrate, phosphate, and other ionic solutes, as well as hydrophobic organic pollutants in soil and water (Taghizadeh-Toosi et al. 2011).

5.6 Implications of Biochar on Environmental Clean-Up

Biochar can be an effective amendment for immobilizing heavy metals in contaminated soils but has variable effects depending on its chemical and physical properties and those of the treated soil. Meta-analysis results showed that across all studies, biochar addition to soils resulted in average decreases of 38, 39, 25 and 17%, respectively, in the accumulation of Cd, Pb, Cu and Zn in plant tissues. The effect of biochar on heavy metal concentrations in plants varied depending on soil properties, biochar type, plant species, and metal contaminants. The largest decreases in plant heavy metal concentrations occurred in coarse-textured soils amended with biochar. Biochar had a relatively small effect on plant tissue Pb concentrations, but a large effect on plant Cu concentrations when applied to alkaline soils. Plant uptake of Pb, Cu, and Zn was less in soils with higher organic carbon contents. Manure-derived biochar was the most effective for reducing Cd and Pb concentrations in plants as compared to biochars derived from other feedstocks. Biochar having a high pH and used at high application rates resulted in greater decreases in plant heavy metal uptake (Chen 2018).

The use of biochar for the removal of organic and heavy-metal contaminants from aqueous media is a relatively new and promising water and wastewater treatment technology. The presence of cellulose, hemicelluloses, lipids, sugars and proteins in agricultural residue feedstocks provides a variety of functional groups. These functional groups can be physically activated upon pyrolysis and by further steamer CO_2 treatment, to improve their ability to adsorb contaminants (Inyang et al. 2011).

The type and concentration of surface functional groups on biochar plays an important role in the adsorption capacity of the biochar. The carbon-structured matrix, the high degree of porosity, surface area, and a strong affinity for non-polar substances such as PAHs, dioxins, furans, and other compounds enable it to play a vital role as a surface sorbent for in controlling contaminants in the environment as a surface sorbent(Yu et al. 2009).

Biocharactsasasuper-sorbent for the removal of both organic and inorganic contaminants in soil and water. Considering the wide variety and availability of cheap feedstocks for biochar production, the use of biochar could be a cheaper remediation technology option for Pb adsorption than activated charcoal (Shang et al. 2012).

The surface of biochar can contain abundant and abundance of chemically reactive groups, (OH, COOH and ketones) that bestow biochar with an immense potential to adsorb heavy metals and toxic substances, such as aluminum (Al) and

Table 5.4 Effect of biochar application on the bioavailability of heavy metals in soils

Feedstock	Production temperature	Contaminant	Effect
Cotton stalks	450 °C	Cd	Reduction of the bioavailability of Cd in soil by adsorption or co-precipitation (Zhou et al. 2008)
Hardwood biochar	400 °C	As	Significant reduction of as in the foliage of *Miscanthus* (Hartley et al. 2009)
Eucalyptus	550 °C	As, Cd, Cu, Pb, Zn	A decrease in As, Cd, Cu, and Pb in maize shoots (Namgay et al. 2010)
Orchard prune residue	500 °C	Cd, Cr, Cu, Ni, Pb, Zn	Significant reduction of the bioavailable Cd, Pb, and Zn, with Cd showing the greatest reduction; an increase in the pH, CEC, and water-holding capacity (Fellet et al. 2011)
Chicken manure & green waste	550 °C	Cd, Cu, Pb	Significant reduction of Cd, Cu, and Pb accumulation by Indian mustard (Park et al. 2013)
Chicken manure	550 °C	Cr	Enhanced soil Cr (VI) reduction to Cr (III) (Choppala et al. 2012)
Sewage sludge	500 °C	Cu, Ni, Zn, Cd, Pb	Significant reduction in plant availability of the metals studied (Mendez et al. 2014)
Rice straw	300–400 °C	Cu, Pb, Cd	Significant reduction in concentrations of free Cu, Pb, and Cd in contaminated soils; identification of functional groups on biochar with high adsorption affinity to Cu (Jiang et al. 2012)
Quail litter	500 °C	Cd	Reduction of the concentration of Cd in the physic nut; greater reduction with the higher application rates (Suppadit et al. 2012)
Oakwood	400 °C	Pb	Bioavailability reduction by 75.8%; bio accessibility reduction by 12.5% (Ahmad et al. 2012)
Peanut shell & wheat straw	350–500 °C	Cd and Pb	5% PBC addition lowered Cd and Pb concentrations in grains by 22.9 and 12.2%, while WBC addition lowered them by 29.1 and 15.0%, respectively (Xu et al. 2017)

manganese (Mn) in acid soils, and arsenic (As), cadmium (Cd), copper (Cu), nickel (Ni) and lead (Pb) in heavy metal contaminated soils (Singh et al. 2017a, b, c, d, e, 2018; Tiwari et al. 2018; Yuan et al. 2011). Due to the dissociation of oxygen-containing functional groups, biochars mostly carry net negative charges on their surfaces. So, it can be utilized as low-cost adsorbents for the removal of organic contaminants and heavy metal cations from water (Qiu et al. 2008). Several other researchers reported the efficiency of heavy metal removal using biochar with derived from rice husk (Liu and Zhang 2009), corn straw (Chen et al. 2011), peanut straw (Tong et al. 2011), olive pomace (Pellera et al. 2012) as well as oak wood and bark (Xue et al. 2012). Most of the heavy metals were adsorbed on the biochar surface using through inorganic constituents in the biochar (Table 5.4). Organic contaminants were either removed due to sorption or interaction with functional groups and surface charges.

Several studies report the removal of heavy metals including As(III, V) using iron-impregnated magnetic biochars (Wang et al. 2015), Cr(VI) using zinc and chitosan-modified biochars (Gan et al. 2015 and Huang et al. 2016), Pb (II), Cu(II), and Cd(II) using a $KMnO_4$-treated wood biochar (Wang et al. 2015), and Hg(II) using a graphene-treated biochar; Pb(II) and As(V) is the most studied of these heavy metals. Adsorption capacities for Pb (II) of 4.9–367.6 mg^{-1} were reported for a ZnS-biochar (Yan et al. 2015). Removal of phosphorus has also been reported in several studies using oxides of Ca, Mg, and Al-modified biochars (Liu et al. 2016). The use of catalytic and degradative nanoparticles, such as nanoscale zerovalent ions (Yan et al. 2015) and graphitic C_3N_4 have been reported to remove several organic chemicals. The use of nanocomposites is clearly a promising technology for the treatment of aqueous media, but it is in its infancy and requires a lot of further research, especially regarding the re-use, desorption, and disposal of these metal-attached nanocomposites.

5.7 Conclusions

Agricultural residues and municipal yard wastes can be a significant burden on the environment. Nutrients contained in the wastes may cause eutrophication of surface waters or pollution of groundwaters. Landfills of municipal green wastes may generate large quantities of greenhouse gases. But all these substances can be useful when managed properly. Biochar production is an intelligent way of recycling organics for soil amendment and reduces environmental pollution. Across the results of global biochar experiments clearly revealed that the application of different sources and quantity of biochar in different soil types showed positive improvement of soil health; positive, negative and no effect on greenhouse gas emission reduction. The better and poor crop growth and high and low yield response of biochar were noticed in different biochar experiments. Heavy metal abatement potential of biochar also varied among the biochar and heavy metals. The wider variation of biochar responses in the soil system mainly due to the feedstock types, pyrolysis method, and temperature. So prioritize and standardize the production technique, characterization and application of different sources biochars in different crops under varied soil types are very imperative to mitigate the vulnerability of climate change and sustain soil health in a different ecosystem (Fig. 5.6).

- The opportunities for carbon sequestration and the reduction of greenhouse gas emissions have not been explored in different ecosystems under field conditions, but they are potentially significant under changing climate.
- Further, it is to be studied in detail for promoting biochar as a greening approach to the environment as well as human health. Published data for the effect of biochar on trace gas emission is extremely limited under field conditions but has a potentially great impact on the net benefit of a biochar strategy.

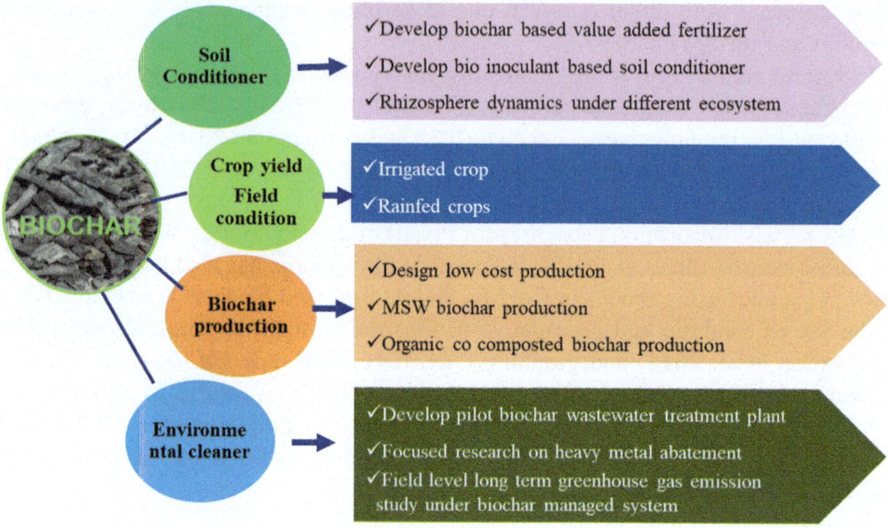

Fig. 5.6 Future Thrust Area in Biochar Research

- Good predictive models will be necessary for this to be reflected in future accounting for biochar projects.
- Municipal solid waste disposal through biochar production one of the viable options and it has to be studied in detail about production, characterization, and standardization for different crops.
- Long term effect of biochar application on soil health to be ascertained and its effect on crop yield to be studied in a long-term experiment under changing climate.

References

Ahmad M, Lee SS, Yang JE, Ro HM, Lee YH, Ok YS (2012) Effects of soil dilution and amendments (mussel shell, cow bone and biochar) on Pb availability and hytotoxicity in military shooting range soil. Ecotoxicol Environ Saf 79:225–231

Alhashimi HA, Aktas CB (2017) Life cycle environmental and economic performance of biochar compared with activated carbon: a meta-analysis. Resour Conserv Recycl 118:13–26

Berek AK, Hue N, Ahmad A (2011) Beneficial use of biochar to correct soil acidity. Food Provider Hanai Ai 9:1–3

Berihun T, Tolosa S, Tadele M, Kebede F (2017) Effect of biochar application on growth of garden pea (*Pisum sativum* L.) in acidic soils of Bule Woreda Gedeo Zone Southern Ethiopia. Int J Agron 2017:6827323

Brady NC, Weil RR (2008) The nature and properties of soils, 14th edn. Pearson- Prentice Hall, Upper Saddle River, p 990

Brown R (2009) Biochar production technology. In: Lehmann J, Joseph S (eds) Biochar for environmental management: science and technology. Earthscan, London, pp 127–139

Bruun E, Muller-Stover D, Ambus P, Hauggaard-Nielsen H (2011) Application of biochar to soil and N_2O emissions: potential effects of blending fast-pyrolysis biochar with anaerobically digested slurry. Eur J Soil Sci 62(4):581–589

Bryan E, Deressa TT, Gbetibouo GA, Ringler C (2009) Adaptation to climate change in Ethiopia and South Africa: options and constraints. Environ Sci Pol 12:413–426

Chan KY, Van Zwieten L, Meszaros I, Downie A, Joseph S (2007) Agronomic values of green waste biochar as a soil amendment. Aust J Soil Res 45:629–634

Chan KY, Zwieten VL, Meszaros I, Downie A, Joseph S (2008) Using poultry litter biochars as soil amendments. Aust J Soil Res 46:437–444

Chen D (2018) Effects of biochar on availability and plant uptake of heavy metals – a meta-analysis. J Environ Manag 222:76–85

Chen B, Chen Z, Lv S (2011) A novel magnetic biochar efficiently sorbs organic pollutants and phosphate. Bioresour Technol 102(2):716–723

Cheng CH, Lehmann J, Thies J, Burton SD, Engelhard MH (2006) Oxidation of black carbon by biotic and abiotic processes. Org Geochem 37:1477–1488

Chia CH, Downie A, Munroe P (2015) Biochar for environmental management: science, technology, and implementation. Routledge, Abingdon, pp 89–110

Choppala GK, Bolan NS, Megharaj M, Chen Z, Naidu R (2012) The influence of biochar and black carbon on reduction and bioavailability of chromate in soils. J Environ Qual 41:1175–1184

De Gryze S, Cullen M, Durschinger L, Lehmann J, Bluhm D (2010) Evaluation of the opportunities for generating carbon offsets from soil sequestration of biochar. In: Lehmann J, et al (eds) An issues paper commissioned by the Climate Action Reserve, final version. Climate Action Reserve

Demirbas A, Gonenc A (2002) An overview of biomass pyrolysis. Energy Sources 24:471–482

Duku MH, Gu S, Hagan EB (2011) Biochar production potential in Ghana-A review. Renew Sust Energ Rev 15(8):3539–3551

Ericksen PJ, Ingram JSI, Liverman DM (2009) Food security and global environmental change: emerging challenges. Environ Sci Pol 2:373–377

Fellet G, Marchiol L, Delle Vedove G, Peressotti A (2011) Application of biochar on mine tailings: effects and perspectives for land reclamation. Chemosphere 83:1262–1297

Fernandez-Lopez M, Marisa C, Almeida R, Pereira C, Iolanda Ribeiro L (2015) Life cycle assessment of swine and dairy manure: Pyrolysis and combustion processes. Bioresour Technol 182:184–192

Freibauer A, Rounsevell MA, Smith P, Verhagen A (2004) Carbon sequestration in the agricultural soils of Europe. Geoderma 122(1):1–23

Fresco LO (2009) Challenges for food system adaptation today and tomorrow. Environ Sci Pol 12:378–385

Gan C, Liu Y, Tan X, Wang S, Zeng G, Zheng B, Li T, Jiang Z, Liu W (2015) Effect of porous zinc-biochar nano composites on Cr (VI) adsorption from aqueous solution. RSC Adv 5(44):35107–35115

Gaunt JL, Cowie AL (2009) Biochar, greenhouse gas accounting, and emissions trading. In: Lehmann J, Joseph S (eds) Biochar for environmental management: science and technology. Earthscan, London, pp 317–340

Genesio L, Miglietta F, Baronti S, Vaccari FP (2015) Biochar increases vineyard productivity without affecting grape quality: results from a four years field experiment in Tuscany. Agric Ecosyst Environ 201:20–35

Han G, Lan J, Chen Q, Yu C, Bie S (2017) Response of soil microbial community to application of biochar in cotton soils with different continuous cropping years. Sci Rep 7:10184

Hartley W, Dickinson NM, Riby P, Lepp NW (2009) Arsenic mobility in brownfield soils amended with green waste compost or biochar and planted with Miscanthus. Environ Pollut 157:2654–2662

Hossain MK, Strezov V, Chan KY, Ziolkowski A, Nelson PF (2011) Influence of pyrolysis temperature on the production and nutrient properties of wastewater sludge biochar. J Environ Manag 92(1):223–228

Huang XX, Liu YG, Liu SB, Tan XF, Ding Y, Zeng GM, Zhou YY, Zhang MM, Wang SF, Zheng BH (2016) Effective removal of Cr (VI) using b-cyclo dextrine chitosan modified biochars with sorption/reduction bifunctional roles. RSC Adv 6:94

Hussain Z, Khan N, Ullah S, Liaqat A, Nawaz F, Khalil AUR, Ali Shah J, Junaid M, Ali M (2017) Response of mung bean to various levels of biochar, farmyard manure and nitrogen. World J Agric Sci 13(1):26–33

International Biochar Initiative(2008) Biochar sustainability and security in a changing climate. IBI conference, Newcastle, United Kingdom, September 8–10, 2008

Inyang M, Gao B, Ding W, Pullammanappallil P, Zimmerman AR, Cao X (2011) Enhanced lead sorption by biochar derived from anaerobically digested sugarcane bagasse. Sep Sci Technol 46(12):1950–1956

Jiang J, Xu R, Jiang T, Li Z (2012) Immobilization of Cu (II), Pb (II) and Cd (II) by the addition of rice straw derived biochar to a simulated polluted Ultisol. J Hazard Mater 229–230:145–150

Kannan P, Ponmani S, Prabukumar G, Swaminathan C (2016) Effect of biochar amendment on soil physical, chemical and biological properties and groundnut yield in rainfed Alfisol of semi-arid tropics. Arch Agron Soil Sci 6:1293–1310

Katterer T, Roobroeck D, Andren O, Kimutai G, Karltun E, Kirchmann H, Nyberg G, Vanlauwe B, deNowina KR (2019) Biochar addition persistently increased soil fertility and yields in maize-soybean rotations over 10 years in sub-humid regions of Kenya. Field Crop Res 235:18–26

Khodadad CLM, Zimmerman AR, Green SJ, Uthandi S, Foster JS (2011) Taxa-specific changes in soil microbial community composition induced by pyrogenic carbon amendments. Soil Biol Biochem 43:385–392

Kimetu J, Lehmann J, Ngoze S, Mugendi D, Kinyangi J, Riha S, Verchot L, Recha J, Pell A (2008) Reversibility of soil productivity decline with organic matter of differing quality along a degradation gradient. Ecosystems 11:726–739

Koomen E, Kuhlman T, Groen J, Bouwman A (2005) Simulating the future of agricultural land use in the Netherlands. Tijdschr Econ Soc Ge 96:218–224

Kour D, Rana KL, Yadav N, Yadav AN, Rastegari AA, Singh C, Negi P, Singh K, Saxena AK (2019) Technologies for biofuel production: current development, challenges, and future prospects. In: Rastegari et al (eds) Prospects of renewable bioprocessing in future energy systems, biofuel and biorefinery technologies. Springer, Cham, pp 1–50

Kumar A, Singh JS (2017) Cyanoremediation: A green-clean tool for decontamination of synthetic pesticides from agro-and aquatic ecosystems. In: Singh JS, Seneviratne G (eds) Agro-environmental sustainability: Vol (2) managing environmental pollution. Springer, Cham, pp 59–83

Kumar A, Kaushal S, Saraf S, Singh JS (2017) Cyanobacterial biotechnology: an opportunity for sustainable industrial production. Clim Change Environ Sustain 5(1):97–110

Kumar A, Kaushal S, Saraf S, Singh JS (2018) Microbial bio-fuels: a solution to carbon emissions and energy crisis. Front Biosci (Landmark) 23:1789–1802

Laird DA (2008) The charcoal vision: a win-win scenario for simultaneously producing bioenergy, permanently sequestering carbon, while improving soil and water quality. Agron J 100:178–181

Lal R (2004) Agricultural activities and the global carbon cycle. Nutr Cycl Agroecosyst 70(2):103–116

Lal R (2010) Managing soils for warming earth in a food-insecure and energy-starved world. J Plant Nutr Soil Sci 173:4–15

Lehmann J (2007) Bio-energy in the black. Front Ecol Environ 5(7):381–387

Lehmann J (2009) Biological carbon sequestration must and can be a win-win approach. Clim Chang 97(3):459–463

Lehmann J, Joseph S (2009) Biochar for environmental management: an introduction. In: Lehmann J, Joseph S (eds) Biochar for environmental management: science and technology. Earthscan, London

Lehmann J, Rondon M (2006) Biochar soil management on highly weathered soils in the humid tropics. In: Uphoff N et al (eds) Biological approaches to sustainable soil systems. CRC Press, Boca Raton, pp 517–530

Lehmann J, Pereira da Silva J, Steiner C, Nehls T, Zech W, Glaser B (2003) Nutrient availability and leaching in an archaeological Anthrosol and a Ferralsol of the Central Amazon basin: fertilizer, manure and charcoal amendments. Plant Soil 249:343–357

Lehmann J, Gaunt J, Rondon M (2006) Biochar sequestration in the terrestrial ecosystems-a review. Mitig Adapt Strat Glob Change 11(2):395–419

Lehmann J, Matthias R, Janice T, Caroline M, William H, David C (2011) Biochar effects on soil biota – a review. Soil Biol Biochem 43:1812–1836

Liang B, Lehmann J, Solomon D, Kinyangi J, Grossman J, O Neill B, Skjemstad JO, Thies J, Luizao FJ, Petersen J (2006) Black carbon increases cation exchange capacity in soils. Soil Sci Soc Am J 70:1719–1730

Lin XW, Zie ZB, Zheng JY, Liu Q, Bei QC, Zhu JG (2015) Effects of biochar application on greenhouse gas emissions, carbon sequestration and crop growth in coastal saline soil. Eur J Soil Sci 66:329–338

Liu Z, Zhang FS (2009) Removal of lead from water using biochars prepared from hydrothermal liquefaction of biomass. J Hazard Mater 167(1):933–939

Liu P, Ptacek CJ, Blowes DW, Landis RC (2016) Mechanisms of mercury removal by biochars produced from different feedstocks determined using X-ray absorption spectroscopy. J Hazard Mater 308:233–242

Lorenz K, Lal R, Preston CM, Nierop KGJ (2007) Strengthening the soil organic carbon pool by increasing contributions from recalcitrant aliphatic bio (macro) molecules. Geoderma 142(1):1–10

Major J, Rondon M, Molina D, Riha SJ, Lehman J (2010) Maize yield and nutrition for 4 years after biochar application to a Colombian savanna oxisol. Plant Soil 333:117–128

Matthews JA (2008) Carbon negative biofuels. Energy Policy 36:940–945

McCarl BA, Peacocke C, Chrisman R, Kung CC, Sands R (2009) Economics of biochar production, utilization and greenhouse gas offsets. In: Lehmann J, Joseph S (eds) Biochar for environmental management: science and technology. Earthscan, London, pp 341–357

McHenry MP (2009) Agricultural bio-char production, renewable energy generation and farm carbon sequestration in Western Australia: certainty, uncertainty, and risk. Agric Ecosyst Environ 129:1–7

Mendez A, Paz-Ferreiro J, Araujo F, Gasco G (2014) Biochar from pyrolysis of de-inking paper sludge and its use in the treatment of a nickel polluted soil. J Anal Appl Pyrolysis. https://doi.org/10.1016/j.jaap.2014.02.001

Mohan D, Pittman CU, Steele PH (2006) Pyrolysis of wood/biomass for bio-oil: a critical review. Energy Fuel 20:848–889

Mukherjee A, Lal R, Zimmerman AR (2014) Effects of biochar and other amendments on the physical properties and greenhouse gas emissions of an artificially degraded soil. Sci Total Environ 487:26–36

Namgay T, Singh B, Singh BP (2010) Influence of biochar application to soil on the availability of As, Cd, Cu, Pb, and Zn to maize (*Zea mays* L.). J Aust Soil Res 48:638–647

Noguera D, Rondon M, Laossi KR, Hoyos V, Lavelle P, de Carvalho MHC, Barot S (2010) Contrasted effect of biochar and earthworm on rice growth and resource allocation in different soils. Soil Biol Biochem 42:1017–1027

Oya A, Iu WG (2002) Deodorization performance of charcoal particles loaded with orthophosphoric acid against ammonia and trimethylamine. Carbon 40(9):3891–3899

Park J, Choppala H, Lee GH, Bolan SJ, Chung JW, Edraki M (2013) Comparative sorption of Pb and Cd by biochars and its implication for metal immobilization in soil. Water Air Soil Pollut 224:1711. https://doi.org/10.1007/s11270-013-1711-1

Pellera FM, Giannis A, Kalderis D, Anastasiadou K, Stegmann R, Wang J, Gidarakos E (2012) Adsorption of Cu (II) ions from aqueous solutions on biochars prepared from agricultural by-products. J Environ Manag 96(1):35–42

Purakayastha TJ (2010) Effect of biochar on the yield of different crops. IARI. Annual Report 2010-11, Indian Agricultural Research Institute, New Delhi, p 55

Qiu Y, Cheng H, Xu C, Sheng SD (2008) Surface characteristics of crop-residue-derived black carbon and lead (II) adsorption. Water Res 42(3):567–574

Quayle W (2010) Biochar potential for soil improvement and soil fertility. IREC Farm Newsl 182:22–24

Reichenauer TG, Panamulla S, Subasinghe S, Wimmer B (2009) Soil amendments and cultivar selection can improve rice yield in salt-influenced (tsunami-affected) paddy fields in Sri Lanka. Environ Geochem Health 31:573–579

Rondon M, Ramirez A, Hurtado M (2004) Charcoal additions to high fertility ditches enhance yields and quality of cash crops in the Andean hillsides of Colombia. CIAT Annual Report 2004, Cali, Colombia

Shackley S, Sohi S, Haszeldine S, Manning D, Masek O (2009) Biochar, reducing and removing CO2 while improving soils: a significant and sustainable response to climate change

Shafie S, Mohd Salleh MA, Lau Lek Hang M (2012) Effect of pyrolysis temperature on the biochar nutrient and water retention capacity. J Purity Util React Environ 1(6):293–307

Shang G, Shen G, Wang T, Chen Q (2012) Effectiveness and mechanisms of hydrogen sulfide adsorption by camphor-derived biochar. J Air Waste Manage Assoc 62(8):873–879

Simon S, Wiegmann K (2009) Modelling sustainable bioenergy potentials from agriculture for Germany and Eastern European countries. Biomass Bioenergy 33:603–609

Singh JS (2013) Anticipated effects of climate change on methanotrophic methane oxidation. Clim Change Environ Sustain 1(1):20–24

Singh JS (2014) Cyanobacteria: a vital bio-agent in eco-restoration of degraded lands and sustainable agriculture. Clim Change Environ Sustain 2:133–137

Singh JS (2015) Biodiversity: current perspective. Clim Change Environ Sustain 3(1):71–72

Singh JS (2016) Microbes play major roles in ecosystem services. Clim Change Environ Sustain 3:163–167

Singh JS (ed) (2019) New and future developments in microbial biotechnology and bioengineering: microbes in soil, crop and environmental sustainability. Elsevier, Amsterdam

Singh JS, Boudh S (2016) Climate change resilient crops to sustain Indian agriculture. Clim Change Environ Sustain 5:97–110

Singh JS, Singh DP (eds) (2019) New and future developments in microbial biotechnology and bioengineering: microbial biotechnology in agro-environmental sustainability. Elsevier, United States

Singh BP, Hatton BJ, Singh B, Cowie A, Kathuria A (2010a) Influence of biochars on nitrous oxide emission and nitrogen leaching from two contrasting soils. J Environ Qual 39:1224–1235

Singh B, Singh BP, Cowie AL (2010b) Characterization and evaluation of biochars for their application as a soil amendment. Soil Res 48(7):516–525

Singh C, Tiwari S, Boudh S, Singh JS (2017a) Biochar application in management of paddy crop production and methane mitigation. In: Singh JS, Seneviratne G (eds) Agro-environmental sustainability: Vol-2: managing environmental pollution. Springer, Cham, pp 123–146

Singh C, Tiwari S, Boudh S, Singh JS (2017b) Biochar application in management of paddy crop production and methane mitigation. In: Singh JS, Seneviratne G (eds) Agro-environmental sustainability: Vol (2) managing environmental pollution. Springer, Cham, pp 123–145

Singh C, Tiwari S, Singh JS (2017c) Impact of rice husk biochar on nitrogen mineralization and methanotrophs community dynamics in paddy soil. Int J Pure Appl Biosci 5(5):428–435

Singh JS, Koushal S, Kumar A, Vimal SR, Gupta VK (2017d) Book review: microbial inoculants in sustainable agricultural productivity- Vol II: Functional application. Front Microbiol 7:2015
Singh C, Tiwari S, Singh JS (2017e) Application of biochar in soil fertility and environmental management: a review. Bull Environ Pharmacol Life Sci 6(12):07–14
Singh C, Tiwari S, Gupta VK, Singh JS (2018) The effect of rice husk biochar on soil nutrient status, microbial biomass and paddy productivity of nutrient poor agriculture soils. Catena 171:485–493
Singh C, Tiwari S, Singh JS (2019) Biochar: a sustainable tool in soil 2 pollutant bioremediation. In: Bharagava RN, Saxena G (eds) Bioremediation of industrial waste for environmental safety. Springer, Dordrecht, pp 475–494
Singh JS, Kumar A, Singh M (2019a) Cyanobacteria: a sustainable and commercial bioresource in production of bio-fertilizer and bio-fuel from waste waters. Environ Sustain Indic 3–4:100008
Singh MK, Rai PK, Rai A, Singh S, Singh JS (2019b) Poly-β-hydroxybutyrate production by the cyanobacterium *Scytonema geitleri* Bharadwaja under varying environmental conditions. Biomolecules 9(198):1–10
Spokas K, Koskinen WC, Baker JM, Reicosky DC (2009) Impacts of woodchip biochar additions on greenhouse gas production and sorption/degradation of two herbicides in a Minnesota soil. Chemosphere 77(4):574–581
Steiner C, Teixeira W, Lehmann J, Nehls T, de Macedo J, Blum W, Zech W (2007) Long term effects of manure, charcoal and mineral fertilization on crop production and fertility on a highly weathered Central Amazonian upland soil. Plant Soil 291:275–290
Steiner C, Das KC, Garcia M, Forster B, Zech W (2008) Charcoal and smoke extract stimulate the soil microbial community in a highly weathered xanthic Ferralsol. Pedobiologia 51:359–366
Suppadit T, Kitikoon V, Phubphol A, Neumnoi P (2012) Effect of quail litter biochar on the productivity of four new physic nut varieties planted in cadmium-contaminated soil. Chilean J Agric Res 72:125–132
Taghizadeh-Toosi A et al (2011) Biochar incorporation into pasture soil suppresses in situ nitrous oxide emissions from ruminant urine patches. J Environ Qual 40(2):468–476
Tiwari AP, Singh JS (2017) Plant growth promoting rhizospheric *Pseudomonas aeruginosa* strain inhibits seed germination in *Triticum aestivum* (L) and *Zea mays* (L). Microbiol Res 8(7233):73–79
Tiwari S, Singh C, Singh JS (2018) Land use changes: a key ecological driver regulating methanotrophs abundance in upland soils. Energy Ecol Environ 3(6):355–371
Tiwari S, Singh C, Boudh S, Rai PK, Gupta VK, Singh JS (2019a) Land use change: a key ecological disturbance declines soil microbial biomass in dry tropical uplands. J Environ Manag 242:1–10
Tiwari S, Singh C, Singh JS (2019b) Wetlands: a major natural source responsible for methane emission. In: Upadhyay AK et al (eds) Restoration of wetland ecosystem: a trajectory towards a sustainable environment. Springer, Singapore, pp 59–74
Tong X, Li JY, Yuan JH, Xu RK (2011) Adsorption of Cu (II) by biochars generated from three crop straws. Chem Eng J 172(2):828–834
Uchimiya M, Lima IM, Thomas Klasson K, Chang S, Wartelle LH, Rodgers JE (2010) Immobilization of heavy metal ions (CuII, CdII, NiII, and PbII) by broiler litter-derived biochars in water and soil. J Agric Food Chem 58(9):5538–5544
Van Zwieten L, Singh BP, Joseph S, Kimber S, Cowie A, Chan KY (2009) Biochar and emissions of non-CO_2 greenhouse gases from the soil. In: Lehmann J, Joseph S (eds) Biochar for environmental management: science and technology. Earthscan, London, pp 227–249
Van Zwieten L, Kimber S, Morris S, Chan KY, Downie A, Rust J, Joseph S, Cowie A (2010) Effects of biochar from slow pyrolysis of paper mill waste on agronomic performance and soil fertility. Plant Soil 327:235–246
Verheijen F, Jeffery S, Bastos AC, van der Velde M, Diafas I (2010) Biochar application to soils: a critical scientific review of effects on soil properties. Processes and Functions, European Commission Joint Research Centre for Scientific and Technical reports 51–68

Vimal SR, Singh JS (2019) salt tolerant pgpr and fym application in saline soil paddy agriculture sustainability. Clim Change Environ Sustain 7(1):23–33

Vimal SR, Gupta J, Singh JS (2018) Effect of salt tolerant *Bacillus* sp. and *Pseudomonas* sp. on wheat (*Triticum aestivum* L.) growth under soil salinity: a comparative study. Microbiol Res 9(1):1–14

Vitkova J, Kondrlova E, Rodny M, Surda P, Horak J (2017) Analysis of soil water content and crop yield after biochar application in field conditions. Plant Soil Environ 63(12):569–573

Wang C, Deng H, Zhao F (2015) The remediation of chromium (VI)-contaminated soils using microbial fuel cells. Soil Sedim Contamin 1:1–12

Warnock DD, Lehmann J, Kuyper TW, Rillig MC (2007) Mycorrhizal responses to biochar in soil e concepts and mechanisms. Plant Soil 300:920

Woolf D (2008) Biochar as a soil amendment: a review of the environmental implications of black carbon (charcoal) in terrestrial and aquatic ecosystems. Science of the total of green waste biochar as a soil amendment. Aust J Soil Res 45:629–634

Xu C, Chen H, Xiang Q, Zhu H, Wang S, Zhu Q, Huang D, Zhang YZ (2017) Effect of peanut shell and wheat straw biochar on the availability of Cd and Pb in a soil–rice system. Environ Sci Pollut Res. https://doi.org/10.1007/s11356-017-0495

Xue Y et al (2012) Hydrogen peroxide modification enhances the ability of biochar (hydrochar) produced from hydrothermal carbonization of peanut hull to remove aqueous heavy metals: batch and column tests. Chem Eng J 200:673–680

Yaghoubi P, Reddy KR (2011) Characteristics of biochar-amended soil cover for landfill gas mitigation. Pan-Am CGS Geotechnical Conference, 2011

Yan LL, Kong L, Qu Z, Li L, Shen GQ (2015) Magnetic biochar decorated with ZnS nano crytals for Pb (II) removal. ACS Sustain Chem Eng 3:125–132

Yanai Y, Toyota K, Okazaki M (2007) Effects of charcoal addition on N_2O emissions from soil resulting from rewetting air-dried soil in short-term laboratory experiments. Soil Sci Plant Nutr 53:181–188

Yang Y, Wei Z, Zhang X, Chen X, Yue D, Yin Q, Xiao L, Yang L (2014) Biochar from Alternanthera philoxeroides could remove Pb (II) efficiently. Bioresour Technol 171:227–232

Yang Q, Han F, Chen Y, Yang H, Chen H (2016) Greenhouse gas emissions of a biomass-based pyrolysis plant in China. Renew Sust Energ Rev 53:1580–1590

Yu XY, Ying GG, Kookana RS (2009) Reduced plant uptake of pesticides with biochar additions to soil. Chemosphere 76(5):665–671

Yuan JH, Xu RK, Zhang H (2011) The forms of alkalis in the biochar produced from crop residues at different temperatures. Bioresour Technol 102(3):3488–3497

Zackrisson O, Nilsson MC, Wardle DA (1996) The key ecological function of charcoal from wild fire in the Boreal forest. Oikos 77:10–19

Zhang A, Cui L, Pan G, Li L (2010) Effect of biochar amendment on yield and methane and nitrous oxide emissions from a rice paddy from Tai Lake plain, China. Agric Ecosyst Environ 139(4):469–475

Zhang W, Mao S, Chen H, Huang L, Qiu R (2013) Pb (II) and Cr (VI) sorption by biochars pyrolyzed from the municipal wastewater sludge under different heating conditions. Bioresour Technol 147:545–552

Zhou JB, Deng CJ, Chen JL, Zhang QS (2008) Remediation effects of cotton stalk carbon on cadmium (Cd) contaminated the soil. Ecol Environ 17:1857–1860

Chapter 6
Recent Development in Bioremediation of Soil Pollutants Through Biochar for Environmental Sustainability

Gulshan Kumar Sharma, Roomesh Kumar Jena, Surabhi Hota, Amit Kumar, Prasenjit Ray, Ram Kishor Fagodiya, Lal Chand Malav, Krishna Kumar Yadav, Dipak Kumar Gupta, Shakeel A. Khan, and S. K. Ray

Abstract Soil contamination due to heavy metals has become a great concern nowadays. The main reasons for soil contamination are both natural as well as anthropogenic. Natural processes like volcanic eruption, weathering of rocks, landslides and soil erosion while anthropogenic involves several activities like smelting, mining, application of agrochemicals (pesticides, herbicides and fertilizers) and industrial wastes. Heavy metals pollution has a direct influence on the fertility of agricultural soils. The removal of heavy metals from soil is very difficult as it stored in the environment for a long time, because of its persistent nature. Several in-situ bioremediation technologies are used for removal of heavy metals from the environment. Out of that in-situ biochar application is one of the prominent technologies for remediation of heavy metals and it was found to be effective in reducing the mobility of heavy metals in soils. Biochar effectively adsorbs heavy metals and

---

G. K. Sharma (✉) · R. K. Jena · S. Hota · P. Ray · S. K. Ray
National Bureau of soil Survey and Land Use Planning, RC, Jorhat, Assam, India

A. Kumar
Host Plant Section, Central Muga Eri Research & Training Institute, Jorhat, Assam, India

R. K. Fagodiya
Division of Irrigation and Drainage Engineering, CSSRI, Karnal, Haryana, India

L. C. Malav
National Bureau of soil Survey and Land Use Planning, RC, Nagpur, Maharashtra, India

K. K. Yadav
Institute of Environment and Development Studies, Bundelkhand University, Jhansi, India

D. K. Gupta
ICAR-Central Arid Zone Research Institute, Regional Research Station, Pali Marwar, Rajasthan, India

S. A. Khan
Centre for Environmental Science and Climate Resilient Agriculture, IARI, New Delhi, India

J. S. Singh, C. Singh (eds.), *Biochar Applications in Agriculture and Environment Management*, https://doi.org/10.1007/978-3-030-40997-5_6

decreases bioavailability and toxin-induced stress to the biotic component of soil. In this chapter, the emphasis has been given on heavy metal pollution and types of biochar used for remediation of heavy metals from the soil and water.

Keywords Biochar · Bioremediation · Pollutants · Contaminated soil · Heavy metal

6.1 Introduction

Heavy metals contamination in soil and water has become an alarming issue in the recent times due to accelerated anthropogenic activities with increasing industrialization and decreasing land holdings leading to various health hazards as well as their adverse effect on the quality of soil and water (Singh et al. 2013; Bhatia et al. 2015; Matta et al. 2018; Gupta et al. 2019). Heavy metals are generally found in the soil environment in traces which are a result of weathering of rocks and minerals over the earth crust. These heavy metals become pollutant only when they cross a certain threshold limit due to anthropogenic activities which accelerate the release and addition of these metals into the environment, soils and water bodies. The sources of the heavy metals pollutions may be fertilizers, pesticides, bio solids, manures, metal mining, waste water, milling process, industrial wastes and airborne sources like dust or smoke from storage areas or waste dumps (Wuana and Okieimen 2011). The major heavy metal (loid)s pollutants commonly found in the contaminated environment are lead (Pb), chromium (Cr), arsenic (As), zinc (Zn), cadmium (Cd), copper (Cu), mercury (Hg), and nickel (Ni) (GWRTAC 1997). The common source of heavy metals includes cars, mining activities such as smelting, mine spoils and trailing, antiseptics, self-cleaning ovens, plastics, solar panels, mobile phones and particle accelerators (Koduru et al. 2017; Hubner et al. 2010). In addition to that the potential sources pyro-metallurgical industries, automobile exhausts, fossil fuel combustion are also the main sources of heavy metals (Lottermoser 2010; Prasad 2001), industries such as plastics, textiles, microelectronics, wood preservatives, refineries, agrochemicals (excessive use of fertilizers and pesticides) and waste disposal from sewage treatment plant, leachate from landfill and fly ash disposal some of the chief sources of the heavy metals (Singh and Kumar 2006; Kumar et al. 2016). Although, heavy metals content in domestic municipal sewage wastewater is mainly below threshold level (Sharma et al. 2014; Malla et al. 2015; Khan et al. 2019). The details about the sources of heavy metals and respective anthropogenic activities are given in Table 6.1.

Various research has been carried out all over the world towards remediation of the heavy metal contamination. This remediation may be by physical, chemical or biological means. Physical remediation is effective but cost-intensive and poses a risk of secondary diffusion while chemical methods cause secondary pollution to

Table 6.1 Anthropogenic sources of heavy metals

S. No.	Heavy metals	Anthropogenic activities
1	Chromium (Cr)	Mining, industrial coolants, chromium salts manufacturing, leather tanning
2	Lead (Pb)	Lead acid batteries, paints, e-waste, smelting operations, coal-based thermal power plants, ceramics, bangle industry
3	Mercury (Hg)	Chlor-alkali plants, thermal power plants, fluorescent lamps, hospital waste (broken thermometers, barometers, sphygmomanometers), electrical appliances, etc.
4	Arsenic (As)	Geogenic/natural processes, smelting operations, thermal power plants, fuel burning
5	Copper (Cu)	Mining, electroplating, smelting operations
6	Cadmium (Cd)	Zinc smelting, waste batteries, e-waste, fuel combustion
7	Molybdenum (Mo)	Spent catalyst
8	Zinc (Zn)	Smelting, electroplating
9	Beryllium (Be)	Alloy (with Cu), electrical insulators in power transistors, moderator or neutron deflectors in nuclear reactors
10	Cobalt (co)	Metallurgy (in super alloys), ceramics, glasses, paints
11	Iron (Fe)	Cast iron, wrought iron, steel, alloys, construction, transportation, machine-manufacturing
12	Mercury(Hg)	Extracting of metals by amalgamation, mobile cathode in the chloralkali cell for the production of NaCl and Cl_2 from brine, electrical and measuring apparatus, fungicides, catalysts, pharmaceuticals, dental fillings, scientific instruments, rectifiers, oscillators, electrodes, mercury vapour lamps, X-Ray tubes, solders
13	Manganese (Mn)	Production of ferromanganese steels, electrolytic manganese dioxide for use in batteries, alloys, catalysts, fungicides, antiknock agents, pigments, dryers, wood preservatives, coating welding rods
14	Molybdenum (Mo)	Alloying element in steel, cast irons, non-ferrous metals, catalysts, dyes, lubricants, corrosion inhibitors, flame retardants, smoke represents, electroplating
15	Nickel (Ni)	As an alloy in the steel industry, electroplating, Ni/Cd batteries, arc-welding, rods, pigments for paints and ceramics, surgical and dental protheses, moulds for ceramic and glass containers, computer components, catalysts
16	Lead (Pb)	Antiknock agents, tetramethyl lead, lead-acid batteries, pigments, glassware, ceramics, plastic, in alloys, sheets, cable sheathings, solder, ordinance, pipes or tubing
17	Antimony (Sb)	Type-metal alloy, electrical applications, Britannia metal, pewter, Queen's metal, Sterline, in primers and tracer cells in munition manufacture, semiconductors, flameproof pigments and glass, medicines for parasitic diseases, nauseant, expectorant, combustion.
18	Selenium (Se)	glass industry, semiconductors, thermoelements, photoelectric and photo cells, and xerographic materials, inorganic pigments, rubber production, stainless steel, lubricants, dandruff treatment

(continued)

Table 6.1 (continued)

S. No.	Heavy metals	Anthropogenic activities
19	Stannum (Sn)	Tin-plated steel, brasses, bronzes, pewter, dental amalgam, stabilizers, catalysts, pesticides
20	Titanium (Ti)	White pigments, UV-filtering agents, nucleation Agent for glass ceramics, alloy in aeronautics Tl: Used for alloys (with Pb, Ag, or Au) with special properties, electronics industry, infrared optical systems, catalyst, deep temperature thermometers, low melting glasses, semiconductors, supra conductors
21	Vanadium (V)	Steel production, Alloys, catalyst

Singh and Kumar (2006), Kumar et al. (2016, 2019), Lottermoser (2010), Prasad (2001), Koduru et al. (2017), Hubner et al. (2010)

the environment. Biological methods, though safe to use but time consuming (Yang et al. 2019). Hence, in the last decade, the application of biochar has become the trend in addressing heavy metal contamination. Biochar based remediation are cost-effective as well as quick in action and raw material for biochar production (biomass) is easily available as biomass contributes 35% of the entire energy requirement of the world and it is the third-largest source of energy (Armynah et al. 2018). This chapter gives an insight into the action of biochar towards remediation of the heavy metal contamination.

6.2 Characteristics of Biochar

Biochar derived from two words, bio and char. Char means partially burnt to become blackened. So literally, biochar means charred biomass. It is carbonaceous or carbon-rich solid obtained by thermochemical conversion of biomass-derived from various sources like manure, leaves or wood at a temperature range of 300–1000 ° C (Xie et al. 2015) in a partial or anaerobic condition. The products of this conversion may be heat, power, fuel and/or chemicals. The thermochemical processes involved in the production of biochar are pyrolysis, gasification, hydrothermal carbonization and microwave carbonization. Biochar when produced from woody biomass, called as charcoal. Biochar is a recalcitrant and stabilized organic C compound. Biochar is produced from a varied range of sources or biomass. Many studies have been carried out towards the characterization of biochars derived from different sources. Biochar can be produced from forest species like acacia, gmelina, eucalyptus, pine (Suárez et al. 2017; Zhang et al. 2018) and animal manures like yak manure (Zhang et al. 2018) and chicken manure (Domingues et al. 2017) and from agricultural field crop by-products like coffee husk, sugarcane bagasse and rice husk (Armynah et al. 2018; Domingues et al. 2017). Biochar also has a range of application as wide as its sources. It is used in agriculture for fertility enhancement (Domingues et al. 2017) and amendment of soil acidity (Zhang et al. 2018), C sequestration (Domingues et al. 2017), wastewater treatment and environmental

Table 6.2 Characterization of biochar

Index	Value
Specific surface area (m^2/g)	6.86
Micropore area (m^2/g)	0.17
Total pore volume (mm^3/g)	22.29
Micropore volume (mm^3/g)	0.02
Ash (wt%)	42.25
pH	10.09
C (wt%)	48.45
H (wt%)	1.78
N (wt%)	1.47
S (wt%)	0.78

Brewer et al. (2009), Chen et al. (2014)

application for bioremediation. The application of biochar for any specific purpose depends upon its physicochemical properties. Physiochemical properties of biochar largely depend upon the type of feedstock from which it has been derived and the operation temperature conditions (Suárez et al. 2017; Yang et al. 2019; Armynah et al. 2018; Jindo et al. 2014; Zhang et al. 2018; Domingues et al. 2017). In most of the studies, the pyrolysis temperature has been maintained in the range of 300–800 ° C (Jindo et al. 2014; Zhang et al. 2018). Many studies have used pyrolysis technique for production of biochars (Yang et al. 2019; Jindo et al. 2014; Zhang et al. 2018; Domingues et al. 2017). Other than that, the process like flash carbonization and torrefaction have also been used (Yang et al. 2019).

6.2.1 *Characterization of Biochar*

For determination of morphological and physical properties like surface area and pore size distribution of biochar, Breuner Emmet and Teller (BET) physio-sorption method is mostly used (Suárez et al. 2017; Armynah et al. 2018; Jindo et al. 2014; Zhang et al. 2018; Nartey and Zhao 2014; Askeland et al. 2019). For surface characteristics and elemental analysis, Scanning Electron Microscope (SEM) and Energy Dispersive X-ray (EDX) can be used (Jindo et al. 2014; Jechan et al. 2017; Bouraoui et al. 2015). Chemical bonding and structural properties of biochar can be determined by Fourier Transform Infrared (FTIR) Spectroscopy (Nartey and Zhao 2014; Askeland et al. 2019). Contaminant and heavy metals in biochars are determined by Atomic Absorption Spectrometer (AAS) and Inductively Coupled Plasma Mass Spectrometer (ICP MS) (Askeland et al. 2019). For analysis of structural elements C, H and O the CHN analyzer can be used. Biochars can also be characterized and differentiated based on elemental ratios like H/C, O/C or C/N.

The specific surface area of biochar (6.86 m^2/g), micropore area (0.17 m^2/g), total pore volume (22.29 mm3/g), and micropore volume (0.02 mm^3/g). Ash content and pH were 42.25% and 10.09, respectively. Content of C element (48.45%), H (1.78%), N (1.47%), and S (0.78%) elements as shown in Table 6.2.

Table 6.3 The removal of heavy metals by biochar and their effects

Contaminants	Biochar type	Source	Effects	References
As and Cu	Hardwood	Soil	Mobilization due to enhanced pH and DOC	Beesley et al. (2010)
As, Cr, Cd, Cu, Ni, Pb, and Zn	Sewage sludge (500–550 °C)	Soil	Immobilization of As, Cr, Ni, and Pb due to rise in soil pH. Mobilization of Cu, Zn, and Cd due to highly available concentrations in biochar	Khan et al. (2013)
Cd and Zn	Hardwood	Soil	Immobilization due to enhanced pH	Beesley et al. (2010)
Cd, Cu, and Pb	Chicken manure and green waste (550 °C)	Soil	Immobilization due to partitioning of metals from exchangeable phase to less bioavailable organic-bond fraction	Park et al. (2011)
Cu and Pb	Oakwood	Soil	Complexation with phosphorus and organic matter	Karami et al. (2011)
Pb	Dairy manure (450 °C)	Soil	Immobilization by Hydroxyl-pyromorphite formation	Cao et al. (2011)
Pb	Oakwood (400 °C)	Soil	Immobilization by rise in soil pH and adsorption on biochar	Ahmad et al. (2012)
Pb	Rice straw	Soil	Non-electrostatic adsorption	Jianga et al. (2012)
Pb, Cu and Zn	Broiler litter (300 and 600 °C)	Soil	Stabilization of Pd and Cu	Uchimiya et al. (2012)
Cd (II)	Pine Residue 600 °C	Water	Reduction of Cd content in water by adsorption	Park et al. (2019)
Cr (VI)	Eucalyptus Bark	Soil and water	Removal from soil and ground water by sorption, electrostatic attraction, aqueous reduction by dissolved oraganic matter (DOM)	Choudhary et al. (2017)
Zn(II)	Wheat straw, pine needles (350 °C and 550 °C)	Soil	Immobilization in soil by OH^-, CO_3^{2-} and Si released from biochar.	Qian et al. (2016)
Pb (II)	Maple wood (500 °C)	Water	Removal from water by adsorption by oxygen containing surface functional group	Wang et al. (2018)
Cd(II) and Pb(II)	Dairy manure (300 °C)	Soil	Removal from soil by precipitation as carbonate minerals complexation with surface functional carboxyl and hydroxyl groups	Chen et al. (2019)
Cr (VI)	Sewage Sludge (300 °C)	Water	Removal for aqueous solution by adsorption through electrostatic interactions between Cr(VI) anion and the positively charged functional groups on the surface of biochar	Agrafioti et al. (2014)
As(III), As(V)	Empty oil palm fruit bunch and rice husk	Soil	Surface complexes were formed between As(III) and As(V) and the functional groups (hydroxyl, carboxyl, and C–O ester of alcohols) of the two biochars	Samsuri et al. (2013)

6.2.2 Morphological Characteristics

The surface area and porosity of final biochar product depends on feedstock type and operation temperature. With increase in temperature, the surface area and total pore volume of the biochar increases and the particle size decreases for woody biochar (Suliman et al. 2017), biochar from agricultural products (Suárez et al. 2017) or from animal manures (Zhang et al. 2018). With increase in surface area the CEC of biochar also increases. Porosity and surface area of biochar material are important parameters for their use as remediation of contaminants.

6.2.3 Composition of Biochar

At low temperatures the biochar yield is higher and with increasing temperature the yield of biochar has been reported to decrease while ash and C contents have found to be increase (Jindo et al. 2014; Domingues et al. 2017). The increase in total C content of woody biochar at high temperatures may be contributed to higher degree of polymerization resulting to more condensed C structure (Lehman and Joseph 2009). Therefore, an increase in reaction temperature leads to an increase in carbon content, and decrease in the hydrogen and oxygen content (Jindo et al. 2014). In contrast, animal manure such as chicken manure (Domingues et al. 2017) and yak manure (Zhang et al. 2018) derived biochars show decrease in C content with increasing temperature which might be due to the high labile nature of organic compounds found in animal waste which are rapidly lost before biochar formation (Domingues et al. 2017). The different types of biochar on the basis of feedstock with removal of heavy metal from soil and water is given in Table 6.3. C and O are dominant elements present in biochar derived from any feedstock. Other than C and O the elemental composition varies according to feed stock and the percentage of elements also varies with operation temperatures. Apart from C and O biochar consists mostly of basic elements Ca, Mg, K (Armynah et al. 2018). For example, biochar derived from rice plant by products have Silicon as a dominant element (Armynah et al. 2018; Nwajiaku et al. 2018). With increase in pyrolysis temperature, the content of the basic elements Ca, Mg, K increases (Armynah et al. 2018; Zhang et al. 2018; Nwajiaku et al. 2018). Some studies also show that animal manure derived biochars contain traces of Fe, Cu, Mn and Zn (Zhang et al. 2018).

The amount of volatile compounds also gradually decreases with increasing pyrolysis temperature. The lignin containing biomass or woody biomass show a large change in volatile component with increasing temperature, compared to non woody species because of lignin decomposition at higher temperatures (Jindo et al. 2014).

6.2.4 *Acidity/Alkalinity*

The pH of biochar is an important property with regards to its use as soil amendment or as remediation of contaminated soils (Zhang et al. 2018). Because the changes in pH greatly influence some of the soil processes such as precipitation of minerals, nitrogen mineralization and ion exchange (Novak et al. 2016; Dai et al. 2018; Fidel et al. 2017). The resultant pH of biochar is dependent upon the temperature and type of feedstock. Generally, the pH value of biochar increase with increasing pyrolysis temperature (Jindo et al. 2014; Zhang et al. 2018; Suliman et al. 2017). Increase in pH at higher temperature may be due to increase in concentration of inorganic elements which are not pyrolysed and basic oxides formed at the surface under high temperature of pyrolysis (Zhang et al. 2018). The other reason may be the detachment of functional group during pyrolysis which is principally acidic in nature, such as hydroxyl, formyl or carboxyl groups (Weber and Quicker 2018; Tiwari et al. 2019a, b; Singh et al. 2019; Kour et al. 2019). Therefore removal of acidic functional group during pyrolysis shifted the pH to more basic condition. Hence, an increased pH-value is mainly contributed to increasing degree of carbonization (Quicker et al. 2016).

6.3 Mechanism of Biochar Interaction with Heavy Metals in Soil

The main reason behind application of biochar in soil is increasing the population of hydrocarbon-degrading microbes (Karppinen et al. 2017). Heavy metals cannot be degraded completely from the soil and water, but it can be transformed from more toxic to less toxic form. It can also be accumulated in plants and animals. Therefore, there are mainly four commonly used strategies for bioremediation of heavy metals.

1. Passive adsorption of ionic cations on the surface of biochar from the soil
2. Precipitation of anions like phosphate, carbonates, silicate, and chloride
3. Complexation with functional groups on the surface of biochar
4. Nutrients release such as N, P, K, Ca, Mg, S etc.

The processes passive adsorption, precipitation and complexation can reduce the bioavailability of heavy metals in the soil and further reduce the phytotoxicity to plants, on the other hand the nutrient release mechanism can increase the availability of nutrients to plants and microorganism in rhizosphere. The different mechanism of bioremediation and phytoremediation are given in Fig. 6.1. The mechanism of biochar application in phytoremediation is shown in Fig. 6.2. The mechanism of biochar application in phytoremediation is shown in Fig. 6.1. Phytoremediation includes the phytostabilization, phytoextraction, phytodegradation and phytovolatilization. Phytostabilization and phytoextraction are the commonly used

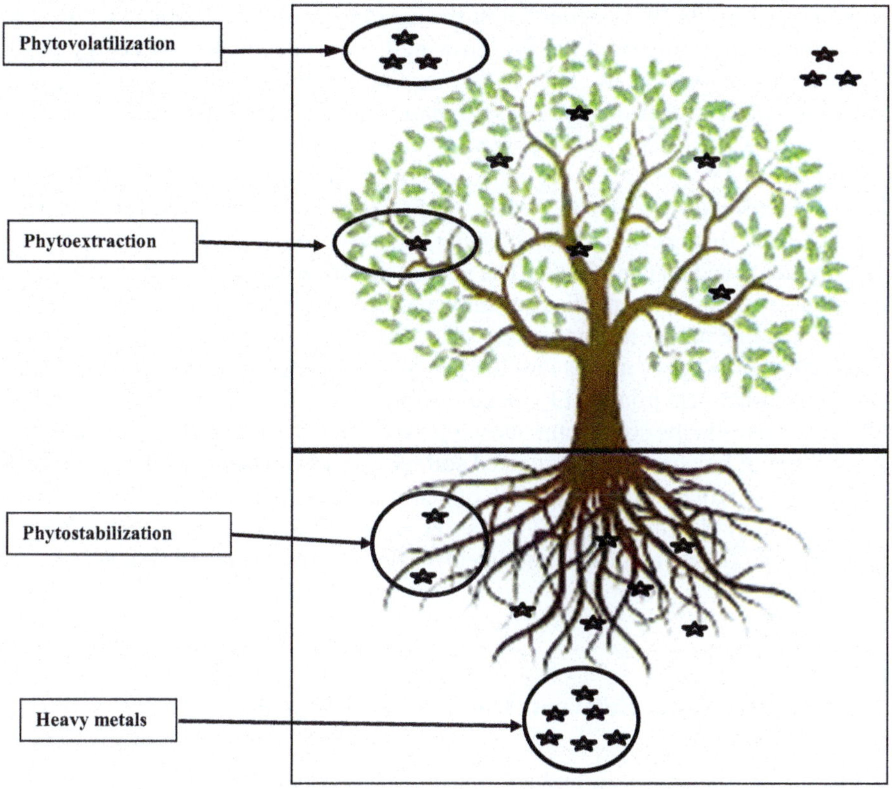

Fig. 6.1 Mechanism of remediation of heavy metals absorbed from soil in plants

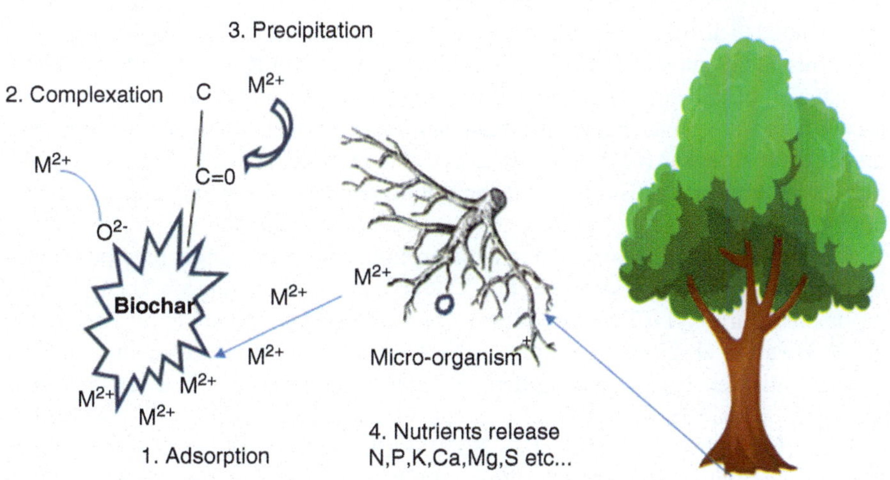

Fig. 6.2 Mechanism of combining effects of phytoremediation and biochar for bioremediation of heavy metal contaminated soil. Cationic metal ions expressed as M^{2+}. (Source: Wenjie et al. 2018)

technologies for the bioremediation of the contaminated agricultural soils. In phytoextraction, generally used the fast-growing hyper-accumulators or energy plants to uptake the heavy metals. The biochar amended phytoremediation from agricultural soils was performed using *Brassica napus* (Houben et al. 2013), *Anthyllis vulneraria* and *Noccaea rotundifolium* (Fellet et al. 2014), *Brassica juncea* (Rodrıguez-Vila et al. 2014), *Spinacia oleracea* and *Brassica napus* (Kelly et al. 2014), *Amaranthus tricolor* (Lu et al. 2015), *Clitocybe maxima* (Wu et al. 2016), *Cassia alata* (Huang et al. 2018), *Solanum nigrum* (Li et al. 2019) and *Lycopersicon esculentam* (Bandara et al. 2017). The impact of various dose of biochar and different plant species in bioremediation are mentioned in Table 6.4.

The alkaline nature and negatively charged surface of biochar help it to adsorb and retaining the heavy metals ions. Inspite of this, biochar also provide support and favourable micro environment for benificial microbes, nutrient supply to plant roots and contributes in the soil strcture development. Namgay et al. (2010) reported that application of biochar in contaminated soil reduces the availability of heavy metals which lead to reduction of absorption of heavy metal to plants. Apart from that biochar also increases the soil pH (Novak et al. 2009) hence might have improved sorption of these metals, consequently decreasing their bioavailability for plant uptake.

6.4 Interactions Between Biochar and Metals

These metals occur as their cations, anions or complexes. The form of occurrence depends upon the reaction (pH) (Park et al. 2019) and the aeration status (aerobic or anaerobic) of the environment. The mechanism of interaction of these metal (loid)s with biochar depend upon the form of their occurrence and the biochar characteristics (Ahmed et al. 2014) like surface area, porosity, content of exchangeable cations or anions and elemental composition of biochar which ultimately depends upon the feedstock type and operation temperature. Ion-exchange, electrostatic attraction and precipitation are predominant mechanisms for the remediation of inorganic contaminants by biochar (Ahmed et al. 2014). The increase in soil pH due to application of biochar also results in the immobilization of the metal cations (Egene et al. 2018).

Cd interacts by mechanisms of chemisorption, cation exchange, precipitation by carbonate (CO_3^{2-}) and phosphate (PO_4^{3-}) and complexation with surface functional groups like carboxyl (COO^-) and hydroxyl (OH^-) groups (Singh et al. 2017a, b, c, 2018; Tiwari et al. 2018; Park et al. 2019; Chen et al. 2019). In its compounds Cd occurs as divalent Cd (II). Adsorption of $Cd^{2+}$is reported to be dominated by boundary layer diffusion and very small amount of intraparticle diffusion or adsorption due to diffusion by pores (Park et al. 2019). At the surface, the interaction is mainly by exchange with cations of Ca, Mg and K but intra particle interaction is due to precipitation by PO_4^{3-}(Park et al. 2019; Chen et al. 2019). Pb^{2+} follows similar sorption mechanism as $Cd^{2+}$and interacts with biochar by mechanisms of

Table 6.4 Effects of biochar application on bioremediation and phytoremediation of contaminated soils

Feedstock	Biochar dose	Contaminants	Experiment type	Mechanism/Effect	Microbial/Plant species	References
Corncob biochar	0%, 1% and 5%	Cd	Pot experiment	Phyto extraction	*Solanum nigrum* L	Li et al. (2019)
Pruning residues Manure biochr	0%, 1.5%, and 3%	Cd and Pb	Pot experiment	Phytostabilization	*Anthyllis vulneraria* and *Noccaea rotundifolium,*	Fellet et al. (2014)
Miscanthus biochar	1%, 5% and 10%,	Cd, Zn and Pb	Greenhouse pot experiment	Phytostabilization	*Brassica napus* L.	Houben et al. (2013)
Sewage sludge biochar	0.4%, 1%, and 3%	Pb, Zn, Cu, Cd and As	100-day pot experiment	Immobilization	*Cassia alata* L.	Huang et al. (2018)
Pine wood biochar	0%, 10%, 20%, and 30%	Cd and Al	Pot experiment	Phytostabilization	*Spinacia oleracea and Brassica napus*	Kelly et al. (2014)
Hordeum vulgare straw biochar	0%, 3% and 6%	Cd and Cu	Pot experiment	Mycoextraction	*Clitocybe maxima*	Wu et al. (2016)
Biochar with compost	1%, 2%, 4% and 8% biochar in 20% 40% 80% and 100 % amendment	Co, Cu and Ni	Greenhouse experiments	Phytostabilization	*Brassica juncea* L.	Rodrıguez-Vila et al. (2014)
Woody biochar	1%, 2.5% and 5%	Cr, Ni and Mn	–	Phytostabilization	*Lycopersicon esculentam* L.	Bandara et al. (2017)

chemisorptions, cation exchange, precipitation by carbonate (CO_3^{2-}), phosphate (PO_4^{3-}) and silicate (SiO_4^{2-}) (Wang et al. 2018; Chen et al. 2019). Pb is released as Pb^{2+}, metal oxyanion complexes and its oxides and hydroxides in to the soil and water environment. But most stable forms are Pb^{2+} and lead hydroxy complexes (Wuana and Okieimen 2011). However, due to easy hydrolysis of Pb at low pH, biochar has been reported to have a higher affinity for Pb than Cd (Chen et al. 2019). In aqueous solution, the sorption of these divalent metal cations is pH dependant due to their strong tendency to be hydrated in aqueous solution (Dong et al. 2017). According to a study, adsorption of Cd increased with increasing pH up to 5 and maximum adsorption occurred at pH from 5–8 and the adsorption decreased gradually from pH 9 (Park et al. 2019). Similar behaviour has been reported for Pb adsorption with increasing pH. This behaviour is due to formation of hydroxide or carbonate complexes at higher pH (Park et al. 2019; Wang et al. 2018).

The dominant and toxic form of Cr found at contaminated sites is Cr (VI) and major species are chromate (CrO_4^{2-}) and dichromate ($Cr_2O_7^{2-}$). At reduced condition it may occur as Cr (III) (Wuana and Okieimen 2011). The mechanism of Cr (VI) removal by biochar is governed by sorption due to electrostatic attraction of anionic Cr (VI) by protonated biochar surface, where it is reduced to Cr (III) by carboxylic and phenolic hydroxyl group. Other mechanism is sorption and reduction mediated by surface organic complexes (Cr (III) - DOM complex) of biochar, and direct aqueous reduction by dissolved organic matter (DOM), derived from biochar (Choudhary et al. 2017).

Arsenic is a metalloid contaminant. In aerobic environments the dominant form is As (V) and it occurs as arsenate ion. Cr and as occur as their anions (H_3AsO_4, $H_2AsO_4^-$, $HAsO_4^{2-}$ and AsO_4^{3-}) and hence, the positively charged biochar surfaces are important for their adsorption. Under reducing condition, dominant form is As (III). As may interact with biochar by anion exchange at surface with anions like PO_4^{2-}, surface complexation with carboxylic and phenolic groups, adsorption on surface by metal bridging and DOM mediated reduction and physical adsorption (Vithanage et al. 2017).

6.5 Impacts of Biochar on Agriculture Productivity

The application of biochar has resulted in to improve in soil fertility, increase crop productivity, and water holding capacity of soil etc. as given below.

6.5.1 *Soil Property and Crop Productivity*

The excess use of fertilizer lead to leaching of excess nutrients, runoff losses of nutrients especially nitrogen to nearby water bodies, reduce soil fertility, acidification of soil, etc. On the other hand application of biochar in soil increases the organic

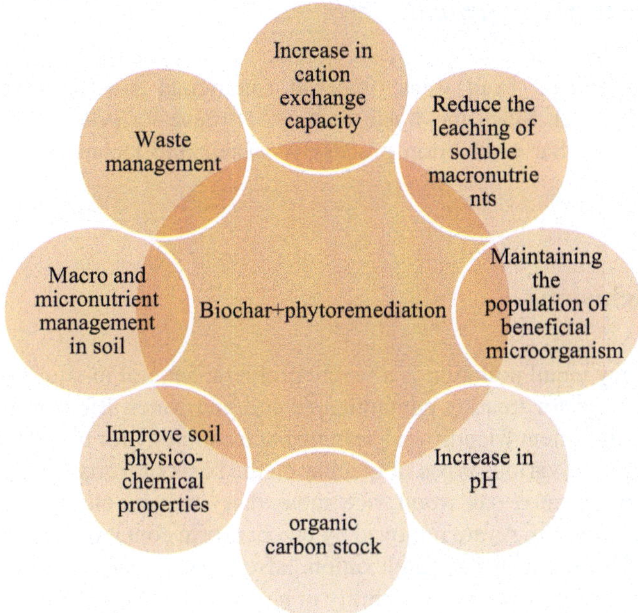

Fig. 6.3 A summary of the potential positive effects achieved by combining phytoremediation and biochar in heavy metal pollution remediation

carbon stock, helpful for maintaining the population of beneficial microorganism, improves physico-chemical properties, cation exchange capacity of soil, reduce the leaching of soluble macronutrients and therefore enhancing the soil fertility and crop yield. Biochar application also improves the water holding capacity of soil by improving soil quality in terms of physical chemical and biological properties of soil. An overview of combining biochar application with phytoremediation is given in Fig. 6.3.

6.5.2 Carbon Sequestration

As biochar is rich source of organic carbon, so have a potential role in carbon sequestration in soil. Biochar degrade slowly in soil, hence play an important role in built up of organic carbon in soil. Carbon sequestration improves the organic carbon status of soil. As organic carbon is tightly bound to soil particle, results in to less emission of CO_2 to the atmosphere from agriculture. Therefore biochar has an effective role in mitigating greenhouse gases emission and climate change.

6.5.3 *Waste Management*

Biochar is produced from waste material (municipal sewage waste, biomass, chicken waste etc.). Since the waste material is reuse for production of biochar, therefore the biochar production process is sustainable technique and economically viable.

6.6 Conclusions

Globally, heavy metal pollution is a severe problem. Several in-situ bioremediation methods are used for treating contaminated soil and water. Biochar application is one of the environment friendly and sustainable technology for treating heavy metal contaminated soil. Various types of biochar on the basis of feedstock used are hardwood, chicken manure, oak wood, rice straw, dairy manure etc. The heavy metals such as Cr, Zn, Pb, Cu, As are removed by biochar form contaminated soil and water through mechanism such as complexation, adsorption, precipitation and stabilization. In spite of decreasing bioavailability of heavy metal to soil, biochar application also improve soil physicochemical properties, organic carbon content, increase in pH and CEC of soil, maintaining the population of beneficial microorganism and waste management. Hence, it can be concluded that biochar application to contaminated soil and water plays a dual role of controlling soil and water pollution and improving the soil fertility.

References

Agrafioti E, Kalderis D, Diamadopoulos E (2014) Arsenic and chromium removal from water using biochars derived from rice husk, organic solid wastes and sewage sludge. J Environ Manag 133:309–314

Ahmad M, Lee SS, Yang JE, Ro HM, Lee YH, Ok YS (2012) Effects of soil dilution and amendments (mussel shell, cow bone, and biochar) on Pb availability and phytotoxicity in military shooting range soil. Ecotoxicol Environ Saf 79:225–231

Ahmad M, Rajapaksha AU, Lim JE, Zhang M, Bolan N, Mohan D, Ok YS (2014) Biochar as a sorbent for contaminant management in soil and water: a review. Chemosphere 99:19–33

Armynah B, Djafar Z, Piarah WH, Tahir D (2018) Analysis of chemical and physical properties of Biochar from Rice husk biomass. J Phys Conf Ser 979(1)

Askeland M, Clarke B, Paz-Ferreiro J (2019) Comparative characterization of biochars produced at three selected pyrolysis temperatures from common woody and herbaceous waste streams. Peer J 7:1–20

Bandara T, Herath I, Kumarathilaka P, Seneviratne M, Seneviratne G, Rajakaruna N, Vithanage M, Ok YS (2017) Role of woody biochar and fungal-bacterial co-inoculation on enzyme activity and metal immobilization in serpentine soil. J Soils Sed 17:665–673

Beesley L, Jiménez EM, Eyles JLG (2010) Effects of biochar and green waste compost amendments on mobility, bioavailability and toxicity of inorganic and organic contaminants in a multi-element polluted soil. Environ Pollut 158:2282–2287

Bhatia A, Singh SD, Kumar A (2015) Heavy Metal Contamination of Soil, Irrigation Water and Vegetables in Peri-Urban Agricultural Areas and Markets of Delhi. Water Environment Research 87(11):2027–2034

Bouraoui Z, Jeguirim M, Guizani C, Limousy L, Dupont C, Gadiou R (2015) Thermogravimetric study on the influence of structural, textural and chemical properties of biomass chars on CO_2 gasification reactivity. Energy 88:703–710

Brewer CE, Schmidt-Rohr K, Satrio JA, Brown RC (2009) Characterization of biochar from fast pyrolysis and gasification systems. Environ Prog Sustain Energy 28(3):386–396

Cao X, Ma L, Liang Y, Gao B, Harris W (2011) Simultaneous immobilization of lead and atrazine in contaminated soils using dairy-manure biochar. Environ Sci Technol 45:4884–4889

Chen Z, Li K, Pu L (2014) The performance of phosphorus (P)-doped activated carbon as a catalyst in air-cathode microbial fuel cells. Bioresour Technol 170:379–384

Chen ZL, Zhang JQ, Huang L, Yuan ZH, Li ZJ, Liu MC (2019) Removal of Cd and Pb with biochar made from dairy manure at low temperature. J Integr Agric 18(1):201–210

Choudhary B, Paul D, Singh A, Gupta T (2017) Removal of hexavalent chromium upon interaction with biochar under acidic conditions: mechanistic insights and application. Environ Sci Pollut Res 24(20):16786–16797

Dai ZM, Zhang XJ, Tang C, Nwajiaku (2018) Potential role of biochars in decreasing soil acidification – a critical review. Sci Tot Environ 581–582:601–611

Domingues RR, Trugilho PF, Silva CA, De Melo ICNA, Melo LCA, Magriotis ZM, Sánchez-Monedero MA (2017) Properties of biochar derived from wood and high-nutrient biomasses with the aim of agronomic and environmental benefits. PLoS One, 12(5):1–19

Dong Li H, Da Silva, De Oliveira EB, Chen LM, Ma LQ (2017) Mechanisms of metal sorption by biochars: biochar characteristics and modifications. Chemosphere 178:466–478

Egene CE, Van Poucke R, Ok YS, Meers E, Tack FMG (2018) Impact of organic amendments (biochar, compost and peat) on Cd and Zn mobility and solubility in contaminated soil of the Campine region after three years. Sci Tot Environ 626:195–202

Fellet G, Marmiroli M, Marchiol L (2014) Elements uptake by metal accumulator species grown on mine tailings amended with three types of biochar. Sci Total Environ 468–469:598–608

Fidel RB, Laird DA, Thompson ML (2017) Characterization and quantification of biochar alkalinity. Chemosphere 167:367–373

Gupta N, Yadav KK, Kumar V, Kumar S, Chadd RP, Kumar A (2019) Trace elements in soil-vegetables interface: Translocation, bioaccumulation, toxicity and amelioration - A review. Science of The Total Environment 651:2927–2942

GWRTAC (1997) Remediation of metals-contaminated soils and groundwater. Tech. Rep. TE-97-01, GWRTAC, Pittsburgh, Pa, USA, GWRTAC-E Series

Houben D, Evrard L, Sonnet P (2013) Beneficial effects of biochar application to contaminated soils on the bioavailability of Cd, Pb and Zn and the biomass production of rapeseed (Brassica napus L.), Biomass and Bioenergy, https://doi.org/10.1016/j.biombioe.2013.07.019

Huang L, Li Y, Zhao M, Chao Y, Qiu R, Yang Y, Wang S (2018) Potential of Cassia alata L., coupled with biochar for heavy metal stabilization in multi-metal mine tailings. Int J Environ Res Public Health 15(3):494

Hübner R, Astin KB, Herbert RJ (2010) "Heavy metal"—time to move on from semantics to pragmatics. J Environ Monit 12:1511–1514

Jechan L, Ki-Hyun K, Eilhann K (2017) Biochar as a catalyst. Renew Sust Energy Rev 77:70–79

Jianga J, Xua R, Jianga T, Li Z (2012) Immobilization of Cu(II), Pb(II) and Cd(II) by the addition of rice straw derived biochar to a simulated polluted Ultisol. J Hazard Mater 229–230:145–150

Jindo K, Mizumoto H, Sawada Y, Sanchez-Monedero M, Sonoki T (2014) Physical and chemical characterization of biochars derived from different agricultural residues. Biogeosciences 11(23):6613–6662

Karami N, Clemente R, Moreno-Jiménez E, Lepp NW, Beesley L (2011) Efficiency of green waste compost and biochar soil amendments for reducing lead and copper mobility and uptake to ryegrass. J Hazard Mater 191:41–48
Karppinen EM, Stewart KJ, Farrell RE, Siciliano SD (2017) Petroleum hydrocarbon remediation in frozen soil using a meat and bonemeal biochar plus fertilizer. Chemosphere 173:330–339
Kelly CN, Peltz CD, Stanton M, Rutherford DW, Rostad CE (2014) Biochar application to hardrock mine tailings: soil quality, microbial activity, and toxic element sorption. Appl Geochem 43:35–48
Khan S, Chao C, Waqas M, Arp HPH, Zhu YG (2013) Sewage sludge biochar influence upon rice (Oryza sativa L.) yield, metal bioaccumulation and greenhouse gas emissions from acidic paddy soil. Environ Sci Technol 47:8624–8632
Khan SA, Sharma GK, Malla FA, Kumar A, Rashmi GN (2019) Microalgae based biofertilizers: A biorefinery approach to phycoremediate wastewater and harvest biodiesel and manure. Journal of Cleaner Production 211:1412–1419
Koduru JR, Shankar S, More NS, Shikha, Lingamdinne LP, Singh J (2017) Toxic metals contamination in environment toxicological effects and bioremediation approaches for environmental cleanup in book environmental pollutants and their bioremediation approaches. https://doi.org/10.1201/9781315173351-8
Kour D, Rana KL, Yadav N, Yadav AN, Rastegari AA, Singh C, Negi P, Singh K, Saxena AK (2019) Technologies for biofuel production: current development, challenges, and future prospects AA Rastegari et al., Prospects of renewable bioprocessing in future energy systems, Biofuel and Biorefinery Technologies 10, Cham, Springer, pp 1–50
Kumar SS, Malyan S, Kumar A, Bishnoi NR (2016) Optimization of Fenton's oxidation by Box-Behnken design of response surface methodology for landfill leachate. J Mater Environ Sci 7:4456–4466
Kumar A, Chaturvedi AK, Yadav K, Arunkumar KP, Malyan SK, Raja P, Kumar R, Khan SA, Yadav KK, Rana KL, Kour D, Yadav Yadav N, Yadav AN (2019) Fungal phytoremediation of heavy metal- contaminated resources: Current scenario and future prospects. In: In Recent advancement in white biotechnology through fungi, fungal biology. Springer Nature Switzerland, pp 437–461
Lehmann J, Joseph S (2009) Biochar for environmental management: science and technology. London; Sterling, VA: Earthscan. Litter biochars in small arms range soil: impact of pyrolysis temperature. Agric Food Chem 60:5035–5044
Li K, Yang B, Wang H, Xu X, Gao Y, Zhu Y (2019) Dual effects of biochar and hyperaccumulator Solanum nigrum L. on the remediation of Cd-contaminated soil. Peer J 7:e6631. https://doi.org/10.7717/peerj.6631
Lottermoser BG (2010) Introduction to mine wastes. In: Mine wastes. Springer, Berlin/Heidelberg, pp 1–41
Lu H, Li Z, Fu S, Méndez A, Gascó G, Paz-Ferreiro J (2015) Combining phytoextraction and biochar addition improves soil biochemical properties in a soil contaminated with Cd. Chemosphere 119:209–216
Malla FA, Khan SA, Khan R, Sharma GK, Gupta N, Abraham G (2015) Phycoremediation potential of Chlorella minutissima on primary and tertiary treated wastewater for nutrient removal and biodiesel production. Ecol Eng 75:343–349
Matta G, Kumar A, Kumar A, Naik PK, Kumar A, Srivastava N (2018) Assessment of Heavy Metals Toxicity and Ecological Impact on Surface Water Quality Using HPI in Ganga River. INAE Letters 3(3):123–129
Namgay T, Singh B, Singh BP (2010) Influence of biochar application to soil on the availability of As, Cd, Cu, Pb, and Zn to maize (Zea mays L.). Soil Res 48:638–647
Nartey OD, Zhao B (2014) Biochar Preparation, Characterization, and Adsorptive Capacity and Its Effect on Bioavailability of Contaminants: An Overview. Adv Mater Sci Eng 2014:1–12
Novak JM, Busscher WJ, Laird DL, Ahmedna M, Watts DW, Niandou MAS (2009) Impact of biochar amendment on fertility of a south eastern coastal plain soil. Soil Sci 174:105–112

Novak JM, Ippolito JA, Lentz RD (2016) Soil health, crop productivity, microbial transport, and mine spoil response to biochars. Bioenergy Res 9:454–464

Nwajiaku IM, Olanrewaju JS, Sato K, Tokunari T, Kitano S (2018) Change in nutrient composition of biochar from rice husk and sugarcane bagasse at varying pyrolytic temperatures. Int J Recycling Org Waste Agric 7(4):269–276

Park JH, Choppala GK, Bolan NS, Chung JW, Chuasavathi T (2011) Biochar reduces the bioavailability and phytotoxicity of heavy metals. Plant Soil 348:439–451

Park JH, Wang JJ, Kim SH, Kang SW, Jeong CY, Jeon JR, Seo DC (2019) Cadmium adsorption characteristics of biochars derived using various pine tree residues and pyrolysis temperatures. J Colloid Sci 553:298–307

Prasad MNV (2001) Metals in the environment: analysis by biodiversity. Marcel Dekker, New York

Qian T, Wang Y, Fan T, Fang G, Zhou D (2016) A new insight into the immobilization mechanism of Zn on biochar: the role of anions dissolved from ash. Sci Rep 6:33630

Quicker P, Weber B, Herstellung K, Eigenschaften, Verwendung V (2016) Biomass ekarbonisaten. Springer Vieweg, Wiesbaden. https://doi.org/10.1007/978-3-658-03689-8

Rodrıguez-Vila A, Covelo EF, Forj'an R, Asensio V (2014) Phytoremediating a copper mine soil with Brassica juncea L, compost and biochar. Environ Sci Pollut R 21:11293–11304

Samsuri W, Sadegh-Zadeh F, Seh-Bardan BJ (2013) Sorption of As (III) and As (V) by Fe coated biochars and biochars produced from empty fruit bunch and rice husk. J Environ Chem Eng 1:981–988

Sharma GK, Khan SA, Fayaz MA, Gupta N (2014) Nutrient sequestration and phycoremediation of sewage wastewater by selective microalgae. Green Farming. 5:623–626

Singh S, Kumar M (2006) Heavy metal load of soil, water and vegetables in peri-urban Delhi. Environ Monit Assess 120:79–91

Singh J, Rawat KS, Kumar A (2013) Mobility of Cadmium in sewage sludge applied soil and its uptake by Radish (Raphanus sativus L.) and Spinach (Spinacia oleracea L.). Int J Agric Food Sci Technol 4(4):291–296

Singh C, Tiwari S, Boudh S, Singh JS (2017a) Biochar application in management of paddy crop production and methane mitigation. In: Singh JS, Seneviratne G (eds) Agro-environmental sustainability: managing environmental pollution, 2nd edn. Springer, Cham, pp 123–146

Singh C, Tiwari S, Singh JS (2017b) Impact of rice husk biochar on nitrogen mineralization and methanotrophs community dynamics in Paddy soil. Int J Pure App Biosci 5(5):428–435

Singh C, Tiwari S, Singh JS (2017c) Application of biochar in soil fertility and environmental management: a review. Bull Environ Pharmacol Life Sci 6(12):07–14

Singh C, Tiwari S, Gupta VK, Singh JS (2018) The effect of rice husk biochar on soil nutrient status, microbial biomass and paddy productivity of nutrient poor agriculture soils. Catena 171:485–493

Singh C, Tiwari S, Singh JS (2019) Biochar: a sustainable tool in soil 2 pollutant bioremediation. In: Bharagava RN, Saxena G (eds) Bioremediation of Industrial Waste for Environmental Safety. Springer, Singapore, pp 475–494

Suárez-Hernández L, Ardila-A AN, Barrera-Zapata R (2017) Morphological and physicochemical characterization of biochar produced by gasification of selected forestry species. Rev Fac Ing 26(46):123–130

Suliman W, Harsh JB, Abu-Lail NI, Fortuna AM, Dallmeyer I, Garcia-Pérez M (2017) The role of biochar porosity and surface functionality in augmenting hydrologic properties of a sandy soil. Sci Tot Environ 574:139–147

Tiwari S, Singh C, Singh JS (2018) Land use changes: a key ecological driver regulating methanotrophs abundance in upland soils. Energy Ecol Environ 3(6):355–371

Tiwari S, Singh C, Boudh S, Rai PK, Gupta VK, Singh JS (2019a) Land use change: a key ecological disturbance declines soil microbial biomass in dry tropical uplands. J Environ Manag 242:1–10

Tiwari S, Singh C, Singh JS (2019b) Wetlands: a major natural source responsible for methane emission AK Upadhyay et al. , Restoration of wetland ecosystem: a trajectory towards a sustainable environment, Springer, Singapore pp 59–74

Uchimiya M, Bannon DI, Wartell LH, Lima IM, Klasson KT (2012) Lead retention by broiler litter biochars in small arms range soil: impact of pyrolysis temperature. Agric Food Chem 60:5035–5044

Vithanage M, Herath I, Joseph S, Bundschuh J, Bolan N, Ok YS, Rinklebe J (2017) Interaction of arsenic with biochar in soil and water: a critical review. Carbon 113:219–230

Wang Q, Wang B, Lee X, Lehmann J, Gao B (2018) Sorption and desorption of Pb (II) to biochar as affected by oxidation and pH. Sci Tot Environ 634:188–194

Weber K, Quicker P (2018) Properties of biochar. Fuel 217:240–261

Wenjie S, Sha Z, Chunming S (2018) Impact of biochar on the bioremediation and phytoremediation of heavy metal(loid)s in soil, Advances in Bioremediation and Phytoremediation, Naofumi Shiomi, IntechOpen https://doi.org/10.5772/intechopen.70349

Wu B, Cheng G, Jiao K, Shi W, Wang C, Xu H (2016) Mycoextraction by Clitocybe maxima combined with metal immobilization by biochar and activated carbon in an aged soil. Science of The Total Environment 562:732–739

Wuana RA, Okieimen FE (2011) Heavy metals in contaminated soils: a review of sources, chemistry, risks and best available strategies for remediation. ISRN Ecol:1–20

Xie T, Reddy KR, Wang C, Yargicoglu E, Spokas K (2015) Characteristics and applications of biochar for environmental remediation: a review. Crit Rev Environ Sci Technol 45(9):939–969

Yang X, Zhang S, Ju M, Liu L (2019) Preparation and modification of biochar materials and their application in soil remediation. Appl Sci 9(7):1365

Zhang J, Huang B, Chen L, Li Y, Li W, Luo Z (2018) Characteristics of biochar produced from yak manure at different pyrolysis temperatures and its effects on the yield and growth of highland barley. Chem Speciat Bioavail 30(1):57–67

Chapter 7
Role of Biochar in Carbon Sequestration and Greenhouse Gas Mitigation

Dipak Kumar Gupta, Chandan Kumar Gupta, Rachana Dubey, Ram Kishor Fagodiya, Gulshan Sharma, Keerthika A., M. B. Noor Mohamed, Rahul Dev, and A. K. Shukla,

Abstract Global warming and associated climate change are becoming a threat to almost all the ecosystems on the earth. According to the intergovernmental panel on climate change (IPCC) special report 2019, the global mean surface (land and ocean) temperature has been increased by 0.87 °C while mean of land surface air temperature has increased by 1.53 °C since 1850–2015. Climate change is affecting food security and human life due to warming, changing precipitation patterns, and the greater frequency of some extreme events. The main cause of global warming is the continuous increase in the atmospheric concentration of greenhouse gases (GHGs) like CO_2, CH_4, N_2O and fluorinated gases due to several anthropogenic activities. Therefore, reducing the increasing concentration of GHG is necessary to slow down global warming and climate change. Among several options of greenhouse mitigation, application of biochar into the soil is gaining popularity due to several advantages over other options. Biochar is a highly stable form of carbon derived from pyrolysis of biomass at relatively low temperatures. Application of biochar into the soil has been reported to provide multiple benefits like increase in

D. K. Gupta (✉) · Keerthika A. · M. B. Noor Mohamed · A. K. Shukla
ICAR-Central Arid Zone Research Institute, Regional Research Station, Pali Marwar, Rajasthan, India
e-mail: deepak.gupta@icar.gov.in

C. K. Gupta
Department of Agro-meteorology and Environmental Science, Birsa Agricultural University, Ranchi, Jharkhand, India

R. Dubey
ICAR-Research Complex for Eastern Region, Patna, Bihar, India

R. K. Fagodiya
Division of Irrigation and Drainage Engineering, CSSRI, Karnal, Haryana, India

G. Sharma
ICAR-National Bureau of Soil Survey and Land Use Planning, Regional Center, Jorhat, Assam, India

R. Dev
ICAR-Vivekananda Parvatiya Krishi Anusandhan Sansthan, Almora, Uttarakhand, India

J. S. Singh, C. Singh (eds.), *Biochar Applications in Agriculture and Environment Management*, https://doi.org/10.1007/978-3-030-40997-5_7

crop yield, nutrient and water use efficiency and several environmental benefits. Recalcitrant nature, relatively higher carbon content and easily available feedstock make biochar a highly sustainable and quick option for carbon sequestration into the soil. Biochar application into the soil not only helps in carbon sequestration but also provides a better option for managing agricultural residues. The application of biochar has also reported for reducing a considerable amount of methane and nitrous oxide emission from the agricultural field due to its priming effect on the soil. Biochar yield, physical properties, and carbon content varies with the type of feedstock and pyrolysis condition. Therefore, the rate of carbon sequestration and mitigation of greenhouse gas is also highly variable, however, the biochar application ultimately leads to a positive contribution towards climate change mitigation. However, most of the reported benefits are confined to laboratory and field trial at institute level, widespread adoption of biochar on farmer's field is still lacking. In the present chapter, all the aspects of biochar towards carbon sequestration and greenhouse mitigation have been well discussed.

Keywords Carbon sequestration · Climate change mitigation · Greenhouse gas · Global warming · Crop residue

7.1 Introduction

The global mean surface (land and ocean) temperature has been increased by 0.87 °C (0.75 °C–0.99 °C) while mean land surface air temperature has increased by 1.53 °C (1.38 °C–1.68 °C) since 1850 to 2015 (IPCC 2019). Global warming is resulting in changes in the global climate system and almost all the natural and human systems in many countries. According to the intergovernmental panel on climate change (IPCC) special report 2019, global warming has resulted in an increased frequency, intensity and duration of heat-related events; change in precipitation patterns and greater frequency of some extreme events in most land regions. Yields of some crops (e.g., maize and wheat) in many lower-latitude regions are being affected negatively by observed climate changes. The main cause of global warming is a continuous increase in the atmospheric concentration of greenhouse gases (GHG) like CO_2, CH_4, N_2O and fluorinated gases (F-gases: hydrofluorocarbons, perfluorocarbons, sulfur hexafluoride, and nitrogen trifluoride) due to several anthropogenic activities. The total GHG emission has increased from 27 Gt CO_2 eq/year in 1970 to 49 Gt CO_2 eq/year in 2010 (IPCC2014a, b). Out of these greenhouse gases, CO_2 is a major greenhouse gas responsible for about 72% of global warming followed by CH_4 (20%), N_2O (5%) and F-gases (3%) (Fig. 7.1).

Despite several climate change mitigation measures, annual GHG emission is increasing on an average rate of 2.2% (1.0 Gt CO_2 equivalent) per year from 2000

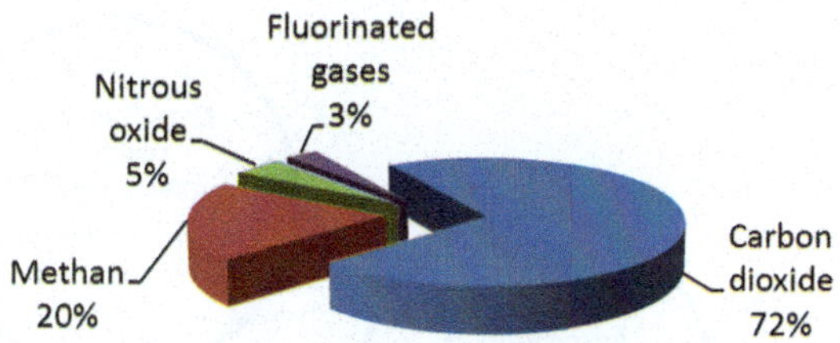

Fig. 7.1 The contribution of greenhouse gases in total global warming (IPCC 2014a, b)

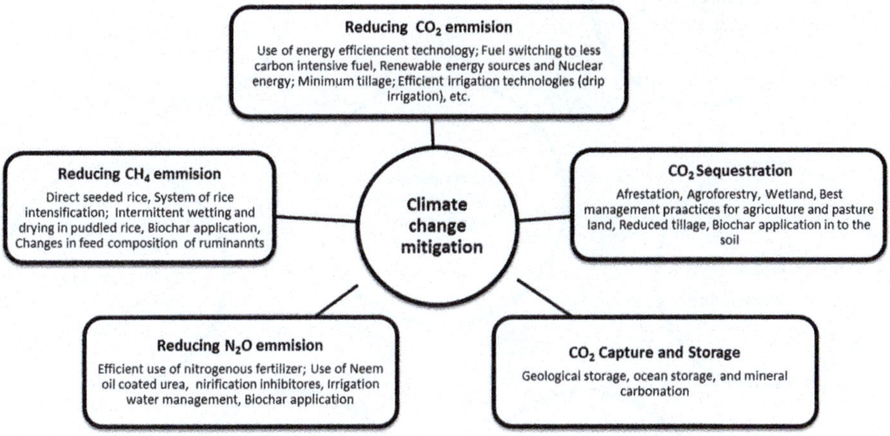

Fig. 7.2 Strategies for climate change mitigation

to 2010 compared to 1.3% (0.4 Gt CO_2 equivalent) per year from 1970 to 2000 (IPCC 2014a, b). It is expected that future emission of GHG will cause warming and long-lasting changes in all components of the climate system. Limiting total human-induced warming to less than 2 °C relative to the period 1861–1880 is necessary to stop the devastating effect of climate change. Without any additional mitigation, the global mean surface temperature has been predicted to increase between 3.7 °C and 4.8 °C in 2100 as compared to pre-industrial levels. Limiting warming to less than 2 °C would require to keep cumulative CO_2 emissions since 1870 to remain below about 2900 Gt CO_2 (IPCC 2014a, b). However, about 1900 Gt CO_2 had already been emitted up to 2011. Therefore, reduction in the atmospheric concentration of GHG is an urgent global need to arrest global warming and climate change.

Global warming can be reduced by the simultaneous effort of reducing CO_2, CH_4, N_2O, and F-gases emission at the source and capturing already emitted atmospheric CO_2 and storing in the long-lived pool. Therefore, identifying technologies that reduce anthropogenic GHG emission at the source as well as technologies that capture and store already emitted GHG become necessary to mitigate climate change. Several suggested strategies for the mitigation of climate change have been depicted in Fig. 7.2.

Among these, strategies that led to carbon sequestration in the soil as well as reduce CH_4 and N_2O emission should be considered as one of the most potent

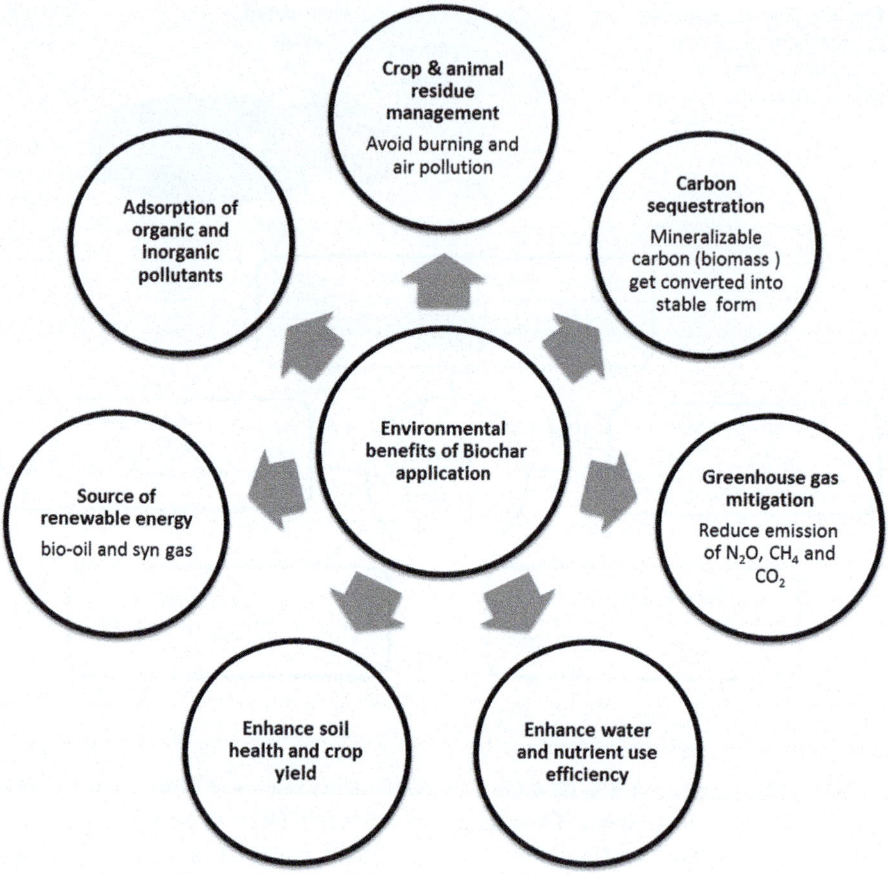

Fig. 7.3 Environmental benefits of biochar application in to the soil

options. The conventional methods like agroforestry, afforestation and soil management are being promoted for soil carbon sequestration. However, increasing the levels of soil organic carbon by conventional agricultural management can take many years (Denman et al. 2007). Recently, interest has grown to sequester atmospheric carbon into biochar followed by its application to the soil. Soil application of biochar is being frequently reported as a potential option for climate change mitigation through carbon sequestration and other agricultural and environmental benefits. Application of biochar into the soil not only sequester carbon but also reduce emission of N_2O and CH_4 from soil, provide suitable option for management of agricultural and forestry wastes, enhancement of soil sustainability, reduction in fertilizer requirements, production of renewable energy and several other environmental benefits (Waters et al. 2011; Lehmann and Joseph 2009; Jeffery et al. 2011; Wang et al. 2016; Majumder et al. 2019). The reported potential benefits of biochar in environmental management have been depicted in Fig. 7.3. Therefore, the present chapter will discuss the role of biochar in carbon sequestration and greenhouse gas mitigation.

7.2 Carbon Dioxide and its Impact on Global Warming

Among different greenhouse gases responsible for global warming, CO_2 has the least global warming potential. Therefore, the global warming potential of other GHGs is determined with respect to CO_2. It means CO_2 absorbs less heat per molecule than the other greenhouse gases like methane and nitrous oxide. However, higher atmospheric concentration and a long life cycle in the atmosphere make CO_2 the most responsible GHG for global warming. It contributes about 72% of the total global warming induced by anthropogenic activities (IPCC 2014a, b). Before the industrial revolution (the mid-1700s), the global average atmospheric concentration of CO_2 was about 280 ppm and it has reached 407.4 ± 0.1 ppm in 2018 (Dlugokencky et al. 2019). The major sectors responsible for CO_2 emission are fossil fuel combustion and industrial processes. CO_2 emissions from fossil fuel combustion and industrial processes contributed about 78% of the total GHG emission increase from 1970 to 2010 (IPCC 2014a, b). Therefore, rising atmospheric CO_2 concentration can be reduced by two processes (a) reducing emission by using energy-efficient technology; switching to renewable energy like biodiesel, ethanol, wind energy, solar power, reduced tillage, and efficient irrigation system and (b) capturing and storing emitted CO_2 into the long-lived pool.

7.3 Role of Biochar in Carbon Sequestration

Plant assimilates CO_2 by the process of photosynthesis and stores it into the above and below-ground biomass as well as supply liters to the soil for soil organic carbon buildup. This process of CO_2 assimilation and storage by the plant into the long-lived reservoir of plant biomass and soil organic carbon is known as carbon sequestration. However, the duration of C-storage in biomass is highly variable. It varied from a few years in the annual crop to 5–60 years in the agroforestry system and a few hundred years in the forests. Soil organic carbon buildup by natural process is very slow and also prone to lose under intensive tillage practices and soil erosion. Therefore, land-use systems which lead to a large amount of carbon assimilation within a relatively short period and it's storage for long period are suitable options for reducing the concentration of atmospheric CO_2 and mitigating global warming and climate change.

Among different options of soil carbon sequestration, biochar application to soil is being considered as the most promising option for long term storage of carbon sequestered in biomass. It is reported that natural biochar sink in Australia is sequestering about 21 million tons of carbon dioxide annually (Graetz and Skjemstad

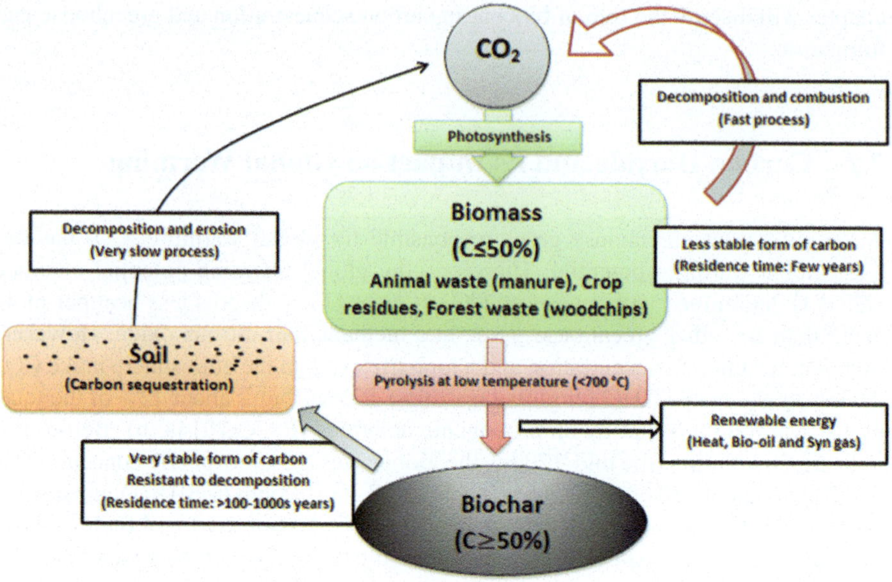

Fig. 7.4 Process of carbon sequestration by biochar

2003). Preliminary calculations suggests that if 2.5% of the world's agricultural land produces biochar (ideally from wastes) for application into topsoil then atmospheric CO_2 levels might be reduced to pre-AD 1752 level by 2050 (Jacquot 2008). According to Lehmann et al. 2006, globally, up to 12% of all anthropogenic land-use change emissions can be offset annually if slash-and-burn agriculture is replaced by slash-and-char systems. Further, it has been also suggested that, if other greenhouse gas emissions from soils after application of biochar are not elevated, and if emissions associated with production and transport of biochar and/or its feedstocks do not off-set the sequestered C, then the overall greenhouse effect will be abated by application of biochar into the soil (Roberts et al. 2010). This high potential of biochar for carbon sequestration is being advocated mainly due to its high carbon content and very stable form. The overall mechanism of carbon sequestration by biochar can be understood in Fig. 7.4.

The mechanism responsible for the higher potential of long term carbon storage by biochar application into the soil can be grouped in following properties of biochar

(i) Biochar is a stable and rich form of carbon
(ii) It leads to a relatively higher rate of carbon sequestration
(iii) It leads to agricultural and forestry waste management

7.3.1 Biochar: A Stable and Rich Form of Carbon

There are two intrinsic properties of biochar that make it a very potential option for long term carbon sequestration in the soil. First is, its stability i.e. resistance to biotic and abiotic decay into the soil and second is relatively higher carbon content as compared to biomass. The estimated C-residence time of biochar in the soils has been reported to range between hundreds to thousands of years (Liang et al. 2008; Kuzyakov et al. 2009; Major et al. 2010; Zimmerman 2010). Wang et al. (2016) through a meta-analysis of 24 studies reported 108 days and 556 years mean residence time of labile and recalcitrant biochar resulting in only a small part of biochar bioavailable and remaining 97% contribute directly to long-term C sequestration in soil. The initial rate of biochar decomposition is relatively faster and gets decreased with time. The median rate of biochar decomposition in this meta-analysis was found to be 0.0046% day^{-1}. The initial fast rate of decomposition is mainly due to the decomposition of the labile condensed fraction of biochar. This initial fast rate of decomposition disappears after 2 years and maintained at a very low level over prolonged periods (Kuzyakov et al. 2014; Fang et al. 2014).

The primary reason for the higher stability of biochars in soils is their chemical recalcitrance i.e. resistant to microbial decomposition (Liang et al. 2008) which is due to the presence of aromatic structures. An increase in the aromaticity of organic matter leads to a more recalcitrant nature of the organic matter. Biochar is produced by the process of pyrolysis of biomass in an oxygen-limited environment at relatively low temperatures. During pyrolysis, the biomass undergoes devolatilization and the solid portion gets enriched in carbon. The H and O are preferably removed over C and the H/C and O/C ratios tend to decrease as biomass undergoes its transformation into biochar. This process results in very low H/C and O/C ratios in biochar as compared to the original biomass feedstock (Nsamba et al. 2015). The ratio of H/C and O/C is an indicator of the degree of aromaticity, carbonization, maturation and the stability of biochar (Lehmann and Joseph 2009). According to Krull et al. (2009), decreasing the H/C ratio in biochars indicates an increasing aromatic structure in the biochar. As pyrolysis temperature increases, the turbostratic layering inside of biochar increases in orderliness, the mass percentage of the fused aromatic C thereby increases, the produced biochar is thus often low in easily degradable C but high in recalcitrant C (Nguyen et al. 2010). The value of H/C ratios greater than 0.7 indicates a low biochar quality and pyrolysis deficiencies and the value of the O/C ratio greater than 0.4 indicates lower biochar stability (EBC 2012). The International Biochar Initiative (IBI) recommends a maximum value of 0.7 for the molar H/C ratio (Nsamba et al. 2015, Mary et al. 2016) to distinguish biochar from biomass that has not been or only somewhat thermo-chemically altered.

On the weight basis, biochar contains a high percentage of carbon as compared to original feedstock, however, it varied with the type of feedstock and pyrolysis condition. The carbon content has been reported to vary 29–50% in rice straw derived biochar and 70–85% in apple and oak tree branch at pyrolysis temperature ranging from 400–800 °C (Jindo et al. 2014). Billa et al. (2019) reported 57.59%

carbon in rice husk derived biochar to 93.38% in Cassava residue derived biochar at the standard procedure for Biochar production. Lee et al. (2013) reported more than 84% carbon in biochar obtained from Bagasse, Cocopeat, Paddy straw, Palm kernel shell, Wood stem and Wood bark at constant pyrolysis temperature of 500 °C. Windeatt et al. (2014) reported 90.6, 88.6, 54.5, 93.9, 75.3, 83.2, 71.8 and 82.6% carbon in biochar obtained from Palm shell, Sugarcane bagasse, Rice husk, Coconut shell, Wheat straw, Cotton stalk, Olive pomace, and Coconut fiber. Purakayastha et al. (2015) reported 66% carbon in maize biochar, followed by 64% in pearl millet biochar, 64% in wheat biochar and 60% in rice biochar.

Charcol is also mineralized in soil otherwise the earth's surface would be converted into charcoal within a period of time. Shneour (1966) successful oxidized artificial graphitic 14C to 14CO_2 in the presence of soils with high microbial activity. Thermodynamically, the abiotic oxidation of elemental C to CO_2 is a strongly exothermic reaction ($\Delta H = -94{,}052$ kJ). However, under environmental conditions, this process is extremely slow (Shneour 1966). Therefore, biochar is much more stable than uncharred organic matter i.e. biomass, and that this difference is the relevant measure for its ability to prevent carbon from being returned rapidly to the atmosphere.

7.3.2 *Carbon Storage Potential of Biochar*

Biochar is a rich source and a very stable form of carbon obtained from the pyrolysis of biomass. Biochar production itself do not sequester carbon from the atmosphere, however, it leads to the transformation of carbon sequestered in biomass into more stable form i.e. biochar as well as helps in enhancing soil organic carbon sequestration. It has been reported that conversion of biomass carbon to biochar leads to storage of about 50% of the initial carbon compared to the low amounts of carbon retained after burning (3%) and biological decomposition (less than 10–20% after 5–10 years) (Lehmann et al. 2006). Windeatt et al. (2014) reported 21.3–32.5% carbon retained in biochar obtained from pyrolysis of Palm shell, Sugarcane bagasse, Rice husk, Coconut shell, Wheat straw, Cotton stalk, Olive pomace and Coconut fiber at 600 °C. The balance of carbon stored in the biochar, as the mass fraction of the carbon remaining in biochar from the original feedstock carbon, was between 45% for olive pomace and 57% for coconut fiber. The average value of carbon stored from the original feedstock carbon was 51%. Therefore, the biochar itself represents a carbon stock that once added to soil tends to persist for a long time. Further, biochar additions can also interact with the native organic matter already present in soils, and either stimulate or reduce the rate of decomposition of the native soil organic matter (Paustian et al. 2019). Zhang et al. (2018) reported a 76.29% increase in SOC after 5 years of wheat straw biochar application in the soil at the rate of 40 t/ha. Both positive and negative effects on native soil organic matter decomposition following biochar addition have been reported (Song et al. 2016;

Wang et al. 2016), but in most cases, these effects on the long-term soil C balance are small (Wang et al. 2016).

However, the net amount of carbon available for storage depends upon the percentage of biochar obtained from the original feedstock and carbon content of biochar. The carbon sequestration potential of the biochars, which is a measure of the amount of the original feedstock carbon that would be retained in biochar for long time periods upon addition to soil, is generally calculated by the following equation (Windeatt et al. 2014).

$$CSB(\%) = \frac{\mathrm{M} \times \mathrm{Ch} \times C_{Ch} \times R_{50}}{(\mathrm{M} \times C_F)}$$

where, CSB is carbon sequestration potential of biochar (%), M is the mass of feedstock (g), Ch is the yield of biochar (%), C_{Ch} is the carbon content of the biochar, R_{50} is the recalcitrance index and C_F is the carbon content of the feedstock.

$$R_{50X} = \frac{T_{50x}}{T_{50\,graphite}}$$

where T_{50x} and $T_{50\ graphite}$ are the temperatures at which 50% of the biochar and graphite are oxidised respectively.

During pyrolysis both mass and carbon content of biomass feedstock decreases due to loss of volatile compounds. The biochar yield and carbon content have an inverse relation with the temperature of pyrolysis. With the increase in pyrolysis temperature, the biochar yield decreases while carbon content increases (Singh et al. 2017a, b, c; Singh et al. 2018; Tiwari et al. 2018; Jindo et al. 2014; Zhang et al. 2017). Increased heating temperature results in higher concentrations of fixed C, total C and stable-C in biochar, as well as higher heating value due to the increased release of volatile compounds (Crombie and Masek 2015). Further, chemical composition and carbon content of feedstock varied and therefore of biochar. It has been reported that biochar carbon content and biochar yield increase with the increase in lignin content of the feedstock (Sun et al. 2017). Biomass feedstock's with relatively high ash contents produced relatively low fixed carbon biochars, which was attributed to the high ash content inhibiting the formation of aromatic carbon forms (Enders et al. 2012; Sun et al. 2017). The lowest carbon content was seen in rice husk biochar which has the highest ash content, conversely, the highest carbon content was seen in the coconut shell biochar which has the lowest ash content.

Table 7.1 Rotation period of major agroforestry trees species in India

S.N.	Tree species	Family	Rotation (Years)	Uses
1.	*Acacia mangium*	Fabaceae	6–20	Pulpwood and timber
2.	*Acacia nilotica*	Fabaceae	30–40	Timber and gum Arabic
3.	*Ailanthus excelsa*	Simaroubaceae	5–20	Matchwood and leaf fodder
4.	*Alnus nepalensis*	Betulaceae	25–30	Timber
5.	*Neolamarckia cadamba*	Rubiaceae	5–10	Pulpwood and timber
6.	*Albizia lebbeck*	Mimosaceae	10–20	Timber
7.	*Casuarina equisetifolia*	Casuarinaceae	4–6	Pulp and paper, boles
8.	*Dalbergia sissoo*	Papilionaceae	10–60	Timber and fodder
9.	*Eucalyptus tereticornis*	Myrtaceae	4–8	Pulp and paper, plywood
10	*Gmelina arborea*	Verbenaceae	4–12	Timber and firewood
11.	*Hardwickia binata*	Fabaceae	20–30	Timber and fodder
12.	*Leucaena leucocephala*	Leguminosae	3–4	Pulp and paper, poles, fodder
13.	*Melia dubia*	Meliaceae	5–15	Plywood
14.	*Millettia pinnata*	Leguminosae	50–60	Oil from seed kernals (biodiesel)
15.	*Populus deltoides*	Salicaceae	6–8	Pulpwood and paper, plywood
16.	*Prosopis cineraria*	Leguminosae	5–30	Timber and fodder
17.	*Salix alba.*	Salicaceae	15–20	Pulpwood, match splints and timber
18.	*Tectona grandis*	Verbenaceae	20–60	Timber
19.	*Terminalia arjuna*	Combretaceae	10–20	Plywood, timber and bark

7.3.3 A Relatively Faster Rate of Carbon Sequestration

The carbon sequestration in biomass and soil by natural process takes a long time. Gupta et al. (2019) reported that sequestration of about 31.6 Mg C ha^{-1} in biomass and 3.32 Mg C ha^{-1} in soil by *Hardwickia binata* based silvopasture system took 30 years. Measured rates of soil C sequestration through the adoption of recommended management practices have been reported to range from 0.05 to 1.0 Mg C ha^{-1} $year^{-1}$(Lal 2004). In India carbon sequestration potential of agroforestry systems has been estimated between 0.25–19.14 Mg C ha^{-1} $year^{-1}$ in biomass and 0.003 to 3.98 Mg C ha^{-1} $year^{-1}$ in soil (Dhyani et al. 2016). The period of biomass carbon sequestration by agroforestry depends on harvesting rotation. The harvesting period of different agroforestry tree species generally ranged from 5–60 years depending on their use (Chaturvedi et al. 2017) (Table 7.1). It means the agroforestry system stores carbon for 5–60 years and after harvesting of the tree, the stored biomass carbon gradually released to the atmosphere. However, the length of the C-locking period also depends on their use like timber, fuelwood and paper, and pulp. As compared to these systems, the application of biochar can lead to a faster rate of sequestration in the soil.

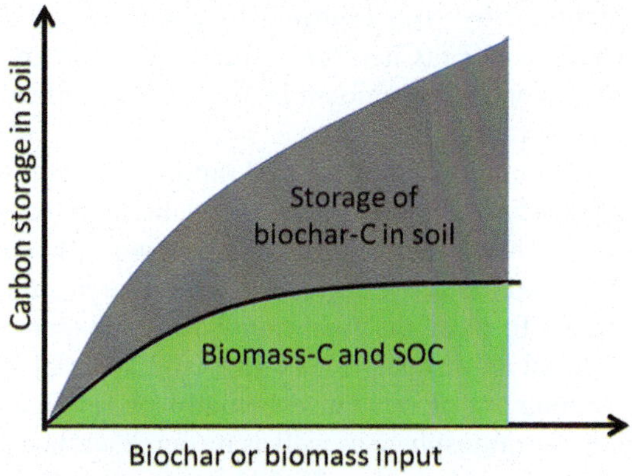

Fig. 7.5 Soil carbon storage after application of biochar and biomass in to the soil. (Modified from Wang et al. 2016)

Most of the study advocates use of agricultural, animal and forest residue as feedstock for the production of biochar. These feedstocks are annually available in large amount and their conversion into biochar takes a relatively short period of time depending upon the capacity of the kiln to produce biomass. After conversion to biochar, they are directly applied to the soil. Application rates of 10, 25, 50 and 100 t ha^{-1} were all found to significantly increase crop productivity when compared to controls, which received no biochar addition (Jeffery et al. 2011). The greatest positive effects were seen in biochar application rates of 100 t ha^{-1} (39%) (Jeffery et al. 2011). Biochar carbon is a natural constituent of many soils and soil function is not generally impaired (and maybe enhanced) with the addition of large quantities (e.g., 100 t/ha or more) of biochar. Thus biochar application to the soil can lead to storage of about 10–100 Mg C ha^{-1} in the form of biochar carbon within 2 years. However, the rate of biochar based carbon sequestration further varies with the frequency of biochar application into the soil. As compared to other biological system, each quantity of biochar applied to soil gets continuously accumulated over time with very little loss or decay (Fig. 7.5). Due to its recalcitrance to decomposition in soil, single applications of biochar can provide beneficial effects over several growing seasons in the field (Steiner et al. 2007; Major et al. 2010).

7.3.4 *Easily Available Feedstock for Biochar Production*

The feedstock suggested for biochar production should be sustainably available. Therefore unused agricultural crop residues are suggested as feedstock for biochar production. The commonly reported feedstock used for biochar production are rice straw, wheat straw, rice husk, coconut husk, coconut shell, cotton stalk, olive pomace, palm shell, rice husk, sugarcane bagasse, corn stover, cassava, corncob, coffee husk, groundnut husk, sawdust, poultry litter, paper pulp, wood chips, green waste, wood, peanut hull, pine chip (Jeffery et al. 2011; Windeatt et al. 2014; Billa et al. 2019). The world and Asian countries produced a large amount of crop residue annually thus provide sufficient feedstock for biochar production. In 2013, 5 billion tons of agricultural residues were produced worldwide out of which 47% was produced by Asia followed by America (29%), Europe (16%), Africa (6%) and Oceania (2%) (Cherubin et al. 2018). Most of this biomass is either burnt or discarded and some amount is used as animal feed, manure production, fuel, and soil incorporation.

India produced about 620 million ton crop residue in 2008–2009 out of which about 15.9% residue was burnt on the farm. Rice straw contributed 40% of the total residue burnt followed by wheat straw (22%) and sugarcane trash (20%) (Jain et al. 2014). In India, the disposal of such a large amount of crop residues is a major challenge. To clear the field rapidly and inexpensively and allow tillage practices to proceed unimpeded by residual crop material, the crop residues are burned in situ. Burning of crop residue eventually leads to unlocking of biomass sequestered CO_2 into the atmosphere as well as resource wastage and atmospheric pollution. Burning of crop residues emitted 8.57 Mt. of CO, 141.15 Mt. of CO_2, 0.037 Mt. of SOx, 0.23 Mt. of NOx, 0.12 Mt. of NH_3 and 1.21 Mt. of particulate matter in 2008–2009 (Jain et al. 2014). The burning of rice straw in Punjab and Haryana causes severe air pollution into adjoining states like Delhi. The efficient utilization of agricultural residue is very important for sustainable agricultural production.

Production of biochar from these resources thus provides a sustainable option for efficient management of crop residue, animal waste, forest by-products vis a vis improvement of soil fertility, carbon sequestration and crop growth. Modern biomass pyrolysis technologies can use agricultural and forestry wastes (such as forest residues, mill residues, field crop residues or urban waste to sequester around 30 kg of carbon for each GJ of energy produced (Lehmann et al. 2006). If available crop and forest biomass are converted into biochar at the rate of 35% recovery and product with 70% carbon then about 1402 million tone biochar can be produced per year worldwide. Similarly in India 113.3 million tons of biochar can be produced from crop residue and 250 million tons from wood residue per year (Table 7.2).

Table 7.2 Available crop residue and potential of biochar production (million ton/ha/year)

Biomass	Residue production	Biochar yield (35% recovery)	Biochar carbon content (70%)
World			
Cereals	3607.6	1443.0	1010.1
Legumes	382.1	152.8	107.0
Oil crops	275.2	110.1	77.1
Sugar crop	625.6	250.2	175.2
Tubers	119.6	47.8	33.5
Total residue	5010.1	2004.0	**1402.8**
India			
Crop residues	463.3	162.2	**113.5**
Wood residues	1019.8	356.9	**249.9**

Note: Value of crop residue production has been adopted from Cherubin et al. (2018) and Venkatesh et al. (2018)

7.4 Factors Affecting Role of Biochar in Carbon Sequestration and Greenhouse Gas Mitigation

7.4.1 Biochar Yield and Carbon Content in Biochar

The amount of carbon sequestered in the soil through biochar depends upon biochar and carbon content in the biochar. Both of these are highly variable and depend upon the type of feedstock and pyrolysis condition (Tiwari et al. 2019a, b; Singh et al. 2019; Kour et al. 2019). Therefore, optimum pyrolysis condition that leads to higher biochar yield and carbon content will ensure the higher potential for carbon sequestration.

7.4.2 Rate and Frequency of Application

The soil carbon stored in the form of biochar is directly proportional to the amount and frequency of biochar applied to the soil. The higher the rate of biochar application, the higher the amount of carbon will be stored in the form of biochar in the soil. Further frequency of application i.e. annually or biannually or another time interval that finally leads to the accumulation of biochar in the soil is the main factor. Therefore the optimum rate of application and frequency which has a positive impact on crop yield and environment is necessary to determine.

7.4.3 Methods of Application and Soil Management

Soil applied biochar is also subjected to loss mainly in three different ways: (1) by erosion from the surface, (2) by abiotic and (3) biotic degradation. Biotic and abiotic factors have a limited effect on biochar loss while erosion through water and wind may lead to loss of biochar from applied soil. Therefore methods of application that led to minimum loss will ensure a higher rate of carbon sequestration. Subsurface application has been suggested to minimize the loss of biochar and enhance its storage.

7.4.4 The Negative Effect of Biochar Application

Besides its potential agricultural benefits, biochar may contain inherent contaminants, either introduced by its feedstock (e.g., heavy metals), or co-produced during (improper) pyrolysis (e.g., polycyclic aromatic hydrocarbons) (Hilber et al. 2017). However, the link between biochar's inherent contaminants and toxicity to soil meso and macrofauna remains unclear, with data being often contradictory and influenced by feedstock and pyrolysis conditions.

7.4.5 Farmer Adoption

The ultimate user of biochar for soil application is farmers. Therefore, the rate of farmer adoption is one of the most influential factors for carbon sequestration in the soil through biochar. The latest development leads to a reduction in the production cost (−10 to 30 USD t^{-1}); however, the use of biochar in commercial agriculture remains scarce (Maroušek et al. 2019). While many beneficial reports on biochar trials exist in the scientific literature, laboratory and institute field trial, the practice of applying it to the soil in the commercial farm had just began, and no widely accepted guidelines currently exist (Majro et al. 2010). Further, biochar has not been officially advocated as a measure for carbon sequestration by any regulatory regime or program. Some of the organization like international biochar initiative is working on developing protocol and package and practices of biochar application in agricultural soil. Dickinson et al. (2015) reported that the net present value of biochar application to soils was positive in a sub-Saharan African context but negative in a Northwestern European context, due to a combination of greater production costs and more modest yield benefits in the latter scenario. Therefore, without an understanding of farmers' roles as the main stakeholders in the generation and use of this innovation, the use of biochar is unlikely to be effective.

7.5 Role of Biochar in Mitigation of Methane Emission

7.5.1 Methane Emission and Its Impact on Global Warming

Methane (CH_4) is the second potential greenhouse gas after CO_2 which contributes about 15% of total anthropogenically induced global warming and climate change. Its global warming potential is 28 times higher than CO_2. Its concentration in the atmosphere has increased from 715 ppb in the pre-industrial era to 1863 ppb in 2018. Since 2007–2015, its concentration is increasing at the rate of $+6.9 \pm 2.7$ ppb $year^{-1}$. Therefore it can be said that the contribution of methane in climate change is increasing as compared to CO_2 (when its slowdown is seen in the last 3 years) (Saunois et al. 2016). Methane is produced under anaerobic conditions (submerged soil) through the anaerobic digestion of organic matter by the action of methanogenic bacteria (Fig. 7.6). Around 70–80% of methane emission is of biological origin and remaining from natural sources. Table 7.3 describes the source and sink of methane emission from soil. Natural sources like swamps, marshes, ocean, forest soil, termites, etc. cause 20–30% of methane emission, while around

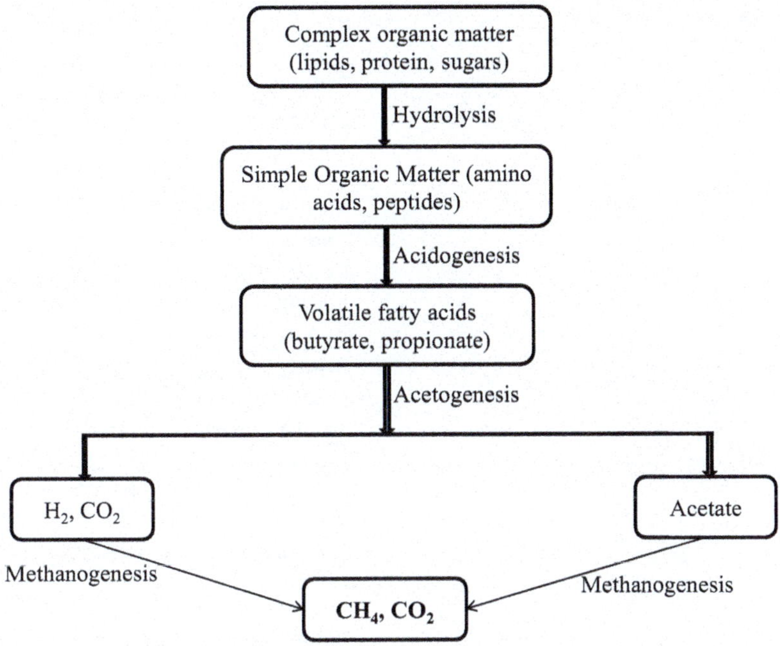

Fig. 7.6 Methanogenesis process of methane emission from soil

70–80% of emission is anthropogenic. Enteric fermentation from ruminants and conventional puddled rice cultivation practices contributes around 20–40% of its emission.

7.5.2 *Biochar and Methane Emission*

Many studies have shown that soil is responsible for around 15–30% of methane emission and most of the anthropogenic emission comes from paddy cultivation (12%) (Xiao et al. 2018; Malyan et al. 2016). Therefore most of the study has been done for the reduction of methane emission from the paddy field. There are several proposed options for reducing methane emission from paddy field. Some of the potential options are direct-seeded rice (Gupta et al. 2016a and 2016b), intermittent wetting and drying (Gupta et al. 2015), system of rice intensification; use of neem oil coated urea (NOCU) (Gupta et al. 2016a) and use of Azolla as biofertilizers (Malyan et al. 2019), etc. Recently it has been reported that soil application of biochar also offers great potential for the reduction of methane.

Aside from impacts on soil C storage, several studies suggest that biochar amendments may decrease soil CH_4 emissions, which would further contribute to greenhouse gas mitigation. Liu et al. (2011) tested the emission rate in rice soil with the incorporation of two biochar i.e. bamboo chips and straw char with control. A

Table 7.3 Sources and sink of methane emission from soil

Source	CH_4 emission (Tg/year)
Natural	
Wetland	127–202
Termite	21–132
Ocean	
Methane hydrates	
Manmade	
Coal mining	77–133
Landfills	
Animal waste	
Sewage	
Enteric fermentation	
Rice cultivation	115–243
Biomass burning	43–58
Others	15–53
SINK	
Atmospheric removal (tropospheric and stratospheric)	510–583
From soil	28–38

Source: Saunois et al. (2016)

reduction of 51.1% and 91.2% of methane in bamboo chips and straw char biochar was found compared to control plots. The reduction in methane emission was attributed to either inhibition of methanogenic bacteria or enhancement in methylotrophic bacteria. Similarly, amending the soil with rice husk biochar @ of 2 and 4% (weight biochar/weight soil) showed 45.2 and 54.9% reduction in methane emission compared to control (Pratiwi and Shinogi 2016). When soil amended with biochar and control (no biochar) were studied in elevated CO_2 and temperature conditions then result showed that cumulative methane emission was much lower with 112.2 mg kg^{-1} compared to 185.4 mg kg^{-1}dry weight soil per season in control (Han et al. 2016). The reduction of methane emission under biochar amended soil was attributed to enhanced methanotrophic activity (especially methanotrophic pmoA gene) which favors rhizospheric activity for methanotrophs. This result confirmed that applying rice straw biochar in soil not only reduces methane activity but at the same time enhances rice productivity (Han et al. 2016). Similarly, many studies show that the application of biochar results in reduced methane emission. Like Rondon et al. (2005) found reduced emission from *Brachiaria humidicola* and soybean plots treated with biochar @ 15 g kg^{-1} soil and 30 g kg^{-1} soil respectively. Similarly, in 2006, Rondon et al. found a reduction in methane emission in tropical soil with the application of wood-based biochar @ 20 t ha^{-1}. Therefore it can be said that methane emission rate varies with soil type, biochar physicochemical properties and agronomic management including water management practices (Zhang et al. 2010).

7.5.3 Factors Affecting Methane Emission from Biochar Applied Soil

Methane flux in the soil is affected by biochar in two ways (i) adsorption of methane on biochar surface (ii) enhanced methanotrophs activity via increased aeration (Jeffery et al. 2016). In the soil, methane is oxidized under aerobic conditions by the action of methanotrophs which takes methane as substrate and converts it into carbon dioxide and water. Additionally, many other things matter like soil physical and chemical properties, biomass type and process of biochar preparation (pyrolysis temperature) also affect the methane emission process. In general, adding biochar helps in enhancing soil fertility which increases methanogenic activity (Feng et al. 2012). The relation of biochar and methane emission is via methanotrophs group of microorganisms. Methanotrophs are gram-negative bacteria whose growth is stimulated through biochar application which ultimately acts as a sink for methane emission. Secondly, biochar increases the porosity of soil which allows new microbial hotspot to develop in this small porous structure where microbes retain methane for their metabolic activity and hence decrease methane emission (Feng et al. 2012). Several biochars and soil factors affect methane emission from soil. Some of them are discussed below:

(i) Water Management: Biochar when added in the puddled field under saturated condition increases methane sink or reduces source emission which means either methanotrophic process enhances or methanogenesis reduced. In the rice field, methane emission is mostly via three ways i.e. diffusion, ebullition, and plant-mediated transport. Plant mediated transport through aerenchyma is the major pathway. The application of biochar increases methane oxidation at anoxic/oxic surface of the plant. Contrary to this, in upland soils where the oxic environment persists, the application of biochar can enhance methane emission as they will provide more substrate for methanogens. Thus the application of biochar in flooded soil is more beneficial than its application in non-flooded soil in reducing methane emission and overall carbon footprint (Jeffery et al. 2016).

(ii) pH: Methane production or oxidation is microbe mediated process where both methanogens and methanotrophs work under a certain optimum conducive environment. Optimum pH for both is 6–8. Most of the biochar is basic ($pH > 7$) in nature which provides a liming effect to the soil. As methanotrophs are more sensitive to aluminium toxicity at lower pH, hence increasing pH via biochar applications helps in enhancing their activity and thus more methane gets sunk (Jeffery et al. 2011).

(iii) Pyrolysis temperature: Biochar produced via high pyrolysis temperature has a higher methane sink due to reduced H: Corganic ratio. Less H: Corganic ratio means more aromaticity. Additionally, it was found that higher temperature pyrolyzed biochar have lesser labile compounds on their surface which provide less area for microbial methanogens for reduction and thus less methane emission (Bruun et al. 2011).

(iv) Feedstocks: In general feedstock composition does not show much difference in methane flux except for biochar produced from sewage sludge. Biochar made from sewage sludge showed enhanced methane sink especially on acidic soil (Cayuela et al. 2014).

(v) Branauer, Emmett, and Teller (BET) surface area: pyrolysis temperature and BET surface area of biochar are positively linked. This indicates that in these types of biochar adsorption of produced methane on its surface occurs which reduces its emission (Jeffery et al. 2016).

7.6 Role of Biochar in Mitigation of Nitrous Oxide Emission

7.6.1 Nitrous Oxide Emission and Its Impact on Global Warming

Nitrous oxide (N_2O) is a potent greenhouse gas (GHG) with 114 years of atmospheric residence time and 265 times higher global warming potential (GWP) than that of CO_2 (IPCC 2014a, b). It contributes about 5% of the total human-induced global warming. The atmospheric concentration of N_2O has reached 350 ppb since the pre-industrial (1750) level (285 ppb). It has steadily increased at a rate of 0.73 ± 0.03 ppb $year^{-1}$ over the last three decades and accounts for approximately 5% of the total greenhouse effect (IPCC 2014a, b). From 1961 to 2010 the global N_2O emission from agriculture has increased by about 3 times from 1.44 Tg to 4.25 Tg (Fagodiya et al. 2017). Besides this, it is also responsible for the destruction of the stratospheric ozone as once it transported to the stratosphere it releases the ozone depleting-chemicals through chlorine or nitrogen oxide catalyzed processes (Ravishankara et al. 2009). A doubling of atmospheric N_2O would cause a 10% decrease in the ozone layer that would increase ultraviolet radiation reaching the earth's surface by about 20%. Agriculture, fossil fuel combustion and industrial process, biomass burning, atmospheric deposition, and human sewage are the major anthropogenic sources of the N_2O emission (Fig. 7.7). Among the anthropogenic sources, agriculture is the largest source which accounted for 67% of total anthropogenic N_2O emission. Out of which 42% is direct emission from nitrogenous fertilizers and manure management; and 25% are indirect emissions from the runoff and leaching of fertilizers (IPCC 2013).

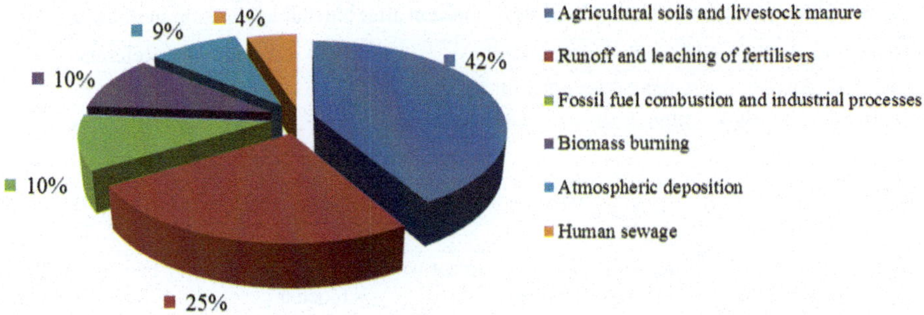

Fig. 7.7 Various sources of nitrous oxide emission

7.6.2 Biochar and Nitrous Oxide Emission

There are several proposed options for reducing N_2O emissions from soil. Some of the potential options are the use of nitrification inhibitors (NIs) and neem oil coated urea (NOCU) (Gupta et al. 2016a, b; Fagodiya et al. 2019) and use of Azolla as biofertilizers (Malyan et al. 2019). Besides increasing soil C storage and reducing CH_4 emissions, biochar application in soil has reported reducing N_2O emission which would further contribute to greenhouse gas mitigation. Recently it has been reported that soil application of biochar also offers great potential for the reduction of N_2O emission from soil mainly due to high porosity, pH, and specific surface area of biochar. Rondon et al. (2005) reported first time in a greenhouse experiment that after biochar application the N_2O emissions were decreased by 50–80%. Since then, biochar gets popularity as a potential option for N_2O mitigation from agricultural soils and since then several studies have been conducted using biochar. Some of these studies are synthesized in Table 7.4. A recent meta-analysis of publication from various field and laboratory study by Borchard et al. (2019) has reported the 38% average reduction of N_2O emissions while Verhoeven et al. (2017) has reported average reductions of 9–12% while an earlier global assessment (Cayuela et al. 2014) suggested greater average reductions of almost 50%, compared to none biochar amended soils.

There are several properties of biochar which are helpful in the reduction of N_2O emissions from agricultural soils. Some of these are porosity, pH, acidic and basic functional groups, specific surface area, and redox properties which are mainly varying with the type of feedstock used for biochar and pyrolysis conditions (Grutzmacher et al. 2018; Cayuela et al. 2014). Once biochar applied to soils, the interaction between the soil and biochar can alter the soil pH, oxygen level, and microbial composition and activity which turn into a reduction in N_2O emission (Edwards et al. 2018; Harter et al. 2014). Biochar application reduced the substrate concentration (NO_3^-, NH_4^+) for nitrification and denitrification reactions which

Table 7.4 Some results of reduction in N_2O emission after biochar application in to the soil

Biochar feedstock	Experimental condition	N_2O reduction	References
Sugarcane straw	Greenhouse pot with wheat crop	71%	
Meta data analysis	Meta-analysis of 88 publication	38%.	Borchard et al. (2019)
Bamboo	Acidic tea field	Biochar@ 0.5% =38% @2% = 61%.	Oo et al. (2018)
Wheat straw and swine manure	Greenhouse pot experiment with five vegetable crops	Wheat straw biochar 36.4–59.1% Swine manure biochar 37.0–49.5%	Fan et al. (2017)
Meta data analysis	Meta data analysis of 30 publication	54%	Cayuela et al. (2014)

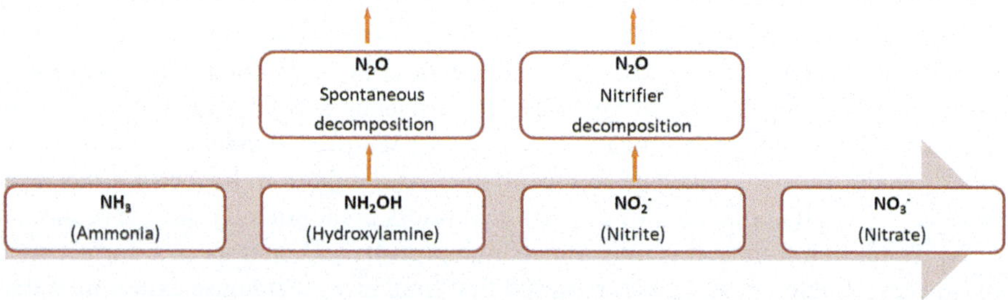

Fig. 7.8 Mechanism of N_2O emission from soil

limits the microbial activity and reduced N_2O emissions (Fig. 7.8) (Zheng et al. 2013). Besides this, it may enhance the soil pH and N_2O reductase concentration into the soils which lead to the final conversion of N_2O into N_2 and thereby reduced N_2O emission (Harter et al. 2014).

7.7 Conclusions

Biochar is a rich form of stable carbon and a suitable option for mitigating climate change through long term carbon storage and reduction in emission of GHG like N_2O and CH_4. Aside from carbon storage, biochar also provides a sustainable solution for managing the large volume of crop residue thus saving them from burning and air pollution. Production of renewable energy and heat during biochar production, increase in nutrient and water use efficiency after soil application further help in mitigating climate change through reduction in emission of CO_2 from fossil fuel combustion. However, yield, characteristics, and rate of biochar application significantly vary in the different study and it depends upon feedstock quality and pyrolysis temperature. Further, most of the reported studies are confined to laboratory or institute level field trial and adoption of biochar for application in agricultural soil by farmers is still lacking. There is a need to develop region and feedstock specific guidelines and policies for biochar production and application in the field to exploit potential benefits for enhancing soil quality and mitigating climate change.

References

Billa F, Angwafo TE, Ngome AF (2019) Agro-environmental characterization of biochar issued from crop wastes in the humid forest zone of Cameroon Samuel. Int J Recycl Org Waste Agric 8:1–13

Borchard N, Schirrmann M, Cayuela ML, Kammann C, Wrag-Mpnnig N, Estavillo JM, Novak J (2019) Biochar, soil and land-use interactions that reduce nitrate leaching and N_2O emissions: a meta-analysis. Sci Total Environ 65:2354–2364

Bruun EW, Hauggaard-Nielsen H, Ibrahim N, Egsgaard H, Ambus P, Jensen PA, Dam-Johansen K (2011) Influence of fast pyrolysis temperature on biochar labile fraction and short-term carbon loss in a loamy soil. Biomass Bioenergy 35(3):1182–1189

Cayuela ML, Van Zwieten L, Singh BP, Jeffery S, Roig A, Sánchez-Monedero MA (2014) Biochar's role in mitigating soil nitrous oxide emissions: a review and meta-analysis. Agric Ecosyst Environ 191:5–16

Chaturvedi OP, Handa AK, Uthappa AR, Sridhar KB, Kumar N, Chavan SB, Rizvi J (2017) Promising agroforestry tree species in India. Jointly published by the Central Agroforestry Research Institute (CAFRI), Jhansi, and the South Asia Regional Programme (based in Delhi, India) of the World Agroforestry Centre, pp 1–190

Cherubin MR, Oliveira DMS, Feigl B et al (2018) Crop residue harvest for bioenergy production and its implications on soil functioning and plant growth: a review. Sci Agric 75(3):255–272

Crombie K, Mašek O (2015) Pyrolysis biochar systems, balance between bioenergy and carbon sequestration. GCB Bioenergy 7:349–361

Denman KL, Brasseur G, Chidthaisong A et al (2007) Couplings between changes in the climate system and biogeochemistry. In: Solomon S, Qin D, Manning M et al (eds) Climate change 2007: the physical science basis. Contribution of working group I to the fourth assessment report of the intergovernmental panel on climate change. Cambridge University Press, Cambridge/New York

Dhyani SK, Ram A, Dev I (2016) Potential of agroforestry systems in carbon sequestration in India. Indian J Agric Sci 86:1103–1112

Dickinson D, Balduccio L, Buysse J, Ronsse F, Van Huylenbroeck G, Prins W (2015) Cost-benefit analysis of using biochar to improve cereals agriculture. GCB Bioenergy 7:850–864

Dlugokencky EJ, Hall BD, Montzka SA, Dutton G, Mühle J, Elkins JW (2019) Atmospheric composition [in *State of the Climate in 2018,* Chapter 2: Global Climate]. Bull Am Meteorol Soc 100(9):S48–S50

EBC (2012) European biochar certificate – guidelines for a sustainable production of biochar. European Biochar Foundation (EBC). Available at: http://www.european-biochar.org/en/download

Edwards JD, Pittelkow CM, Kent AD, Yang WH (2018) Dynamic biochar effects on soil nitrous oxide emissions and underlying microbial processes during the maize growing season. Soil Biol Biochem 122:81–90

Enders A, Hanley K, Whitman T, Joseph S, Lehmann J (2012) Characterization of biochars to evaluate recalcitrance and agronomic performance. Bioresour Technol 114:644–653

Fagodiya RK, Pathak H, Kumar A, Bhatia A, Niveta J (2017) Global temperature change potential of nitrogen use in agriculture: a 50-year assessment. Sci Rep 7:44928. https://doi.org/10.1038/srep44928

Fagodiya RK, Pathak H, Bhatia A, Jain N, Gupta DK, Kumar A, Malyan SK, Dubey R, Radhakrishanan S, Tomer R (2019) Nitrous oxide emission and mitigation from maize–wheat rotation in the upper Indo-Gangetic Plains. Carbon Manag 10(5):489–499

Fan C, Chen H, Li B, Xiong Z (2017) Biochar reduces yield-scaled emissions of reactive nitrogen gases from vegetable soils across China. Biogeosciences 14:2851–2863

Fang Y, Singh B, Singh BP, Krull E (2014) Biochar carbon stability in four contrasting soils. Eur J Soil Sci 65:60–71

Feng Y, Xu Y, Yu Y, Xie Z, Lin X (2012) Mechanisms of biochar decreasing methane emission from Chinese paddy soils. Soil Biol Biochem 46:80–88

Graetz RD, Skjemstad JO (2003) The charcoal sink of biomass burning on the Australian continent. CSIRO atmospheric research technical paper no. 64. CSIRO Atmospheric Research, Aspendale

Grutzmacher P, Puga AP, Bibar MPS, Coscione ARC, Packer AP, de Andrade CA (2018) Carbon stability and mitigation of fertilizer induced N_2O emissions in soil amended with biochar. Sci Total Environ 625:1459–1466

Gupta DK, Bhatia A, Kumar A, Chakrabarti B, Jain N, Pathak H (2015) Global warming potential of rice-wheat cropping system of the Indo-Gangetic Plains. Indian J Agric Sci 85(6):807–816

Gupta DK, Bhatia A, Kumar A, Das TK, Jain N, Chakrabarti B, Pathak H (2016a) Economic analysis of different greenhouse gas mitigation technologies in rice wheat cropping system of the Indo-Gangetic Plains. Curr Sci 110(5):867–874

Gupta DK, Bhatia A, Kumar A, Das TK, Jain N, Tomer R, Malyan S, Fagodiya RK, Dubey R, Pathak H (2016b) Mitigation of greenhouse gas emission from rice-wheat system of the Indo-Gangetic plains: through tillage, irrigation and fertilizer management. Agric Ecosyst Environ 230:1–9

Gupta DK, Bhatt RK, Keerthika A, Noor Mohamed MB, Shukla AK, Jangid BL (2019) Carbon sequestration potential of 30 years old Hardwickia binata Roxb based agroforestry system in hot semi-arid environment of India: an assessment of tree density impact. Curr Sci 116(10):112–116

Han X, Sun X, Wang C, Wu M, Dong D, Zhong T, Thies JE, Wu W (2016) Mitigating methane emission from paddy soil with rice-straw biochar amendment under projected climate change. Sci Rep 6:24731

Harter J, Krause HM, Schuettler S, Ruser R, Fromme M, Scholten T, Kappler A, Behrens S (2014) Linking N_2O emissions from biochar-amended soil to the structure and function of the N-cycling microbial community. ISME J 8:660–674

Hilber I, Bastos AC, Loureiro S, Soja G, Marsz A, Cornelissen G, Bucheli T (2017) The different faces of biochar: contamination risk versus remediation tool. J Environ Eng Landsc Manage 25(02):86–104

IPCC (2013) Climate change 2013: The physical science basis. In: Stocker TF, Qin D. Plattner GK, Tignor M, Allen SK, Boschung J, Nauels A, Xia Y, Bex V, Midgley PM (eds) Contribution of working group I to the fifth assessment report of the Intergovernmental Panel on Climate Change. Cambridge University Press, Cambridge/New York, 1535 pp

IPCC (2014a) Climate change 2014: impacts, adaptation, and vulnerability working group II contribution to the fifth assessment report. Cambridge University Press, Cambridge/New York

IPCC (2014b) Climate change 2014: synthesis report. In: Core Writing Team, Pachauri RK, Meyer LA (eds) Contribution of working groups I, II and III to the fifth assessment report of the Intergovernmental Panel on Climate Change. IPCC, Geneva, 151 pp

IPCC (2019) Summary for policymakers. In: Shukla PR, Skea J, Calvo Buendia E, et al (eds) Climate change and land: an IPCC special report on climate change, desertification, land deg-radiation, sustainable land management, food security, and greenhouse gas fluxes in terrestrial ecosystems (in press)

Jacquot J (2008) Can a kind of ancient charcoal put the brakes on global warming. Pop Mech. December 30, 2008

Jain N, Bhatia A, Pathak H (2014) Emission of air pollutants from crop residue burning in India. Aerosol Air Qual Res 14:422–430

Jeffery S, Verheijen FG, van der Velde M, Bastos AC (2011) A quantitative review of the effects of biochar application to soils on crop productivity using meta-analysis. Agric Ecosyst Environ 144(1):175–187

Jeffery S, Verheijen FG, Kammann C, Abalos D (2016) Biochar effects on methane emissions from soils: a meta-analysis. Soil Biol Biochem 101:251–258

Jindo K, Mizumoto H, Sawada Y, Sanchez-Monedero MA, Sonoki T et al (2014) Physical and chemical characterization of biochars derived from different agricultural residues. Biogeosciences 11:6613–6621

Kour D, Rana KL, Yadav N, Yadav AN, Rastegari AA, Singh C, Negi P, Singh K, Saxena AK (2019) Technologies for biofuel production: current development, challenges, and future prospects. In: Rastegari A et al (eds) Prospects of renewable bioprocessing in future energy systems, Biofuel and biorefinery technologies Vol 10. Springer, Cham, pp 1–50

Krull ES, Baldock JA, Skjemstad JO, Smernik RJ (2009) Characteristics of biochar: organo-chemical properties. In: Lehmann J, Joseph S (eds) Biochar for environmental management: science and technology. Earthscan, Sterling/London, pp 53–66

Kuzyakov Y, Subbotina I, Chen H, Bogomolova I, Xu X (2009) Black carbon decomposition and incorporation into soil microbial biomass estimated by C-14 labeling. Soil Biol Biochem 41(2):210–219

Kuzyakov Y, Bogomolova I, Glaser B (2014) Biochar stability in soil: decomposition during eight years and transformation as assessed by compound-specific 14C analysis. Soil Biol Biochem 70:229–236

Lal R (2004) Soil carbon sequestration to mitigate climate change. Geoderma 123:1–22

Lee Y, Park J, Ryu C, Gang KS, Yang W, Park YK, Jung J, Hyun S (2013) Comparison of biochar properties from biomass residues produced by slow pyrolysis at 500 C. Bioresour Technol 148:196–201

Lehmann J, Joseph S (2009) Biochar for environmental management: science and technology. Earthscan, London

Lehmann J, Gaunt J, Rondon M (2006) Bio-char sequestration in terrestrial ecosystems. Mitig Adapt Strat Glob Chang 11:395–419

Liang B, Lehmann J, Solomon D, Sohi S, Thies JE, Skjemstad JO, FJ L˜a, Engelhard MH, Neves EG, Wirick S (2008) Stability of biomass-derived black carbon in soils. Geochim Cosmochim Acta 72:6096–6078

Liu Y, Yang M, Wu Y, Wang H, Chen Y, Wu W (2011) Reducing CH_4 and CO_2 emissions from waterlogged paddy soil with biochar. J Soils Sediments 11(6):930–939

Major J, Lehmann J, Rondon M, Goodale C (2010) Fate of soil-applied black carbon: downward migration, leaching and soil respiration. Glob Chang Biol 16:1366–1379

Majumdera S, Neogia S, Duttaa T, Powelb MA, Banika P (2019) The impact of biochar on soil carbon sequestration: Meta-analytical approach to evaluating environmental and economic advantages. J Environ Manag 250:109466

Malyan SK, Bhatia A, Kumar A, Gupta DK, Singh R, Kumar SS, Tomer R, Kumar O, Jain N (2016) Methane production, oxidation and mitigation: a mechanistic understanding and comprehensive evaluation of influencing factors. Sci Total Environ 572:874–896

Malyan SK, Bhatia A, Kumar SS, Fagodiya RK, Pugazhendhi A, Duc PA (2019) Mitigation of greenhouse gas intensity by supplementing with Azolla and moderating the dose of nitrogen fertilizer. Biocatal Agric Biotechnol 20:101266. https://doi.org/10.1016/j.bcab.2019.101266

Maroušek J, Strunecký O, Stehel V (2019) Biochar farming: defining economically perspective applications. Clean Techn Environ Policy 21:1389. https://doi.org/10.1007/s10098-019-01728-7

Mary GS, Sugumaran P, Niveditha S et al (2016) Production, characterization and evaluation of biochar from pod (Pisum sativum), leaf (Brassica oleracea) and peel (Citrus sinensis) wastes. Int J Recy Org Waste Agric 5(1):43–53

Nguyen BT, Lehmann J, Hockaday WC, Joseph S, Masiello CA (2010) Temperature sensitivity of black carbon decomposition and oxidation. Environ Sci Technol 44:3324–3331

Nsamba HK, Hale SE, Cornelissen G, Bachmann RT (2015) Sustainable technologies for small-scale biochar production – a review. J Sustain Bioenergy Syst 5:10–31

Oo AZ, Sudo S, Akiyama H, Win KT, Shibata A, Yamamoto A, Sano T, Hirono Y (2018) Effect of dolomite and biochar addition on N_2O and CO_2 emissions from acidic tea field soil. PLoS One 13(2):e0192235. https://doi.org/10.1371/journal.pone.0192235

Paustian K, Larson E, Kent J, Marx E, Swan A (2019) Soil C sequestration as a biological negative emission strategy. Front Clim 1:8. https://doi.org/10.3389/fclim.2019.00008

Pratiwi EPA, Shinogi Y (2016) Rice husk biochar application to paddy soil and its effects on soil physical properties, plant growth, and methane emission. Paddy Water Environ 14(4):521–532

Purakayastha TJ, Kumari S, Pathak H (2015) Characterization, stability, and microbial effects of four biochars produced from crop residues. Geoderma 239–240:293–303

Ravishankara AR, Daniel JS, Portmann RW (2009) Nitrous oxide (N_2O): the dominant ozone-depleting substance emitted in the 21st century. Science 326(5949):123–125

Roberts KG, Gloy BA, Joseph S, Scott NR, Lehmann J (2010) Life cycle assessment of biochar systems: estimating the energetic, economic, and climate change potential. Environ Sci Technol 44:827–833. https://doi.org/10.1021/es902266r

Rondon M, Ramirez JA, Lehmann J (2005) Greenhouse gas emissions decrease with charcoal additions to tropical soils. In: Proceedings of the 3rd USDA symposium on greenhouse gases and carbon sequestration, vol 208, Baltimore, USA

Saunois M, Jackson RB, Bousquet P, Poulter B, Canadell JG (2016) The growing role of methane in anthropogenic climate change. Environ Res Lett 11(12):12

Shneour EA (1966) Oxidation of graphitic carbon in certain soils. Science 151:991–992

Singh C, Tiwari S, Boudh S, Singh JS (2017a) Biochar application in management of paddy crop production and methane mitigation. In: Singh JS, Seneviratne G (eds) Agro-environmental sustainability: Vol-2: managing environmental pollution. Springer, Cham, pp 123–146

Singh C, Tiwari S, Singh JS (2017b) Impact of rice husk biochar on nitrogen mineralization and methanotrophs community dynamics in paddy soil. Int J Pure Appl Biosci 5(5):428–435

Singh C, Tiwari S, Singh JS (2017c) Application of biochar in soil fertility and environmental management: a review. Bull Environ Pharmacol Life Sci 6(12):07–14

Singh C, Tiwari S, Gupta VK, Singh JS (2018) The effect of rice husk biochar on soil nutrient status, microbial biomass and paddy productivity of nutrient poor agriculture soils. Catena 171:485–493

Singh C, Tiwari S, Singh JS (2019) Biochar: a sustainable tool in soil 2 pollutant bioremediation. In: Bharagava RN, Saxena G (eds) Bioremediation of industrial waste for environmental safety. Springer, Cham, pp 475–494

Song X, Pan G, Zhang C, Zhang L, Wang H (2016) Effects of biochar application on fluxes of three biogenic greenhouse gases: a meta-analysis. Ecosyst Health Sustain 2:e01202. https://doi.org/10.1002/ehs2.1202

Steiner C, Teixeira WG, Lehmann J, Nehls T, de Macedo JLV, Blum WEH, Zech W (2007) Long term effects of manure, charcoal and mineral fertilization on crop production and fertility on a highly weathered central Amazonian upland soil. Plant Soil 291:275

Sun X, Shan R, Li X, Pan J, Liu X, Deng R, Song J (2017) Characterization of 60 types of Chinese biomass waste and resultant biochars in terms of their candidacy for soil application. GCB Bioenergy 9:1423–1435

Tiwari S, Singh C, Singh JS (2018) Land use changes: a key ecological driver regulating methanotrophs abundance in upland soils. Energy Ecol Environ 3(6):355–371

Tiwari S, Singh C, Boudh S, Rai PK, Gupta VK, Singh JS (2019a) Land use change: a key ecological disturbance declines soil microbial biomass in dry tropical uplands. J Environ Manag 242:1–10

Tiwari S, Singh C, Singh JS (2019b) Wetlands: a Major natural source responsible for methane emission. In: Upadhyay AK et al (eds) Restoration of wetland ecosystem: a trajectory towards a sustainable environment. Springer, Cham, pp 59–74

Venkatesh G, Gopinath KA, Sammi Reddy K, Sanjeeva Reddy B, Prasad JVNS, Rajeshwar Rao G, Pratibha G, Srinivasarao Ch, Ravindra Chary G, Prabhakar M, Visha Kumari V, Shankar AK Venkateswarlu B (2018) Biochar production and its use in rainfed agriculture: experiences from CRIDA. CRIDA-NICRA Research Bulletin 02/2018. ICAR – Central Research Institute for Dryland Agriculture, Hyderabad, pp 50

Verhoeven E, Pereira E, Decock C, Suddick E, Angst T, Six J (2017) Toward a better assessment of biochar–nitrous oxide mitigation potential at the field scale. J Environ Qual 46:237. https://doi.org/10.2134/jeq2016.10.0396

Wang JY, Xiong ZQ, Kuzyakov Y (2016) Biochar stability in soil: metaanalysis of decomposition and priming effects. Glob Change Biol Bioenergy 8:512–523. https://doi.org/10.1111/gcbb.12266

Waters D, Van Zwieten L, Singh BP, Downie A, Cowie AL, Lehmann J (2011) Biochar in soil for climate change mitigation and adaptation. In: Singh B, Cowie A, Chan K (eds) Soil health and climate change. Soil biology, vol 29. Springer, Berlin/Heidelberg

Windeatt JH, Ross AB, Williams PT, Forster PM, Nahil MA, Singh S (2014) Characteristics of biochars from crop residues: potential for carbon sequestration and soil amendment. J Environ Manag 146:189–197

Xiao Y, Yang S, Xu J, Ding J, Sun X, Jiang Z (2018) Effect of biochar amendment on methane emissions from paddy field under water-saving irrigation. Sustainability 10(5):1371

Zhang A, Cui L, Pan G, Li L, Hussain Q, Zhang X, Zheng J, Crowley D (2010) Effect of biochar amendment on yield and methane and nitrous oxide emissions from a rice paddy from Tai Lake plain, China. Agric Ecosyst Environ 139(4):469–475

Zhang H, Chen C, Gray EM, Boyd SE (2017) Effect of feedstock and pyrolysis temperature on properties of biochar governing end use efficacy. Biomass Bioenergy 105:136–146

Zhang X, Chen C, Chen X, Tao P, Jin Z, Han Z (2018) Persistent effects of biochar on soil organic carbon mineralization and resistant carbon pool in upland red soil, China. Environ Earth Sci 77:177

Zheng H, Wang Z, Deng X, Herbert S, Xing B (2013) Impacts of adding biochar on nitrogen retention and bioavailability in agricultural soil. Geoderma 206:32–39

Zimmerman A (2010) Abiotic and microbial oxidation of laboratory-produced black carbon (biochar). Environ Sci Technol 44:1295–1301

Chapter 8
Biochar Coupled Rehabilitation of Cyanobacterial Soil Crusts: A Sustainable Approach in Stabilization of Arid and Semiarid Soils

Arun Kumar and Jay Shankar Singh

Abstract Cyanobacterial soil crusts (CSCs) are unique microhabitats in desert soil plays a significant role in stabilization of soil surface and provide favourable conditions for the establishment of vascular plants. The CSCs types and its distribution mainly depend up on the locality and climatic factors of the region. They help in retaining soil particles, nutrients, moisture and also add up carbon and nitrogen to the nutrient poor soils. The natural or anthropogenic intervention exerted immense pressure on the crusts community and diversity; leads to disturbed or distressed CSCs. Currently military use of the deserts have destroyed the fragile ecology of these CSCs and delay the time of recovery to reach functional state. To stabilize and rehabilitate the disturbed CSCs, a number of strategies successfully tested and implemented in small scale, some of them are artificial stabilization, resource augmentation and cyanobacterial inoculants. Biochar coupled rehabilitation of CSCs could be effective and sustainable approach for the stabilization of desert soils. Small scale biochar production would be helpful not only reducing the cost of rehabilitation but also help in providing livelihood to the local people.

Keywords Desert soils · Cyanobacterial soil crusts · Rehabilitation · Biochar

A. Kumar
Department of Environmental Science, Guru Nanak Khalsa College, Yamuna Nagar, Haryana, India

J. S. Singh (✉)
Department of Environmental Microbiology, Babasaheb Bhimrao Ambedkar University, Lucknow, Uttar Pradesh, India

J. S. Singh, C. Singh (eds.), *Biochar Applications in Agriculture and Environment Management*, https://doi.org/10.1007/978-3-030-40997-5_8

8.1 Introduction

The deserts are seems to lifeless and unproductive landscapes, as they have arid soils, harsh environment and very sparse vegetation. But they have unique soil surface structure known as cyanobacterial soil crusts (CSCs) that can occur on the surface or just below the surface of soils. Belnap and Gardner (1993) observed that CSCs includes primarily a number of communities such as cyanobacteria, green algae, lichens and mosses which exudates sticky extracellular polysaccharide, helps in binding of soil particles; leads to formation of intimate and living covering on the soil surface.

The CSCs often spread over as the living ground cover in hot, cool, and cold deserts. Further the CSCs can also found in temperate conditions like either in Pine Barrens or vacant area due to reduced plant cover (Belnap and Lange 2003). They play an important role in stabilizing the mobile sand dunes and helps in prevention of soil erosion by water and wind (Danin 1978). They further influence the capture, runoff, infiltration and percolation of rainfall water; improves the water-holding capacity and soil moisture content (Belnap 2003a, b, 2006).

There are various factors such as climate change and human intervention which adversely affected the composition and diversity of crust communities. Nowadays military use of deserts increased the further pressure on already distressed crusts, leads to complete destruction of CSCs. There is also risk of invasion of annual exotic grasses, which increased the chances of summer fires, leads to destruction of crusts. For the rehabilitation and fast recovery of CSCs, mainly three strategies i.e. artificial soil stabilizer, inoculation of cyanobacteria and addition of soil amendments, investigated in a number of studies.

Biochar coupled rehabilitation of CSCs could be a sustainable and feasible method for the stabilization of arid and semiarid soils. Biochar proved to be a very useful soil amendment in agriculture; it not only helps in improving the cation exchange capacity, pH, nutrient contents, plant growth, but also helps in reduction of greenhouse gas (GHG) emissions from the agricultural soils. Further small scale biochar production in regional level could help in better management of arid soils and also supports local livelihood.

This chapter gives a brief account on the structure and formation of CSCs in desert soil conditions and their role in maintaining and regulations of desert ecosystem functions. There are some factors that affecting the CSCs and the strategies for protection and rehabilitation of these CSCs. Further a biochar inoculation based method also discussed for the sustainable and effective approach for the rehabilitation of cyanobacterial soil crusts.

8.2 Cyanobacterial Soil Crusts (CSCs)

Cyanobacterial or biological desert crusts are quite unique and ecological significant microhabitats in the soil of arid areas (Belnap and Lange 2003). These cohesive surface formations on topsoil are mainly started with filamentous growth of cyanobacteria; and consequently expanded through periodic events of moisture availability and capturing mineral particles, either by cyanobacterial filaments or by extracellular slime secreted by cyanobacteria (Belnap and Gardner 1993; Cameron and Blank 1966; Johansen 1993). There would be further succession of other communities of bacteria, fungi, algae, lichens and mosses; made them a unique microhabitat in arid soils.

8.2.1 Formation and Structure

Cyanobacteria are naturally primary colonizers in bare soil of arid regions. Although cyanobacteria are almost present every types of CSCs, but rarely found in CSCs characterized by low pH conditions. Cyanobacteria considered to be one of the earliest inhabitants on planet earth and can thrive in a range of environments including desert soil and rock micro-habitats (Friedmann et al. 1967). Walter et al. (1976) suggested that cyanobacteria seem to be originated over 3 billion years ago, as evidenced in fossil record of marine stromatolites. These marine stromatolites containing large floating cyanobacterial mats considered to oxygenating the atmosphere and responsible for creating the basis of marine food web. Further Horodyski and Knauth (1994) suggested that 1.2 billion-year-old rocks evidenced the appearance of cyanobacteria in terrestrial habitats. Schwartzman and Volk (1989) stated that like the CSCs do currently, cyanobacteria might be hastened the weathering of barren bedrock and played an important role in soil formation which spread across the land. This newly formed soil supported the evolution and establishment of vascular plants and other terrestrial life forms.

Garcia-Pichel and Belnap (1996), Belnap (2003a, b) investigated that large, mobile filamentous cyanobacterial genus such as *Microcoleus vaginatus* (which that preferably live 1–4 mm below the soil surface) firstly inhabited the bare soils and further they can spread on the soil surface upon moisture availability during wet periods. After that, smaller and less mobile cyanobacterial genus such as *Nostoc*, *Scytonema* inhabited either on or just below the soil surface, facilitates the formation of layers of communities in the soils. These cyanobacterial communities constantly secreted out a sticky, polysaccharide outer sheath to the uppermost soil layers, leads to the formation of soil aggregates through binding the soil particles. Further these soil aggregates linked together by cyanobacterial filaments. When cyanobacteria stabilized the soil surface, lichens and mosses colonize according the suitable climate conditions.

The internal structure of CSCs differs due to composition and succession of different crust communities. Cyanobacteria and fungi are primary communities of all crust and provide them most of the cohesive property to CSCs. As cyanobacteria propagate inside the soil, lichens and bryophytes have blanket cover above the soil surface; which keeps underlying soils intact to resist from detachment of soil particles due to raindrops and overland water-flow. Lichens and bryophytes have rhizoptae, rhizinae, and rhizomorphs; act as anchoring structures that could enter in to the soil as deep as 14 mm (Belnap et al. 2003). Beside this, there is protonemata moss which is intermingled throughout the matrix; leads to form a dense, subterranean network that is connected with soil particles (Belnap and Gardner 1993; Belnap 2003a, b).

8.2.2 Distribution and Types of CSCs

The CSCs communities are widely distributed and occurred on every soil types and in almost all the ecosystems where sunlight able to reach the soil surface. Due to low moisture requirements and a high tolerance of extreme temperatures and light, they have ability to survive such conditions which limit the growth of vascular plants (Belnap et al. 2003). They are commonly thriving in low-productivity environments such as hyperarid, arid, semiarid, sub-humid, alpine and polar regions. Further CSCs found to be in limited to more mesic regions such as pine barrens, serpentine soils, temperate steppe. It is evident that tropical evergreen rain forests are the only ecosystems which appeared to lack CSCs (Büdel and Lange 2003).

Among climatic regimes, CSCs may differ in appearance, biomass, and species composition. Due to these differences, CSC's shows distinct external and internal structure; leads to different effect on ecological and hydrologic processes. Belnap et al. (2003) reviewed and proposed various classification schemes of CSCs. There is a classification mainly based on factors that influence runoff, infiltration, and sediment production. According to this, they are primarily categorized into four types:

1. Smooth CSCs-They are primarily found in hot hyper-arid deserts like in Atacama, Sahara deserts; which defined with high PET and absence of soil freezing. In smooth crusts, a thin layer of cyanobacteria and fungi dominated the crusts that can be survived on or just below the soil surface; lichens and mosses pockets rarely found specialized microhabitats. Smooth CSCs are often characterized by very low moisture availability which leads to low biomass and low absorptive of biota; ultimately result in the high porosity and low surface roughness of soil surface.
2. Rugose CSCs- They found in dryland areas like low-elevation Sonoran, Mojave, Australian deserts; which defined with lower PET than hyper-arid deserts and absence of soil freezing. In rugose crusts, a thin layer of cyanobacteria and fungi dominated the crusts but sparse patches of lichens and mosses commonly found

in drier regions of these CSCs. They have comparatively even soil surface. Although with increase in moisture availability in rugose CSCs, lichen and moss cover also increases as well but still have fairly flat soil surface. Overall rugose characterized by low moisture availability, leads to results in moderately low biomass and low absorptive; result in the moderately high soil surface porosity and low surface roughness.

3. Pinnacled crusts- They are found in occur in mid-latitude cool desert like low-elevation Colorado Plateau, mid-latitude China deserts, high-elevation Sonoran and Mojave deserts which defined with lower PET than in hot deserts but soils freezing occurred. In Pinnacled crusts, relatively thick layers of cyanobacteria dominated the crusts with up to 40% lichen and moss cover. Pinnacled crust characterized by remarkably pedicellate mounds, formed due to frost heaving; leads to uplifting. These uplifted mounds further differentially weathered by downward-cutting water. It can be high up to 15 cm with across 4–10 mm thin tip. Unlike smooth and rogose CSCs, they have high biomass & absorptive and high soil surface roughness with comparatively low soil surface porosity.
4. Rolling crusts-They are found in high altitude cold deserts like northern Great Basin, high-latitude deserts which defined with lower PET than pinnacled crusts regions. In rolling crusts, thick layer of lichens and mosses heavily dominated crusts. Unlike pinnacled crusts, soil uplifting due to frost heaving is counteracted by a cohesive and thick encrusted mat. This mat of lichens and mosses makes a roughened, slightly rolling surface that prevents differential downward cutting. Rolling CSCs characterized by high biomass & surface absorptive, with low soil surface porosity and moderate soil surface roughness.

8.2.3 Ecology and Physiology of CSCs

Cyanobacterial soil crusts play a significant role in the biogeochemistry and geomorphology of deserts (Eldridge and Greene 1994; Evans and Johansen 1999). Belnap and Eldridge (2001) investigated that communities of CSCs are almost alike around the world, despite the difference in climates and vegetation types in an area. Some genera such as *Microcoleus vaginatus*, *Psora decipiens*, *Collema tenax*, *Collema coccophorum*, and *Catapyrenium squamulosum* are occurred on almost all the continents. There are also some non-related communities which showed quite similar structures and functions, indicating that CSCs soil surface conditions have produced convergent evolutionary trends within these taxa. There are around hundreds of cyanobacterial and eukaryotic green algal species which found to be in cyanobacterial soil crusts (Evans and Johansen 1999). They mostly distributed in the upper soil layer, as they need sunlight for the photosynthesis. They are responsible for the change in the pH as well as oxygen, ammonium, and nitrate concentrations.

In the deserts, soil surface temperatures reached to very low as–20 °C to very high as over 70 °C. The precipitation is very low and quite sparse. They also face of high radiation as throughout the year. So for the survival of CSCs communities in deserts, they should have the ability to tolerate extreme dehydration. Sometimes the crust communities faced such conditions that dry-weight water content of biomass, might be reduced to extreme low as 5% or less, result in the terminating all metabolic processes (Bewley and Krochko 1982). These abilities can helps the crust communities to withstand extended periods of high heat, strong light, and no water.

Smaller cyanobacterial genera *Nostoc*, *Scytonema*, *Chroococcidiopsis* have large amounts of protective pigments for protection from excess radiation, while large filamentous cyanobacteria *Microcoleus* had no protective pigments, lives beneath the umbrella of smaller cyanobacteria and green algae (Bowker et al. 2002). Intracellular pigmented tissue like carotenoids and xanthophylls are able to reflect and or absorb incoming radiation up to 50–93% from reaching the interior of these communities (Castenholz and Garcia-Pichel 2000). Other taxa like lichens can "roll up" during drying, keeping to protect their photosynthetic pigments from radiation (Büdel and Wessels 1986; Frey and Kürschner 1991). Mosses also have some unique structures which can store and transport the water and also have the revolute (curled-under) leaf margins to reduce water loss through transpiration (Frey and Kürschner 1991).

8.3 Significance and Role of CSCs

The CSCs are unique micro-ecosystems that perform a variety of roles in the formation, stability, and fertility of semi-arid and arid soils. It is clearly evident that undisturbed CSCs shows greater biomass and better ability to perform the various functions than disturbed or damaged crusts at any stages of succession. Besides the discussed below roles of CSCs (Fig. 8.1), there may be another ecosystem services exists that couldn't be investigated.

8.3.1 Dust Entrapment

CSCs have enhanced rough surface and adhesive sheath of polysaccharide, which increased the capability to capture of nutrient-rich dust from the nearby environment. Reynolds et al. (2001) observed that this dust can be able to increase the essential nutrients for the plants such as nitrogen, phosphorus, and potassium, up to fourfold. Due to this, there would be improvement in overall fertility and water holding capability of the soils (Verrecchia et al. 1995). Further undisturbed CSCs have greater capability to capture dust particles as their greater surface roughness as compared to disturbed (flattened) surfaces. Cyanobacteria fibres developed a web-like pattern which not only forms soil aggregates but also responsible for their holding in place.

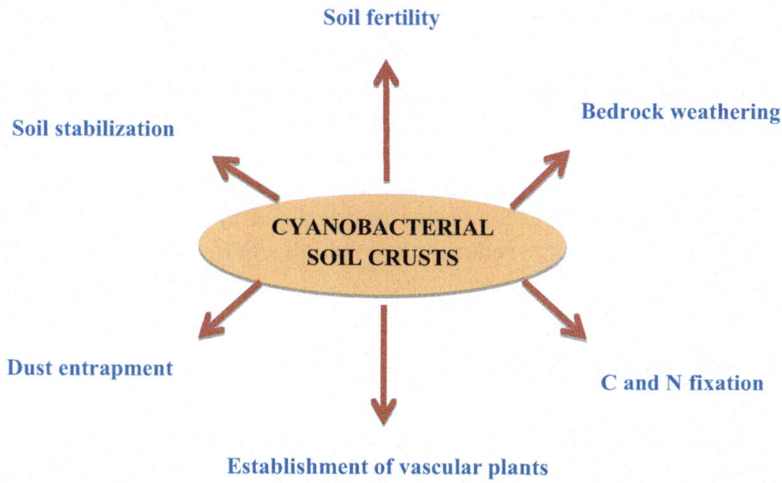

Fig. 8.1 Significance of cyanobacterial soil crusts (CSCs) in arid areas

8.3.2 Bedrock Weathering

Garcia-Pichel and Belnap (1996) observed that many organisms like lichens and cyanobacteria are considered to enhance substrate alkalinity from pH 8 to pH 10.5 in the CSCs of US, Venezuela, and South Africa. Schwartzman and Volk (1989) suggested that as CSCs have greater ability to hold water; lead to enhanced mineral dissolution and freeze–thaw action. Together these two factors can speed up the rate of weathering of bedrock up to 100 times.

8.3.3 Soil Physical Properties

Some crust organisms mainly cyanobacteria secretes extracellular substances (EPS) mainly polysaccharides (Mager and Thomas 2011), which organisms bind soil particles together to form into aggregates. Aggregation is an important aspect for proper functioning of soil and it is responsible to improve soil aeration and infiltration. Aggregate surfaces act as microsites for the most soil organisms and where maximum of the transformations of nutrient occurs (Herrick and Wander 1998). McKenna-Neuman et al. (1996) also suggested that aggregates showed greater resistance to soil erosion.

8.3.4 Soil Stabilization

In arid areas, soils are already nutrient deficient and further more susceptible to erosion (Dregne 1983). CSCs could cover the soil surfaces and provide resistance to wind and water erosion. As CSCs are also contains lichens and mosses which provide a protection cover to these soil surfaces from wind and water erosion; but Belnap and Eldridge (2001) found that as compared to healthy CSCs, disturbed crusts leads to 35 times more sediment loss in high winds or overland water flow.

8.3.5 Soil–Water Interaction

It is evident that water infiltration and soil moisture by CSCs affected by the climate, soil structure, soil texture; and also by the morphology and communities of the crusts. Smooth and rugose CSCs in high potential evapotranspiration areas have lower number of pores and little soil surface roughness that has lower water infiltration. It can be compensated by stored water by communities of CSCs which primarily depends upon rainfall amount. Eldridge et al. (2000) suggested that the phenomenon of better runoff in high PET areas is very important for the survival of heterogeneously distributed downslope plants. And if CSCs are experimentally disturbed in these areas, which accelerate the more localized infiltration; leads to death of downslope plants. While pinnacled and rolling CSCs in lower PET areas have greater soil surface roughness leads to slow movement of water that enhanced the infiltration, supports the better cover of the more homogeneously distributed vegetation found in cooler deserts.

8.3.6 Carbon and Nitrogen Fixation

Beymer and Klopatek (1991), Belnap (2001a) observed that CSCs are more significant in fixing carbon and nitrogen in deserts as there is limited cover of vascular plants and low atmospheric inputs (Peterjohn and Schlesinger 1990; Wullstein 1989). It is estimated that 0.4–2.3 g/m2/year (for cyanobacterial crusts) to 12–37 g/m2/year (for lichen crusts) carbon (Evans and Lange 2001) and 1 kg/ha/year (for cyanobacterial crusts) to 10 kg/ha/year (for lichen crusts) nitrogen (Belnap 2002) fixed by the CSCs. This is the main and significant source of carbon and nitrogen in desert soils (Evans and Ehleringer 1993).

Most of the fixation of carbon and nitrogen occurred during cool season comprises of fall, winter, and spring. Belnap (2001b) suggested that fixed carbon and nitrogen by CSCs generally released upon wetting, it means rainfall facilitates the nutrients and moisture to the desert soils. This released carbon and nitrogen assimilated by vascular plants, fungi, actinomycetes, and bacteria in nearby areas. Mostly cyanobacterial fixation of nitrogen provided by heterocystous cyanobacterial includes *Nostoc*, *Anabaena*, *Calothrix*, *Dicothrix Cylindrospermum*, *Schizothrix*, *Hapalosiphon*, *Nodularia*, *Plectonema*, and *Scytonema*; but some nitrogen fixation also observed in non-heterocystous genera such as *Oscillatoria*, *Lyngbya*, *Phoridium*, *Microcoleus* and *Tolypothrix* (Rogers and Gallon 1988; Harper and Marble 1988; Paerl 1990; Belnap 1996). Despite of free living form cyanobacteria symbiotically fixed nitrogen with lichens i.e. *Nostoc* in *Collema* sp. and *Peltula* sp. and *Scytonema* in *Heppia* sp.

8.3.7 Albedo (Reflective Power)

Belnap (1995) suggested that CSCs can absorbs much of the sunlight and reflect back only half the available sunlight as compared to in uncrusted or disturbed crusted surfaces; leads to reduce surface energy flux about 40 Joules/sec/m2, result in the increase in surface temperature by 10–14 °C. The surface temperature of the soil helps in maintaining many ecosystem functions such as rates of N and C fixation, seed germination, soil water evaporation, nutrient uptake by plants and their growth and microbial activity (Belnap 2003a, b). These ecosystem function and their timing plays an critical role for desert communities and a small change can affect community structure by reducing the species fitness and seedling establishment (Bush and Van Auken 1991). Crawford (1991) observed that many ants, insects and some small mammals segmented their surroundings according to foraging times and burrowing depths and they are regulated by surface temperature.

8.3.8 Establishment of Vascular Plants

CSCs cover and establishment of vascular plant quite interrelated especially in arid areas; at lower elevations vascular plant cover increases the cover of CSCs, because of the shade under the plant canopy. But at higher elevations, most of the soil surface occupied by vascular plants and plant litter, that reduce the opportunity for the CSCs to colonize the soil surface. Further CSCs morphology can influence establishing patterns of vascular plants. Belnap and Eldridge (2001) observed that smooth and rugose CSCs not able to retain the seeds and organic matter in space between the plants, whereas pinnacle and rolling CSCs enhance the retention of seeds and organic materials. In many field studies, germination and survival of native plants either increased or unaffected in CSCs as compared to uncrusted, areas. Once vascular plants established in crusted soils and start to growing, they have more biomass and better nutrient uptakes as compared to plants growing in uncrusted soils.

8.3.9 Soil Fertility

Combining all above said benefits, CSCs contributes great to enhance fertility of arid soils. There are numerous ways by which CSCs can improve the soil fertility and to enhance plant nutrient concentrations:

1. Adding C and N to the arid soils;
2. Secreting adhesive, negatively charged polysaccharides which keep retains the positively charged nutrients and stop further leaching loss of such nutrients essential to plants;
3. Producing chelators (ring shaped chemical compounds that bind the metal ions), which helps to keep minerals that available for plants;
4. Regulating soil temperatures and nutrient uptake rates;
5. Increasing dust capture and soil stabilization, which improves fertility and water-holding capacity of the soils; and
6. Facilitating the soil aggregation.

8.4 Factors Affecting CSCs

CSCs affected by many disturbances such as climate change, land use changes and invasion by exotic annual grasses and their associated risk of fire. All the disturbances responsible for the reduction in total crust cover; leads to decrease in soil surface temperature decreased. Lichens and mosses communities either distressed or substituted by more disturbance-tolerant cyanobacterial communities and soil surfaces are flattened. Further loss of lichens and mosses affected the soil fertility and stability, because of less extrusion of polysaccharide materials, less fixation of C & N, less entrapment of dust and other surface particles, less secretion of chelators and growth factors are, lesser nutrient uptake rates and there is a reduction in number and diversity of soil food web communities.

8.4.1 Land Use Changes

Compression and shear forces like animal hooves, human feet, tank treads, or off-road vehicle tires, thrashed soil crusts; it is more devastated when soil crusts are more dry as most the times crust are in dry state. Crusts are shattered in to pieces of crust, they can be either blow or wash away by the wind or water flow. If the pieces of crusts buried in the soil, they cannot survive as they need light to photosynthesize.

Direct human impact is the dominant force which is more responsible for the simplification and/or the destruction of CSCs. Nowadays deserts are used for the recreation, energy development, livestock grazing, habitation, and military exercises (Brooks and Pokshishevsky 1986), which undoubtedly led to a large scale devastation of lichen–moss cover and their associated ecosystem functions. Due to this slow recovery rate, CSCs cover and diversity decreases, leads to relatively permanent, less diverse and inefficient crusts. Pimm (2001) suggested that as human use of rangelands increased over the time, there are phenomenon of increasing the size and frequency of global dust storms also increased.

8.4.2 Invasion of Annual Exotic Grasses

Disturbances and devastation of CSCs leaves the vacant spaces, where exotic annual grasses easily expanded and their associated fire risk could be responsible for the crust cover and biodiversity loss. As annual exotic grasses starts to occupy the plant interspaces which once acquired by the CSCs and their diverse communities; gradually substituted by a bunch of cyanobacterial and annual moss species. Further absence of limited fires and growth of annual grasses supports the increase in rodent numbers and their burrows; probably responsible for this compositional shift. Unlike in the well-developed CSCs and less growth of annual grasses, limited fires generally expanded from shrub to shrub, leaves soil crusts unaffected between them. However excessive growth of annual grasses might be enhance size and frequency of wildfires, which now burned large areas, including the CSCs between plants; result in the death of soil crusts. This further prevents settlement of perennial lichens and mosses, leaving arid soils dominated by cyanobacteria and annual mosses.

8.4.3 Temperature and Precipitation

Cayan (1996) suggested that climatic alternations in deserts such as higher temperatures, greater summer precipitation, and drier-than-normal winters, responsible for affecting the structure and function of CSCs. It is very often for survival of mosses and lichens, water loss in respiration compensated by water gain in photosynthesis. But at higher temperature, soils of CSCs loose moisture faster (Jeffries et al. 1993); responsible for imbalance in respiratory loss and photosynthetic gain, leads to dryness of crust organisms. This stunted the further growth of crust communities, makes CSCs carbon deficits in the summer time.

Precipitation facilitates the soil wetting and temperature of soil surface also dropped, providing the conditions for CSCs to become metabolically active and perform better physiological functioning. As excess rain in summertime leads to flooding, wash away the soil crusts and less rain in wintertime leads to soil drying. This makes the CSCs carbon and nitrogen deficient. Due to inadequate carbon and nitrogen, CSCs will be less able to avoid or repair any disturbance; leads to increased mortality of more susceptible communities like lichens and mosses or even changing in distribution patterns. Belnap and Eldridge (2001) suggested that the current distribution pattern of lichen and moss in the deserts of US, Australia and central Asia, indicated towards this scenario; as diversity lichen and mosses reduces sharply with increase in temperature and summer rainfall.

8.4.4 UV Radiation

Most of the CSCs communities are quite susceptible to the UV radiation, which increases mortality through affecting the growth, motility, photosynthesis, nitrogen fixation and their uptake, photo-movements and cell differentiation (Castenholz and Garcia-Pichel 2000). UNEP/WMO (2002) predicted the risk of more UV radiation as ozone layer is thinning, which could be avoided through recovery of ozone layer due to decrease in chlorofluorocarbon (CFC) production and replacement with other alternatives. But this recovery might be slowed due to volcanic eruptions, airplane exhaust, and/or the renewed manufacture of ozone depleting substances. As CSCs in some deserts experienced so less days of rain, to rehydration and amplify their biomass; they are more vulnerable to UV radiation. Due to UV radiation, CSCs always faces severe damage, leaves limited time to acquire the carbon necessary to repair and produce new tissue. The condition of high UV radiation, less rain and higher temperature further worsen this situation.

8.4.5 Elevated CO_2 Concentration

Although increasing the atmospheric CO_2 levels might be help in increase the primary production of crust communities. But it would further limit the growth of crust communities. Lange et al. (1999) observed that CO_2 levels limit the photosynthesis in soil lichens as rates at ambient CO_2 levels are reaches to the maximum of 70–80%. Moore et al. (1999) stated that higher plants gradually slowed their processes or down-regulated, upon experiencing long-term exposure of elevated CO_2. In case of CSCs communities there is no substantial and comparable data regarding the response of crust communities to elevated CO_2. Unlike free-living and lichenized green algae in crusts, cyanobacteria have the intracellular CO_2 concentration mechanisms, which help to overcome the situation of, altered photosynthetic. Due to absence of intracellular CO_2 concentration mechanism, elevated CO_2 conditions might be favourable for green algae and lichenized green algae over cyanobacteria and cyano-lichens. So elevated CO_2 concentration induced the growth of some communities to increase in the more cover, leads to change in species composition of higher plant communities. Melillo et al. (1993) observed that enhanced water availability further increased the growth of such communities and result in the significant increase in net primary productivity in arid areas. Smith et al. (1987) further suggested that elevated CO_2 might be influence the competitive balance between higher plants and favoured the growth of invasive annual grasses, leads to reduction in crust cover and diversity.

8.4.6 *Recovery Rates*

Belnap and Eldridge (2001) underlined the factors responsible for recovery of soil crusts such as climate, nature of the soil & location, disturbances & their characteristics, inoculant availability and how recovery is addressed. It is observed that CSCs could be recover from disturbance in fairly quick (20 years) low PET arid areas, but it is tremendously slow (≥1000 years) in high PET arid areas. It is evident that CSCs are faster recovered in fine-textured soils as compared to coarse soils, as their low stability & fertility and poor water-holding capacity. Further stability of soils influence the recovery of crusts, as stable areas having low slopes, low wind deposition of sand, and/or embedded rocks showed better recovery than less stable areas having steep slopes, high sand deposition, and/or unstable rocks.

Due to severe or more frequent disturbance which are enough to disrupt already recovering CSCs, further recovery of crust communities slowed; if communities crumpled but stand still in their place. Although all the cyanobacterial communities vanished and blown by the disturbances. But large, highly mobile filamentous cyanobacteria such as *Microcoleus* survived even after burial and became the first colonizers of unstable soils. After larger cyanobacteria stabilized the crust soils; smaller and less mobile cyanobacteria starts to colonize. It is followed by lichens and mosses.

8.5 Rehabilitation of CSCs

There are many approaches which can be applied for the rehabilitation and stabilization of CSCs (Bowker 2007; Strong et al. 2013; Chock et al. 2019). These approaches are unique and diverse; and further adapted from various fields; related to restoration, ecology and agriculture. These approaches can classify into three major categories: (a) Artificial soil stabilization; (b) Resource augmentation; (c) Inoculation.

8.5.1 *Artificial Soil Stabilization*

In this method, soil surface is stabilized through the use of some artificial medium; which indirectly facilitates the successful rehabilitation of CSCs. There are some mediums such as polyacrilimide, coarse litter (such as straw), and vascular plants; that are successfully applied in the soil surface stabilization. Polyacrilimide (PAM) is a synthetic polymer, which effectively stabilize the soil surface and improves the soil moisture and nutrients availability. Further their application has no negative effect on chlorophyll fluorescence or nitrogenase activity of transplanted *Collema* (Collemataceae) lichens (Davidson et al. 2002).

Another medium straw has been effectively implied and examined in the dune stabilization and CSCs rehabilitation (Fearnehough et al. 1998; Hu et al. 2002; Li et al. 2004). In this approach, straw is vertically buried into soil spaced 1 m apart in lines and lines should make grid pattern. Sometimes, there are plantations of vascular plants along with these lines. Due to this, a succession of CSCs takes place; firstly cyanobacteria which are followed by chlorophytes, and in last mosses are colonized to form a cohesive and diverse CSCs. A number of researchers like van de Ancker et al. (1985), Maxwell and McKenna-Neuman (1994), Danin (1996, 1998) suggested that this approach would be helpful in that arid and semi-arid area where CSCs and vegetation have the capability to naturally stabilize the dunes. But the only problem with this approach is need of considerable economic incentive for the labour resources to execute and maintain it on large area.

Last one is introduction of vascular plants and grasses to stabilize the soil surface (Danin et al. 1998). Native and exotic plant species are more suitable for the stabilization and rehabilitation of CSCs (Aradottir et al. 2000). Danin (1978) suggested that due to plantation of trees in sandy area, wind velocity decreases; leads to succession of shrubs and CSCs and further development of more productive and diverse community. Aradottir et al. (2000) stated that fertilization coupled plantation of grasses could be very useful in highly eroded and unstable soils.

8.5.2 Resource Augmentation Approaches

In this approach, nutrients and moisture conditions are modified for the promotion of CSCs establishment in disturbed areas. However these approaches are not much explored. Singh (1950) investigated during India's monsoon season that earthen water catchments support the cyanobacterial growth and helpful in improving highly alkaline infertile soils in to suitable soils for agriculture. Belnap and Warren (1998), Maestre and Cortina (2002), Bowker et al. (2005) observed that the growth and stabilization of CSCs is favoured by somewhat cooler, shaded and wetter microsites. Although Davidson et al. (2002) reported that additional watering could have negative effect on transplanted lichens as it responsible for soil surface erosion. But in broad perspective, transplanted lichens showed more growth in mesic and cool microaspects of small, upraised elevations and mosses showed better growth in depressions in the CSCs surface (Maestre et al. 2001; Csotonyi and Addicott 2004). Tongway and Ludwig (1996), Maestre and Cortina (2004) suggested that brush piles could be useful to generate favourable microsites for the germination of vascular plants. Although brush piles likely favoured vascular plants more than CSCs. But applying that concept woody debris could be used to facilitate partial shade and mesic conditions for the stabilization CSCs.

Amendments such as minerals (Mn, Zn, and Mg), fertilizers (P, K and NPK) and biochar could be helpful in promoting the growth of cyanobacteria and chlorophytes in CSCs (Qiu and Gao 1999; Aradottir and Arnalds 2001; Elmarsdottir et al. 2003; Grettarsdottir et al. 2004; Bowker et al. 2006). Davidson et al. (2002) studied the effect of P and K fertilizers separately or combined and observed that addition of P and K had no effect on nitrogenase activity or condition of lichen transplants. There are variable effects of fertilization on chlorophyll fluorescence of the transplants. Qiu and Gao (1999) showed that K promote the photosynthetic recovery of *Nostoc flagelliforme* after desiccation in a laboratory study. Bowker et al. (2005, 2006) found that addition of Mn, Zn, K, and Mg have a positive effect on mosses and lichens in CSCs.

8.5.3 Inoculation-Based Approaches

Benefits of inoculation of cyanobacteria successfully investigated for soil reclamation, bioremediation and agricultural land improvement (Venkataraman 1972; Metting and Rayburn 1983; Ashley and Rushforth 1984; Rao and Burns 1990; Rogers and Burns 1994; Falchini et al. 1996; Singh 2014, 2015; Singh et al. 2016a, b, 2017a, b, c, 2018, 2019a, b; Kumar and Singh 2016, 2017; Tiwari et al. 2018; Kumar et al. 2017, 2018a, b;). Tiedemann et al. (1980) and Acea et al. (2001) investigated the benefits of cyanobacteria inoculation in forested ecosystems, either post-fire or as an N source in a tree plantation and observed that it has helpful in enhancing soil fertility and biological activity.

In relation to drylands, some studies carried out by St. Clair et al. (1986), Belnap (1993), Scarlett (1994), Davidson et al. (2002), Kubecková et al. (2003); which involves the application of crushed CSCs material, dry or in a slurry form, to the disturbed area. Although many studies relating inoculation based rehabilitation, significantly improved the enhanced recovery of CSCs; but full recovery time for the CSC development could be much longer in actual field conditions as compared to short duration and controlled field studies. It is found to be very successful to establish the founder communities of particular taxa through transplanting methods (Scarlett 1994; Bowler 1999). Davidson et al. (2002) observed that inoculation of cyanobacterial have apparently distinctive effects on transplanted *Collema* lichens, which primarily reliant on complex interactions with moisture and nutrient additions (Rossi et al. 2017; Wu et al. 2018).

8.6 Biochar Coupled Rehabilitation/Stabilization of CSCs

Biochar is a black and carbon rich solid material which could be obtained by heating the biomass at between 300 °C–700 °C under limited oxygen conditions; this process also known as pyrolysis (Singh et al. 2017a, b, c, 2018; Lehmann et al.

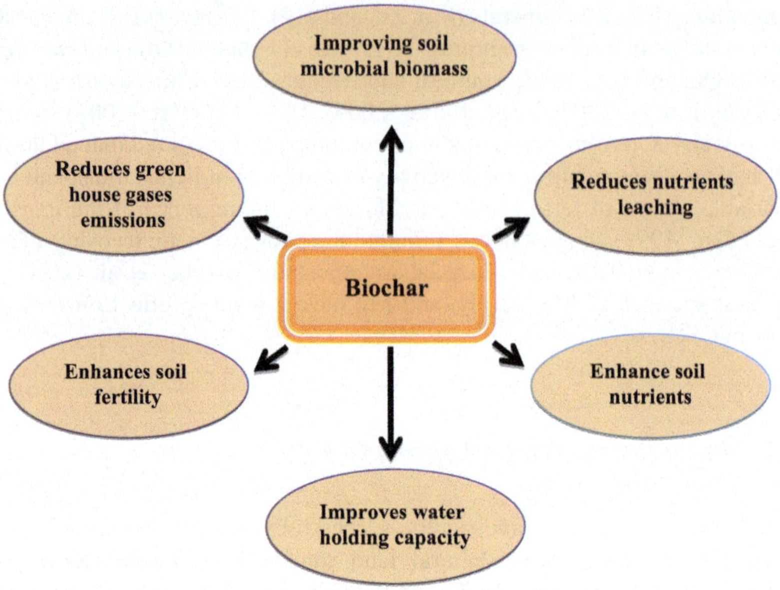

Fig. 8.2 Implications of biochar application in agriculture

2006; Nsamba et al. 2015). It is very helpful in improving soil physical characteristics like soil nutrient retention capacity, water holding capacity and reduced methane & N_20 emissions from soil (Fig. 8.2). Due to porous in structure it can also be helpful in maintaining microbial diversity in the soil (Lehmann et al. 2006; Verheijen et al. 2009; Lehmann 2007; Duku et al. 2011).

Lehmann (2007) investigated that biochar could be able to sink carbon up to 1 $Gtyr^{-1}$, which makes the biochar a attractive solution for the mitigation of climate change (Sohi et al. 2010; Yao et al. 2011). Lehmann et al. (2006), Laird et al. (2009), Lehmann (2007) suggested that it have a significant effect on cation exchange capacity (CEC; 40–80 meq per 100 g) and provide high surface area (51–900 m2•g − 1). Due to this, there is increase in soil pH and water holding capacity, and it also show greater ability to hold and capture micro- and macro- nutrients for the plants.

Considering benefits of biochar as soil amendments, it can be applied for the stabilization of CSCs. Although the previous studies mainly limited to effect of biochar in agricultural soils and emphasised that it improves such as the cation exchange capacity, pH, nutrient contents, plant growth; and also enhance carbon sequestration potential of the amended soils. Further it helps in reducing the greenhouse gas (GHG) emissions from the soils (van Straalen 1998; Gundale and DeLuca 2006; Sharkawi et al. 2006; Asai et al. 2009). Meng and Yuan (2014) investigated the use of biochar in improving the formation of cyanobacterial soil crust on sand under dry conditions and found that biochar have a significant effect on the cyanobacterial growth and sand fixation. Meng and Yuan (2014) conducted a study with application of 2% biochar (produced from the gasification of rice hull) on sandy soils, which undoubtedly enhanced the formation of cyanobacterial soil crust. So it can be concluded that biochar could be coupled with cyanobacteria or algae inoculation to improve or rehabilitation of the CSCs.

Although there are successful but few studies available related to beneficial effects of biochar on the fertility and communities of desert soils. And on quite pilot scale or theoretical way, it could be established that addition of biochar with other approaches recover or rehabilitate the CSCs. However there is need of further research in large scale to find that how biochar helps in rehabilitation of CSCs whether it improving soil physical properties and enhance CSCs formation in arid areas.

8.7 Small Scale Biochar Production

For the small scale biochar production a variety of raw materials such as rice hull, wheat straw, sugarcane bagasse, poultry litter, etc. are used. The pyrolysis reactors depending upon the heating, two methods can be used for the small scale biochar production:

1. Partial combustion- It is the most common pyrolysis method in which raw material combusted with a controlled air flow. But due to a portion of biomass be combusted in this process, it produces low yield of biochar, so they are applied in areas where raw materials are cheap;
2. Carbonization by contact with hot gases-In this method, hot gases from external source provided to the raw material which further converts the biomass into biochar and by-products. Although the costs are increased due to cost associated with heating the required inert gases. But biomass and by-products yield are high which makes the system suitable for medium to large scale production.

Biochar kilns may be simply earthen pits or made up of bricks, concrete or steel and cast iron. Earthen pits or pit kilns are very cost method and has been used from the centuries for the carbonizing woods. Biochar kilns are also made up of bricks or concrete to create a limited oxygen environment. Brick kilns are typically auto thermal and have a long lifespan and further portable as they easily dismantled and moved to be a new location. In last, steel or cast iron can used to make biochar kilns as they the heat easily transferred through walls made from these materials in Table 8.1.

Discarded oil drums also are used to make biochar kilns for the small scale production. Oil drums with both sides intact are most suitable; a big hole at the centre of top side provided for the loading of raw material and many small holes on the bottom side provided for the limited supply of oxygen (Venkatesh et al. 2010; Srinivasarao et al. 2013; Singh et al. 2017a, b, c).

Table 8.1 Biochar physicochemical properties from different raw materials

Raw material	Ash (%)	pH	EC (mS/cm)	C (%)	N (%)	Ca (ppm)	K (ppm)	Mg (ppm)	Si (mg/kg)	P (ppm)
Woodchip	25.4	7.88	0.14	51.9	0.4	0.56	0.21	0.04	–	0.06
Grass	14.7	6.1	–	42.5	1.9	4.3 4	64.8	2.3 4	7.44	2.31
Poultry litter	28.53	23.6	3	38.6	1.37	1.85	0.99	0.19	–	0.35
Rice husk	6.5	6.6	–	41	1.4	250	2604	827	5.8	–
Sugarcane bagasse	11.9–16.4	–	–	60.4–65.3	0.8–1.0	–	–	–	–	–
Wheat straw	5.9	6.76	2770	43.7	0.9	0.18	0.15	–	0.18	0.05

Modified from Mahinpey et al. (2009), Bruun et al. (2012). Jindo et al. (2012), Carrier et al. (2012), Shackley et al. (2012), Yargicoglu et al. (2015), Jouiada et al. (2015), Mohammed et al. (2015), and Singh et al. (2017a, b, c)

8.8 Conclusions

Cyanobacterial soil crusts are very essential in maintaining soil surface stability of desert soils and preventing them from erosion. The role of these crusts in carbon and nitrogen fixation, hydrological properties and surface stability of desert soil well understood. It is well established that extracellular polysaccharide secretion (EPS) from cyanobacteria play a major role in binding soil particles, nutrients and moisture. Further there is need of more *in-situ* studies related to the role of EPS in stabilization of crusts for the better understanding of CSCs.

In absence of CSCs, deserts could more prone to erosion, loss of organic matter, water availability, fine soil particles and nutrient content. Although there are many natural factors like high summer temperature, less rainfall and climate change, which are responsible for degradation of quality of cyanobacterial soil crusts. However there is no match of human intervention in terms of devastating effects that exert more pressure on already distressed crust; result in the complete destruction of crusts or increasing the recovery time for the CSCs. Due to disturbed crusts, exotic annual grasses occupied the vacant spaces and further increased the risk of surface fire; leads to simplifying species composition and flattening the crusts. So there is a need of comprehensive planning to maintain the CSCs in original and diverse condition that they further able to resist the changes caused by either natural or anthropogenic disturbances.

In all rehabilitation measures, cyanobacterial inoculants are seems to be very promising and successfully tested in many studies. But most of the studies carried out in lab conditions, and field applications are not so successful. Further the rehabilitation strategies such as biochar application could be used either separately or with the cyanobacteria inoculants. Biochar could be a game changing option in rehabilitation and stabilization of CSCs in deserts. It not only helps in capturing moisture but also provide microhabitat for microbial activities; helping in nutrient cycling. Small scale biochar production could be sustainable and viable option for the rehabilitation and stabilization of cyanobacterial soil crusts. It provides the livelihood to the locals of that region and also encourages the public participation.

In last, it is not necessary for a particular method to be effective for the rehabilitation and stabilization of all types of CSCs; so there is also need to consider conditions of the region where a particular type of CSCs existed. Further the information and strategies related to rehabilitation of CSCs are still in beginning stage, and the process of learning still going on. Currently the researchers are only focuses on the promoting faster recovery of CSCs in holistic way or of important community within the CSCs. In future, once these technological problems are solved, there would be focus on a particular aspect of CSCs and how this could be rehabilitated to the better recovery of ecosystem functions of interest.

References

Acea MJ, Diz N, Ferna'ndez AP (2001) Microbial populations in heated soils inoculated with cyanobacteria. Biol Fertil Soils 33:118–125

Aradottir AL, Arnalds O (2001) Ecosystem degradation and restoration of birch woodlands in Iceland. In: Wielgolaski FE (ed) Nordic mountain birch ecosystems. UNESCO, Paris and Parthenon Publishing, Carnforth, pp 295–308

Aradottir AL, Svavasdottir K, Jonsson XH, GuCbergsson G (2000) Carbon accumulation in vegetation and soils by reclamation of degraded areas. Icel Agric Sci 13:99–113

Asai H, Samson BK, Stephan HM, Songyikhangsuthor K, Homma K, Kiyono Y, Inoue Y, Shiraiwa T, Horie T (2009) Biochar amendment techniques for upland rice production in Northern Laos 1. Soil physical properties, Leaf SPAD and grain yield. Field Crop Res 111:81–84

Ashley J, Rushforth SR (1984) Growth of soil algae on top soil and processed oil shale from the Uintah Basin, Utah, U.S.A. Reclam Reveg Res 3:49–63

Belnap J (1993) Recovery rates of cryptobiotic crusts: inoculant use and assessment methods. Great Basin Nat 53:89–95

Belnap J (1995) Surface disturbances: their role in accelerating desertification. Environ Monit Assess 37:39–57

Belnap J (1996) Soil surface disturbances in cold deserts: effects on nitrogenase activity in cyanobacterial-lichen soil crusts. Biol Fertil Soils 23:362–367

Belnap J (2001a) Factors influencing nitrogen fixation and nitrogen release in biological soil crusts. In: Belnap J, Lange OL (eds) Biological soil crusts: structure, function, and management. Springer, Berlin, pp 241–261

Belnap J (2001b) Microbes and microfauna associated with biological soil crusts. In: Belnap J, Lange OL (eds) Biological soil crusts: structure, function, and management. Springer, Berlin, pp 167–174

Belnap J (2002) Nitrogen fixation in biological soil crusts from southeast Utah, USA. Biol Fertil Soils 35:128–135

Belnap J (2003a) The world at your feet: desert biological soil crusts. Front Ecol Environ 1(5):181–189

Belnap J (2003b) Comparative structure of physical and biological soil crusts. In: Belnap J, Lange OL (eds) Biological soil crusts: structure, function, and management. Springer, Berlin, pp 177–191

Belnap J (2006) The potential roles of biological soil crusts in dryland hydrologic cycles. Hydrol Process 20:3159–3178

Belnap J, Eldridge D (2001) Disturbance and recovery of biological soil crusts. In: Belnap J, Lange OL (eds) Biological soil crusts: structure, function, and management. Springer, Berlin, pp 363–383

Belnap J, Gardner JS (1993) Soil microstructure in soils of the Colorado Plateau: the role of the cyanobacterium *Microcoleus vaginatus*. Great Basin Nat 53:40–47

Belnap J, Lange O (eds) (2003) Biological soil crusts: structure, function and management, in press. Springer, Berlin

Belnap J, Warren SD (1998) Measuring restoration success: a lesson from Patton's tank tracks. Ecol Bull 79:33

Belnap J, Büdel B, Lange OL (2003) Biological soil crusts: characteristics and distribution. In: Belnap J, Lange OL (eds) Biological soil crusts: structure, function, and management. Springer, Berlin, pp 3–30

Bewley JD, Krochko JE (1982) Desiccation-tolerance. In: Lange OL, Nobel PS, Osmund CB, Ziegler H (eds) Physiological plant ecology II. Springer, Berlin, pp 325–378

Beymer RJ, Klopatek JM (1991) Potential contribution of carbon by microphytic crusts in pinyon-juniper woodlands. Arid Soil Res Rehabil 5:187–198

Bowker MA (2007) Biological soil crust rehabilitation in theory and practice: an underexploited opportunity. Restor Ecol 15(1):13–23

Bowker M, Reed SC, Belnap J, Phillips S (2002) Temporal variation in community composition, pigmentation, and Fv/Fm of desert cyanobacterial soil crusts. Microb Ecol 43:13–25

Bowker MA, Belnap J, Davidson DW, Phillips SL (2005) Evidence for micronutrient limitation of biological soil crusts: importance to arid-lands restoration. Ecol Appl 15:1941–1951

Bowker MA, Belnap J, Davidson DW, Goldstein H (2006) Correlates of biological soil crust abundance: support for a hierarchical conceptual model. J Appl Ecol 43:152–163

Bowler PA (1999) Understory bulbs, grasses, bryophytes and liverworts transplant successfully in coastal sage scrub (California). Ecol Restor 17:82–83

Brooks SI, Pokshishevsky VV (1986) Population problems in the arid and semiarid regions of the world. In: Gerasimov IP (ed) Arid land development and the combat against desertification: an integrated approach. Commission for the United Nations Environment Program, Moscow

Bruun EW, Ambus P, Egsgaard H, Hauggaard-Nielsen H (2012) Effects of slow and fast pyrolysis biochar on soil C and N turnover dynamics. Soil Biol Biochem 46:73–79

Büdel B, Lange OL (2003) Synopsis: comparative biogeography and ecology of soil-crust biota and communities. In: Belnap J, Lange O (eds) Biological soil crusts: structure, function, and management. Springer, Berlin, pp 141–152

Büdel B, Wessels DCJ (1986) Parmelia hueana Gyeln, a vagrant lichen from the Namib Desert, SWA/Namibia. I. Anatomical and reproductive adaptations. Dinteria 18:3–15

Bush JK, Van Auken OW (1991) Importance of time of germination and soil depth on growth of *Prosopis glandulosa* seedling in the presence of a C_4 grass. Am J Bot 78:1732–1739

Cameron RE, Blank GB (1966) Desert algae: soil crusts and diaphanous substrata as algal habitats. JPL Tech Rep 32:1–40

Carrier M, Hardieb AG, Urasa U, Gorgensa J, Knoetze J (2012) Production of char from vacuum pyrolysis of South-African sugar cane bagasse and its characterization as activated carbon and biochar. J Anal Appl Pyrolysis 96:24–32

Castenholz RW, Garcia-Pichel F (2000) Cyanobacterial responses to UV radiation. In: Whitton BA, Potts M (eds) The ecology of cyanobacteria. Kluwer Academic Publishers, Dordrecht, pp 591–611

Cayan DR (1996) Inter-annual climate variability and snowpack in the western United States. J Clim 9:928–948

Chock T, Antoninka AJ, Faist AM, Bowker MA, Belnap J, Barger NN (2019) Responses of biological soil crusts to rehabilitation strategies. J Arid Environ 163:77–85

Crawford CS (1991) The community ecology of macroarthropod detritivores. In: Polis G (ed) Ecology of desert communities. University of Arizona Press, Tucson, pp 89–112

Csotonyi JT, Addicott JF (2004) Influence of trampling-induced microtopography on growth of the soil crust bryophyte *Ceratodon purpureus* in Jasper National Park. Can J Bot 82:1382–1392

Danin A (1978) Plant species diversity and plant succession in a sandy region in the northern Negev. Flora 167:409–422

Danin A (1996) Plants of desert dunes. Springer, Berlin

Danin A, Dor I, Sandler A, Amit R (1998) Desert crust morphology and its relations to microbiotic succession at Mt. Sedom, Israel. J Arid Environ 38:161–174

Davidson DW, Bowker M, George D, Phillips SL, Belnap J (2002) Treatment effects on performance of N-fixing lichens in disturbed soil crusts on the Colorado Plateau. Ecol Appl 12:1391–1405

Dregne HE (1983) Desertification of arid lands. Harwood Academic Publishers, New York

Duku MH, Gu S, Hagan EB (2011) Biochar production potential in Ghana—a review. Renew Sust Energy Rev 15:3539–3551

Eldridge DJ, Greene RSB (1994) Assessment of sediment yield by splash erosion on a semi-arid soil with varying cryptogam cover. J Arid Environ 26:221–232

Eldridge DJ, Zaady E, Shachak M (2000) Infiltration through three contrasting biological soil crusts in patterned landscapes in the Negev, Israel. Catena 40:323–336

Elmarsdottir A, Aradottir AL, Trlica MJ (2003) Microsite availability and establishment of native species on degraded and reclaimed sites. J Appl Ecol 40:815–823

Evans RD, Ehleringer JR (1993) A break in the nitrogen cycle in aridlands? Evidence from 15N of soils. Oecologia 94:314–317

Evans RD, Johansen JR (1999) Microbiotic crusts and ecosystem processes. Crit Rev Plant Sci 18:183–225

Evans RD, Lange OL (2001) Biological soil crusts and ecosystems nitrogen and carbon dynamics. In: Belnap J, Lange OL (eds) Biological soil crusts: structure, function, and management. Springer, Berlin, pp 263–279

Falchini L, Sparvoli E, Tomaselli L (1996) Effect of *Nostoc* (cyanobacteria) inoculation on the structure and stability of clay soils. Biol Fertil Soils 23:346–352

Fearnehough W, Fullen MA, Mitchell DJ, Trueman IC, Zhang J (1998) Aeolian deposition and its effect on soil and vegetation changes on stabilized desert dunes in northern China. Geomorphology 23:171–182

Frey W, Kürschner H (1991) Morphological and anatomical adaptation of the species in terrestrial bryophyte communities along an ecological transect in the Judean Desert, Israel. Bot Jahrb Syst 112:529–552

Friedmann I, Lipkin Y, Ocampo-Paus R (1967) Desert algae of the Negev (Israel). Phycologia 6:185–195

Garcia-Pichel F, Belnap J (1996) Microenvironments and microscale productivity of cyanobacterial desert crusts. J Phycol 32:774–782

Grettarsdottir J, Aradottir AL, Vandvik V, Heegaard E, Birks HJB (2004) Long term effects of reclamation treatments on plant succession in Iceland. Restor Ecol 12:268–278

Gundale MJ, DeLuca TH (2006) Temperature and substrate influence the chemical properties of charcoal in the *Ponderosa* Pine/Douglas-Fire cosystem. Forest Eco Manag 231:86–93

Harper KT, Marble JR (1988) A role for nonvascular plants in management of arid and semiarid rangeland. In: Tueller PT (ed) Vegetation science applications for rangeland analysis and management. Kluwer Academic Publishers, Dordrecht, pp 135–169

Herrick JE, Wander MM (1998) Relationships between soil organic carbon and soil quality in cropped and rangeland soils: the importance of distribution, composition and soil biological activity. In: Lal R, Kimble J, Follett R, Stewart BA (eds) Advances in soil science: soil processes and the carbon cycle. CRC Press, Boca Raton, pp 405–425

Horodyski RJ, Knauth LP (1994) Life on land in the Precambrian. Science 263:494–498

Hu C, Liu Y, Song L, Zhang D (2002) Effect of desert soil algae on the stabilization of fine sands. J Appl Phycol 14(4):281–292

Jeffries DL, Link SO, Klopatek JM (1993) CO_2 fluxes of cryptogamic crusts. I. Response to resaturation. New Phytol 125:163–173
Jindo K, Suto K, Matsumoto K, García C, Sonoki T, Sanchez-Monedero MA (2012) Chemical and biochemical characterisation of biochar-blended composts prepared from poultry manure. Bioresour Technol 110:396–404
Johansen JR (1993) Cryptogamic crusts of semiarid and arid lands of North America. J Phycol 29:140–147
Jouiada M, Al-Nofeli N, Khalifa N, Benyettouc F, Yousef LF (2015) Characteristics of slow pyrolysis biochars produced from rhodes grass and fronds of edible date palm. J Anal Appl Pyrolysis 111:183–190
Kubeckova K, Johansen JR, Warren SD, Sparks RL (2003) Development of immobilized cyanobacterial amendments for reclamation of microbiotic soil crusts. Algol Stud 109:341–362
Kumar A, Singh JS (2016) Microalgae and cyanobacteria biofuels: a sustainable alternate to crop-based fuels. In: Singh JS, Singh DP (eds) Microbes and environmental management. Studium Press Pvt. Ltd, New Delhi, pp 1–20
Kumar A, Singh JS (2017) Cyanoremediation: a green-clean tool for decontamination of synthetic pesticides from agro- and aquatic ecosystems. In: Singh JS, Seneviratne G (eds) Agro-environmental sustainability: Vol-2: managing environmental pollution. Springer, Cham, pp 59–83
Kumar A, Kaushal S, Saraf SA, Singh JS (2017) Cyanobacterial biotechnology: an opportunity for sustainable industrial production. Clim Change Environ Sustain 5(1):97–110
Kumar A, Kaushal S, Saraf SA, Singh JS (2018a) Screening of chlorpyrifos (CPF) tolerant cyanobacteria from paddy field soil of Lucknow, India. Int J Appl Adv Sci Res 3(1):100–105
Kumar A, Kaushal S, Saraf SA, Singh JS (2018b) Microbial bio-fuels: a solution to carbon emissions and energy crisis. Front Biosci-Landmrk 23:1789–1802
Laird DA, Brown RC, Amonette JE, Lehmann J (2009) Review of the pyrolysis platform for coproducing bio-oil and biochar. Biofuels Bioprod Biorefin 3:547–562
Lange OL, Green TGA, Reichenberger H (1999) The response of lichen photosynthesis to external CO_2 concentration and its interaction with thallus water-status. J Plant Physiol 154:157–166
Lehmann J (2007) Bio-energy in the black. Front Ecol Environ 5:381–387
Lehmann J, Gaunt J, Rondon M (2006) Bio-char sequestration in terrestrial ecosystems—a review. Mitig Adapt Strategy Glob Change 11:403–427
Li X, Xiao H, Zhang J, Wang X (2004) Long-term ecosystem effects of sand-binding vegetation in the Tengger Desert, Northern China. Restor Ecol 12:376–390
Maestre FT, Cortina J (2002) Spatial patterns of surface soil properties and vegetation in a Mediterranean semi-arid steppe. Plant Soil 241:279–291
Maestre FT, Cortina J (2004) Insights into ecosystem composition and function in a sequence of degraded semiarid steppes. Restor Ecol 12:494–502
Maestre FT, Bautista S, Cortina J, Bellot J (2001) Potential for using facilitation by grasses to establish shrubs on a semiarid degraded steppe. Ecol Appl 11:1641–1655
Mager DM, Thomas AD (2011) Extracellular polysaccharides from cyanobacterial soil crusts: a review of their role in dryland soil processes. J Arid Environ 75:91–97
Mahinpey N, Murugan P, Mani T, Raina R (2009) Analysis of bio-oil, biogas, and biochar from pressurized pyrolysis of wheat straw using a tubular reactor. Energy Fuel 23:2736–2742
Maxwell CD, McKenna-Neuman C (1994) Photoautotrophs and the microaggregation of sand in a freshwater beach-dune complex: implications for sediment transport by wind. Soil Biol Biochem 26:221–233
McKenna-Neuman C, Maxwell CD, Boulton JW (1996) Wind transport of sand surfaces crusted with photoautotrophic microorganisms. Catena 27:229–247
Melillo JM, McGuire AD, Kicklighter DW, Moore B, Vorosmarty CJ, Schlos AL (1993) Global climate change and terrestrial net primary production. Nature 363:234–240
Meng X, Yuan W (2014) Can biochar couple with algae to deal with desertification? J Sustain Bioenergy Syst 4:194–198

Metting B, Rayburn WR (1983) The influence of a microalgal conditioner on selected Washington soils: an empirical study. Soil Sci Soc Am J 47:682–685
Mohammed IY, Abakr YA, Kazi FK, Yusuf S, Alshareef I, Chin SA (2015) Pyrolysis of napier grass in a fixed bed reactor: effect of operating conditions on product fields and characteristic. Bio Resources 10:6457–6478
Moore BD, Cheng S-H, Sims D, Seemann JR (1999) The biochemical and molecular basis for photosynthetic acclimation to elevated atmospheric CO_2. Plant Cell Environ 22:567–582
Nsamba HK, Hale S, Cornelissen G, Bachmann R (2015) Sustainable technologies for small-scale biochar production—a review. J Sustain Bioenergy Syst 5:10–31
Paerl HW (1990) Physiological ecology and regulation of N_2 fixation in natural waters. Adv Microb Ecol 11:305–343
Peterjohn WT, Schlesinger WH (1990) Nitrogen loss from deserts in the south western United States. Biogeochemistry 10:67–79
Pimm SL (2001) The world according to Pimm. McGraw-Hill, New York
Qiu B, Gao K (1999) Dried field populations of *Nostoc flagelliforme* (Cyanophyceae) require exogenous nutrients for their photosynthetic recovery. J Appl Phycol 11:535–541
Rao DLN, Burns RG (1990) The effect of surface growth of bluegreen algae and bryophytes on some microbiological, biochemical, and physical soil properties. Biol Fertil Soils 9:239–244
Reynolds R, Belnap J, Reheis M, Lamothe P, Luiszer F (2001) Aeolian dust in Colorado Plateau soils: nutrient inputs and recent change in source. Proc Natl Acad Sci 98:7123–7127
Rogers SL, Burns RG (1994) Changes in aggregate stability, nutrient status, indigenous microbial populations, and seedling emergence, following inoculation of soil with *Nostoc muscorum*. Biol Fertil Soils 18:209–215
Rogers L, Gallon J (1988) Biochemistry of the algae and cyanobacteria. Clarendon Press, Oxford
Rossi F, Li H, Liu Y, De Philippis R (2017) Cyanobacterial inoculation (cyanobacterisation): perspectives for the development of a standardized multifunctional technology for soil fertilization and desertification reversal. Earth-Sci Rev 171:28–43
Scarlett N (1994) Soil crusts, germination and weeds—issues to consider. Vic Nat 111:125–130
Schwartzman DW, Volk T (1989) Biotic enhancement of weathering and the habitability of earth. Nature 340:457–460
Shackley S, Carter S, Knowles T, Middelink E, Haefele S, Sohi S, Cross A, Haszeldine S (2012) Sustainable gasification–biochar systems? A case-study of rice-husk gasification in Cambodia, part I: context, chemical properties, environmental and health and safety issues. Energy Policy 42:49–58
Sharkawi H, Yamamoto S, Honna T (2006) Rice yield and nutrient uptake as affected by cyanobacteria and soil amendments—a pot experiment. J Soil Sci Plant Nutr 169:809–815
Singh RN (1950) Reclamation of 'Usar' lands in India through blue green algae. Nature 165:325–326
Singh JS (2014) Cyanobacteria: a vital bio-agent in eco-restoration of degraded lands and sustainable agriculture. Clim Change Environ Sustain 2:133–137
Singh JS (2015) Microbes: the chief ecological engineers in reinstating equilibrium in degraded ecosystems. Agric Ecosyst Environ 203:80–82
Singh JS, Kumar A, Rai AN, Singh DP (2016a) Cyanobacteria: a precious bio-resource in agriculture, ecosystem, and environmental sustainability. Front Microbiol 7:529
Singh JS, Koushal S, Kumar A, Vimal SR, Gupta VK (2016b) Book review: microbial inoculants in sustainable agricultural productivity- Vol. II: Functional Application. Front Microbiol 7:2105
Singh C, Tiwari S, Boudh S, Singh JS (2017a) Biochar application in management of paddy crop production and methane mitigation. In: Singh JS, Seneviratne G (eds) Agro-environmental sustainability: vol-2: managing environmental pollution. Springer, Cham, pp 123–146
Singh C, Tiwari S, Singh JS (2017b) Impact of rice husk biochar on nitrogen mineralization and methanotrophs community dynamics in paddy soil. Int J Pure Appl Biosci 5(5):428–435
Singh C, Tiwari S, Singh JS (2017c) Application of biochar in soil fertility and environmental management: a review. Bull Environ Pharmacol Life Sci 6(12):07–14

Singh C, Tiwari S, Gupta VK, Singh JS (2018) The effect of rice husk biochar on soil nutrient status, microbial biomass and paddy productivity of nutrient poor agriculture soils. Catena 171:485–493

Singh JS, Kumar A, Singh M (2019a) Cyanobacteria: a sustainable and commercial bio-resource in production of bio-fertilizer and bio-fuel from waste waters. Environ Sustain Indic 3–4:100008

Singh C, Tiwari S, Singh JS (2019b) Biochar: a sustainable tool in soil 2 pollutant bioremediation. In: Bharagava RN, Saxena G (eds) Bioremediation of industrial waste for environmental safety. Springer, Cham, pp 475–494

Smith SD, Strain BR, Sharkey TD (1987) Effects of CO_2 enrichment on four Great Basin grasses. Funct Ecol 1:139–143

Sohi S, Krull E, Lopez-Capel E, Bol R (2010) A review of biochar and its use and function in soil. Adv Agron 105:47–82

Srinivasarao C, Gopinath KA, Venkatesh G, Dubey AK, Wakudkar H, Purakayastha TJ, Pathak H, Jha P, Lakaria BL, Rajkhowa DJ, Mandal S, Jeyaraman S, Venkateswarlu B, Sikka AK (2013) Use of biochar for soil health enhancement and greenhouse gas mitigation in India. NICRA Bulletin, Hyderabad, pp 1–51

St. Clair LL, Johansen JR, Webb BL (1986) Rapid stabilization of fire-disturbed sites using a soil crust slurry: inoculation studies. Reclam Reveg Res 4:261–269

Strong CL, Bullard JE, Burford MA, McTainsh GH (2013) Response of cyanobacterial soil crusts to moisture and nutrient availability. Catena 109:195–202

Tiedemann AR, Lopushinsky W, Larsen HJ Jr (1980) Plant and soil responses to a commercial blue-green algae inoculant. Soil Biol Biochem 12:471–475

Tiwari S, Singh C, Singh JS (2018) Land use changes: a key ecological driver regulating methanotrophs abundance in upland soils. Energy Ecol Environ 3(6):355–371

Tongway DJ, Ludwig JA (1996) Rehabilitation of semiarid landscapes in Australia. I. Restoring productive soil patches. Restor Ecol 4:388–397

UNEP/WMO (2002) Executive summary (final): scientific assessment of ozone depletion: 2002. Scientific assessment panel of the montreal protocol on substances that deplete the ozone layer. United Nations, New York

Van den Ancker JAM, Jungerius PD, Mur LR (1985) The role of algae in the stabilization of coastal dune blowouts. Earth Surf Process Landf 10:189–192

van Straalen NM (1998) Evaluation of bio-indicator systems derived from soil arthropod communities. Appl Soil Ecol 9:429–437

Venkataraman GS (1972) Algal biofertilizers and rice cultivation. Today and Tommorrow's, New Delhi

Venkatesh G, Korwar GR, Venkateswarlu B, Gopinath KA, Mandal UK, Srinivasarao C, Grover MT (2010) Preliminary studies on conversion of maize stalks into biochar for terrestrial sequestration of carbon in rainfed agriculture. National Symposium on Climate Change and Rainfed Agriculture, CRIDA, Hyderabad, pp 388–391

Verheijen F, Jeffery S, Bastos AC, Velde MVD, Diafas I (2009) Biochar application to soils—a critical scientific review of effects on soil properties, processes and functions. EUR 24099 EN Office for the Official Publications of the European Communities, Luxemburg, p 149

Verrecchia E, Yair A, Kidron GJ, Verrecchia K (1995) Physical properties of the psammophile cryptogamic crust and their consequences to the water regime of sandy soils, north-western Negev Desert, Israel. J Arid Environ 29:427–437

Walter MR, Bauld J, Brock TD (1976) Microbiology and morphogenesis of columnar stromatolites (Conophyton, Vacerrilla) from hot springs in Yellowstone National Park. In: Walter MR (ed) Stromatolites. Elsevier, Amsterdam, pp 273–310

Wu L, Zhu Q, Yang L, Li B, Hu C, Lan S (2018) Nutrient transferring from wastewater to desert through artificial cultivation of desert cyanobacteria. Bioresour Technol 247:947–953

Wullstein LH (1989) Evaluation and significance of associative dinitrogen fixation for arid soil rehabilitation. Arid Soil Res Rehabil 3:259–265

Yao Y, Gao B, Inyang M, Zimmerman AR, Cao X, Pullammanappallil P, Yang L (2011) Biochar derived from anaerobically digested sugar beet tailings: characterization and phosphate removal potential. Bioresour Technol 102:6273–6278

Yargicoglu EN, Sadasivam BY, Reddy KR, Spokas K (2015) Physical and chemical characterization of waste wood derived biochars. Waste Manag 36:256–268

Chapter 9
Soil Health Management Through Low Cost Biochar Technology

Shaon Kumar Das and Goutam Kumar Ghosh

Abstract The utilization of biochar as an amendment to improve soil health and the environment has been a catalyst for the recent global enthusiasm for advancing biochar production technology and its management. Biochar is simply carbon rich charcoal-like substance which is created by heating biomass (organic matter) in limited oxygen condition, through a process known as pyrolysis. Locally available weed biomass which is not economically important and caused crop loss can be used as an important source of biomass for preparation of biochar. Biochar is able to ameliorate soil acidity as well as it is also able to increase the soil fertility. Biochar reduces leaching of soil nutrients, increases soil structure and pH, reduces dependency on artificial fertilizers, enhances nutrient availability for plants, increases water quality of runoff, reduces toxicity of aluminum to plant roots and microbiota and thus reducing the need for lime, reduces bioavailability of heavy metals, thus works as bioremediation and decreases N_2O and CH_4 emissions from soils, thus further reducing GHG emissions. Employment of biochar as a specialized soil amendment provides a practical approach to address the anticipated problems in the agronomic and environmental sectors. Incorporating huge quantity of biochar into soils provides numerous agricultural benefits, which this special paper examines. But, there is no concrete compilation yet how to apply biochar at farm level. This paper discusses on several factors related to biochar that need to be considered for maximising the soil amelioration and soil quality benefits from the use of biochar.

Keywords Amendment · Biochar · Charcoal · GHG emissions · Soil nutrients

---

S. K. Das (✉)
ICAR-National Organic farming Research Institute, Gangtok, India

G. K. Ghosh
Palli-Siksha Bhavan, Visva-Bharati, Sriniketan, West Bengal, India

J. S. Singh, C. Singh (eds.), *Biochar Applications in Agriculture and Environment Management*, https://doi.org/10.1007/978-3-030-40997-5_9

9.1 Introduction

Biochar is carbon rich charcoal-like substance which is created by heating biomass (organic matter) in a limited oxygen conditions, a process known as pyrolysis. Biochar application in soil has received a growing interest as a sustainable technology to improve highly weathered or degraded soils (Das et al. 2014a, b). It guarantees a long term benefit for soil fertility and productivity. It can enhance plant growth by improving soil physical characteristics (i.e., bulk density, water holding capacity, infiltration, porosity), soil chemical characteristics (i.e., pH, nutrient retention, nutrient availability), and soil biological properties (*i.e.,* microbial biomass carbon), all contributing to an increased crop productivity. The major quality of biochar that makes it attractive as a soil amendment is its highly porous structure which is responsible for improved water retention and increased soil surface area (Das and Avasthe 2015).

9.2 Benefit of Biochar

The major benefits of biochar are impressive because it reduces leaching of soil nutrients, increases soil pH and thus reducing the need for lime, enhances nutrient availability for plants, reduces toxicity of aluminium to plant roots, increases water quality of runoff, reduces dependency on fertilisers, reduces bioavailability of heavy metals and thus works as bioremediation, decreases N_2O and CH_4 emissions from soils, thus further reducing GHG emissions (Mate et al. 2015).

9.2.1 Nutrient Value

Biochar is able to improve soil fertility as well as productivity directly and indirectly as:

(a) *Indirect:* The indirect responses due to biochar application were attributed to either nutrient savings (in term of fertilizers) or improved fertilizer-use efficiency. Biochar being high C/N ratio can immobilize nitrogen which sometimes results in reduced N availability for short duration. This is the ability of biochar to retain applied fertilizer against leaching which results increase in fertilizer use efficiency (Gryze et al. 2010).
(b) *Direct:* Biochar itself contains some amount of nutrients which is available directly to plants. Positive yield responses as a result of biochar application to soils have been reported for a wide range of crops and plants in different parts of the world by improving soil quality (Tiwari et al. 2019a, b; Singh et al. 2019; Kour et al. 2019) with consequent improvement in the efficiency of fertilizer use. From an agronomic perspective it is suggested that biochar could improve

soil health by improving nutrient retention, particularly in coarsely textured soils (Das et al. 2014a, b).

9.2.2 How Can Biochar Help Farmer

Using locally available materials for making biochar could provides an unique opportunity to improve soil fertility for longer period of time to the farmers. Biochar should apply along with other inputs like compost, manure or biopesticides at the same rate every year to realize actual benefits. Application rates of these organic inputs can be reduced when nutrients are combined with biochar because biochar itself contain some nutrient (Major et al. 2009). During conversion of organic residues into biochar farmers can also receive an energy yield by capturing energy given off in the biochar production process. In hilly and desert areas soil loss, weathering and degradation occur at unprecedented rates which causes imbalance in ecosystem properties. Biochar can play a major role in organic agriculture for sustainable soil management by improving existing best management practices, not only to decrease nutrient loss through leaching by percolating water but also to improve soil productivity (Jeffery et al. 2015).

9.2.3 Biochar and Water Availability

Biochar addition in soil increases water holding capacity and plant available water in sandy soils. In dry areas where water quantity and quality is extremely variable, it would contribute a significant benefit. Biochar has a high surface area with increased micro pores and improves the water holding properties of porus sandy soils. Therefore, biochar application for soil water benefits is maximized in sandy soils (Das et al. 2012a, b). Thus, there are enormous benefits of biochar in cropping areas where cost of water is very high such as dry areas.

9.2.4 Effect on Soil pH

Soil pH is an important factor for plant growth because nutrient availability in soils depends on soil pH. Most of the macronutrients are available in neutral soils. In order to neutralize acidic soils, farmers apply thousands of tons of lime to farm soils at great expense. Biochar have an effect on soil pH (Rodríguez-Vila et al. 2014) It can react similarly as agricultural lime do (by increasing soil pH). If a soil has a low cation exchange capacity, it is not able to retain nutrients and the nutrients are often washed out leaching. Biochar in its pores having large surface area develops some negative charges and thus provides more negatively charged sites for cations to be retained when added to soil (Steinbeiss et al. 2009)

9.2.5 *Effect on Soil Physical Properties*

Biochar application improved the saturated hydraulic conductivity of the top soil and xylem sap flow of the rice plant. It increases water holding capacity in sandy soil. Peanut hull biochar have ability to reduce moisture stress in sandy soil. It improves soil physical condition for earthworm populations. Application of 6.6 metric tons cassia biochar/ha is enough to initiate C-accumulation, which reflect in an increase in organic matter and a net reduction in soil bulk density (Das 2014a, b).

9.2.6 *Effect on Soil Chemical Properties*

Biochar contribute some quantity of nutrients in soil through the negative charges that develops on its surfaces. This negative charge can easily buffer acidity in the soil (as does organic matter). Due to its high alkalinity nature it has been demonstrated to reduce aluminium toxicity in acid soils. Application of biochar to acidic soils can avoid significant amounts of direct and indirect costs by avoiding GHG emissions (Hammes and Schmidt 2009). Application of biochar in soil increase soil pH, EC, CEC and decrease exchangeable acidity.

9.2.7 *Effect of Biochar on Soil Biology*

Biochar is able to enhance soil microbial biomass carbon and carbon mineralization. It stimulates the activity of a variety of agriculturally important soil microorganisms and can greatly affect the microbiological properties of soils. The pores in biochar provide a suitable habitat for many microorganisms by protecting them from predation and drying while providing many of their diverse carbon (C), energy and mineral nutrient needs. The intrinsic properties of biochar and its ability to form complex with different soil type, can have an impact on soil-plant-microbe interactions (Hass et al. 2012). Thus, modifications in the soil microbial community can subsequently influence changes in nutrient cycling and crop growth in biochar-amended soil. Biochar application increase Co adsorption which lead to increase local nutrient concentrations for microbial community species and enhanced water retention Dehydrogenase activity and microbial biomass carbon are enhanced due to biochar application in soils (Das and Mukherjee 2012).

9.3 Application of Biochar in Soil

There are different methods for application of biochar in soil like broadcasting, deep banding, band application, spot placement, etc. However, method of biochar application in soil mainly depends on farming system, labour and available machinery. Generally farmers apply biochar in their own field by hand only. But due to prolonged contact with airborne biochar particulates, it is not viable on large-scale considering human health. Broadcasting application needs large amount to cover whole field. Suitable method of application deposits biochar directly into the rhizosphere, and may be viable for perennial cropping systems, and previously established crops. (Jefferym et al. 2011). Deep banding of biochar has been successfully implemented in several wheat fields in Western Australia. Mixing of biochar with composts, manures and other organic input may reduce odours, colour and improve nutrient performance over time due to slower leaching rates (Table 9.1). Mixtures may be applied for uniform topsoil mixing without incorporation (Das and Mukherjee 2011).

9.3.1 Application Rates

Application of biochar in soils is based on its properties like agricultural value from enhanced soils nutrient retention and water holding capacity, carbon sequestration and reduced GHG emissions. There is no specific rate of application of biochar in soil. It depends on many factors including type of biomass used, the types and proportions of various nutrients (N, P, etc.), the degree of metal contamination in the biomass, and also climatic and topographic factors of the land (Jones et al. 2012). It was found that rates between 5–10 t/ha (0.5–1 kg/m^2) have often been found better. Due to variability in biochar materials, nature of crop and soils, farmer should always consider testing several rates of biochar application on a small scale before setting out to apply it on large areas. Even low rates of biochar application can

Table 9.1 Effect of biochar on different soil properties

Factor	Impact
Bulk density	Soil dependent
Soil moisture retention	Upto 25% increase
Liming agent	1 point pH increase
Cation exchange capacity	50% increase
Nutrient use efficiency	10–20% increase
Crop productivity	30–100% increase
CH4 emission	90% decrease
N2O emission	50% decrease
Biological nitrogen fixation	50% increase
Mycorrhizal fungi	30% increase

significantly increase crop productivity assuming if the biochar is rich in nutrients. Biochar application rates sometimes also depend on the amount of dangerous metals present in the original biomass (Das and Mukherjee 2014).

9.4 Soil Health Management

Biochar can act as a soil conditioner by improving soil physical, chemical and biological properties. Benefits from biochar application rates can be maximized only if the soil is rich in nitrogen or if the crops are nitrogen-fixing legumes. Researcher found that application of biochar to soils in a legume-based (e.g. peanut and maize) rotational cropping system, clovers and lucernes is more beneficial. Significant changes in soil quality, including increase in pH, organic carbon and exchangeable cations were observed at higher rates of biochar application, i.e. > 50 t/ha. When mixed with organic matter, biochar can result in enhanced retention of soil water as a result of its pore structure which contributes to nutrient retention because of its ability to trap nutrient rich water within the pores. Biochar is able strongly to adsorb phosphate, even though it is an anion (Knowles et al. 2011). It is reported that the higher BNF with biochar additions is due to greater Mo and B availability. These properties make biochar a unique substance, retaining exchangeable and plant available nutrients in the soil, and offering the possibility of decreasing environmental pollution by nutrients and improving crop yields. Thus, biochar application could provide a new technology for both soil fertility and crop productivity improvement, with potential positive and quantifiable environmental benefits (Kookana 2010).

9.5 Land Restoration/Reclamation

Biochar have received considerable attention in recent years as soil amendment for both sequestering heavy metal contaminants and releasing essential nutrients like sulphur. Biochar are porous with a polar and aromatic surface (Das et al. 2015). They have a high surface to volume ratio and a strong affinity to non-polar substances such as polycyclic aromatic hydrocarbons (PAHs), dioxins, PBDEs, furans (PCDD/Fs), and PCBs. Through the intervention of biochar, groundwater could be protected from the hydrophobic herbicide, insecticide and fungicide. Biochar applications have the potential to absorb pollution by adsorbing ammonia to reduce ammonia volatilization in agricultural soils (Laird et al. 2010)

9.6 Heavy Metal Sorption

The use of biochar to remove contaminants such as organic contaminants or metals is a relatively novel and promising technology. Biochar made from bagasse and other agricultural residues is effective alternative, low-cost environmental sorbents of lead or other heavy metals. Several studies have reported the effective removal of lead by biochar sorbents. Like many other traditional sorbents, the high affinity for lead and other metal ion species bound by biochar may be controlled by other mechanisms as well, including complexation, chelation, and ion exchange. Application of maize stalk biochar is useful to ameliorate chromium (Cr) polluted soils and reduce the amount of carbon produced due to biomass burning (Rajkovich et al. 2012)

9.7 Pathogen and Biochar Interaction

Researchers have reported both increased root colonization and stimulated mycorrhizal fungus spore germination in response to biochar application probably due to improved soil physicochemical properties through enhanced nutrient availability. The efficacy of biochar is dependent on saprophytic fungal activity, which, through their extracellular enzymatic activity and hyphal growth/penetration, can violate the integrity of the material. Citrus wood biochar @1% (w/w) in sandy soil was found to be effective against *Leveillula taurica* (powdery mildew) and *Botrytis cinerea* (grey mold) in pepper and tomato and also in mite *Polyphagotarsonemus latus* in pepper (Das 2014a, b). Beside this, tolerance of asparagus seedlings to *Fusarium oxysporum* is also enhanced by biochar.

9.8 Cattle Feedlot Biochar

Potential sources of organic materials for biochar production include urban green wastes, forestry and crop processing residues as well as animal manures. Biochar made from cattle feedlot manure is an effective soil amendment for improving the productivity in acid soil. This biochar contain high mineral P content which remained as plant available for long period ($\geq$3 years). The increase in P availability led to enhanced P uptake which results in an increase in N uptake and N use efficiency. Manure-based feedstocks tend to have lower carbon content, and higher nutrient/mineral content compared to wood based biochar. Biochar from urban green waste have no harmful effect on pasture productivity (Mohan et al. 2014) Biochar has the capacity to increase soil C accumulation rates in acidic pasture systems. Green waste biochar enhance soil C accumulation at a faster rate than farm manure biochar.

9.9 Crop Production

Biochar applications to soils have shown positive responses for net primary crop production, grain yield and dry matter. Application of wheat straw biochar along with NPK significantly increase the yield of maize in Inceptisol than either crop residue incorporation (CRI) or crop residue burning (CRB). Higher agronomic nitrogen use efficiency was recorded with application of biochar. The combined application of biochar along with organic/inorganic fertilizer has the potential to increase crop productivity, thus providing additional incomes, and may reduce the quantity of inorganic fertilizer use and importation (Kimetu and Lehmann 2010). The impact of biochar application is seen most in highly degraded acidic or nutrient depleted soils. Low biochar application in soil has shown marked impact on various plant species, whereas higher rates seemed to inhibit plant growth. So, moderate additions of biochar are usually beneficial.

9.10 Effect on Upland Rice

Biochar improve saturated hydraulic conductivity of the top soil and the xylem sap flow in upland rice plant. Researchers found that it increased higher grain yields at sites with low P availability and improved the response to N and NP chemical fertilizer treatments (Lehmann et al. 2009). It also reduced leaf SPAD values, possibly through a reduction of the availability of soil nitrogen, indicating that biochar without additional N fertilizer application could reduce grain yields in soils with a low indigenous N supply.

9.11 Effect on Nodulation and Nitrogenise Activity

Biochar addition increase root nodule number, localised N_2 fixation per nodule, nitrogenise activity in legumes, mycorrhizal colonisation and plant-growth promoting organisms in the rhizosphere. Increased nodulation following biochar application could increase sustainable N input into agro ecosystems. Biochar applications also increase nitrogen fixation rates. Increased micronutrient availability (e.g. Mo and B), together with the liming effect on soil pH following biochar application has been proposed as the mechanisms for increased biological N_2 fixation of pot grown beans (Sohi et al. 2010). Symbiotic association between biochar and mycorrhizal association showed that biochar could influence mycorrhizal abundance. Rice biochar showed greater microbial activities than other biochar because of its higher liability (Gaskin et al. 2008).

9.12 Carbon Sequestration

In order to considerably increase long-term C sequestration, biomass has to be converted to a relatively non-degradable form, such as biochar. The biochar is highly resistant to microbial activity, considerably augmenting the recalcitrant fraction of SOC and decreasing emissions of CO_2 from soil. In addition, biochar application was reported to decrease emissions of CH_4, and N_2O from soils. Despite the recalcitrant nature of biochar, about 40% of the total biomass-C of the feedstock is lost during the pyrolysis process, and an additional 10% is mineralized over a few months after biochar application in soil. Nevertheless, the remaining 50% of the total C is relatively stable. The degree of stability of the biochar-C depends on its specifications. While C in biochar produced by high temperature pyrolysis is either recalcitrant or degradable at an extremely slow rate, some of the C in biochar produced under low temperatures is biodegradable. In addition, compared with fallow soils, application of biochar increases rates of CO_2 emissions from the amended soil. This response may be explained by several factors, such as lower bulk density, improved aeration, and higher pH, providing a favorable habitat for soil microorganisms (Novak et al. 2009).

Considering an application rate of between 10 and 100 Mg biochar per hectare and that biochar's C concentration is between 50% and 78%, and assuming a total area of 1411Mha cropland around the world, then the global capacity for storing biochar-C under this landuse is between 7 and 110 Pg. Annual net emissions of carbon dioxide, methane and nitrous oxide could be reduced by a maximum of 1.8 Pg CO_2–C equivalent (CO_2–Ce) per year (12% of current anthropogenic CO_2–Ce emissions), and total net emissions over the course of a century by 130 Pg CO_2–Ce, by utilizing the maximum sustainable technical potential of biochar to mitigate climate change, without endangering food security, habitat or soil conservation. When the use of the process of biochar sequesters more carbon than it emitted, it is carbon negative. Biochar holds 50% of the carbon biomass and it sequesters that carbon for centuries when applied into the soil, removing the CO_2 from the active cycle and thus reduce overall amount of atmospheric CO_2. Plant growth is also enhanced by this process as it absorbs more CO_2 from atmosphere. Overall, these benefits make the biochar process carbon negative as long as biomass production is managed sustainably. Biochar system also needs to be taken into account, *viz.*, emissions resulting from biomass growth, collection, pyrolysis, spreading and transport, to consider it a truly carbon negative. Due to its capability to actively reduce the atmospheric concentrations of greenhouse gases, biochar technology may be considered as geoengineering solution. It may also be considered as a long wave geoengineering option for climate change mitigation as it plays a role into the removal of CO_2 from the atmosphere and enhances the level of long wave radiation leaving from the planet. A biochar system is a carbon sink, where agricultural crops are grown and is subsequently pyrolysed to produce biochar, which is then applied to soil (Novak et al. 2009). In carbon cycle, plants remove CO_2 from atmosphere via photosynthesis and convert it into biomass. But all of that carbon (99%) is returned to atmosphere as CO_2 when plants die and decay, or immediately if biomass is burned as a renewable substitute for fossil fuels. In biochar cycle, half (50%) of that carbon is removed and sequestered as biochar and the rest half (50%) is converted to renewable energy co-products before being returned to the atmosphere. A more efficient way to increase and maintain a high soil organic matter content would be to apply more stable C products such as biochar. Future political agreements may make it profitable for farmers to add biochar to soil. Large amounts of carbon in biochar may be sequestered in the soil for long periods estimated to be hundreds to thousands of years. Terra preta soils suggest that biochar can have carbon storage permanence in soil for many hundreds to thousands of years. Biochar mineralizes in soils in a little fraction and remains in a very stable form which provides it the potential to be a major carbon sink. About 12% of the total anthropogenic carbon emissions by land use change (0.21 Pg C) can be offset annually in soil, if the slash-and-burn system is replaced by the slash-and-char system. Compared with other terrestrial sequestration strategies, such as afforestation or re-forestation, carbon sequestration in biochar increases its storage time. The principal mechanisms operating in soils through which biochar entering the soil is stabilized and increase its residence time in soil are due to formation of interactions between mineral surfaces, intrinsic recalcitrance and spatial separation of decomposers and substrate (Githinji 2013).

9.13 Carbon Credit

Application of higher amounts of biochar to soils may increase the carbon credit benefit to the farmers. Carbon added to the fields in the form of biochar could give farmers C credits that can be sold on a C credit market for additional income. Increasing the C sink in soil will help reduce the amounts of CO_2, CH^4, and N_2O.

9.14 Stability in Soil

Biochar is not a single material, and its characteristics vary depending upon from where and how it is made. Stability of biochar in soil is important in determining environmental benefits because stability determines how long carbon (C) applied to soil as biochar will remain sequestered in soil and contribute to mitigate climate change and how long biochar can provide benefits to soil and water quality (Singh et al. 2017a, b, c, 2018; Tiwari et al. 2018; Das and Mukherjee 2012). Most of the biochar commonly used by the farmer have a small labile (easily decomposed) fraction in addition to a much larger stable fraction. The mean residence time of this stable fraction is estimated to range from 100 to 1000 years.

9.15 Impact on Climate Change

Biochar technology is called as geoengineering solution that has potential to actively reduce the atmospheric concentrations of green house gases. As it results in the removal of CO2 from the atmosphere and increases level of long wave radiation leaving the planet, it is considered as a long wave geoengineering option for climate change mitigation. A biochar system, where agricultural crops are grown, and subsequently pyrolyzed to produce biochar, which is then applied to soil, is a carbon sink. This means CO2 from atmosphere is sequestered as carbohydrates in the growing plant and conversion of the plant biomass to biochar stabilizes this carbon (Keith et al. 2011). The stabilization of carbon in biochar delays its decomposition and ensures that carbon remains locked away from the atmosphere for hundreds to thousands of years. In addition, gases released in the process of creating biochar can be used to make bio fuels. If we want to tackle climate change challenges, we must emphasize the potential of soil to sequester carbon. Sustainable biochar can be used now to combat global warming by holding carbon in soil and by displacing fossil fuel use.

9.16 Safety Concerns

Application of large amounts of biochar to agricultural soils entails significant practical and technical barriers like safe production and use. This risk is similar to other dusts that can become combustible hazards, such as coal, plastics, some metals, foods, and woods. The dust of biochar can spontaneously combust and poses a minor risk when handled, stored, or transported in enclosed spaces (Renner 2007; Sohi et al. 2010; Liu et al. 2012; Nelissen et al. 2012). Some biochar contain toxic materials that are controlled by "permissible exposure limit" standards in many countries. The levels of these toxic materials in the biochar are highly dependent on both the biomass feedstock and its production. So, there is no straightforward permissible exposure limit available for biochar as yet (Ogawa and Okimori 2010).

9.17 Conclusions

Soil amendment with biochar has attracted a fair amount of research interest due to its abundant usage and wide potential, which includes enhancing crop production by improving soil fertility, decreasing greenhouse gas emissions and increasing soil carbon sequestration. Use of biochar in agricultural systems is one viable option that can improve the soil quality, increase carbon sequestration in soil, reduce farm waste. The initial outcomes reveal that biochar application helps in improving soil health and crop productivity. However, to promote the application of biochar as a soil amendment and also as a climate change abatement option, research, development and demonstration on biochar production and application is very vital.

References

Das SK (2014a) Recent development and future of botanical pesticides in India. Popular Kheti 2:93–99

Das SK (2014b) Role of micronutrient in rice cultivation and management strategy in organic agriculture-A reappraisal. Agric Sci 5:765–769

Das SK, Avasthe RK (2015) Biochar as carbon negative in carbon credit under changing climate. Curr Sci 109:1223

Das SK, Mukherjee I (2011) Effect of light and pH on persistence of flubendiamide. Bull Environ Contam Toxicol 87:292–296

Das SK, Mukherjee I (2012) Effect of moisture and organic manure on persistence of flubendiamide in soil. Bull Environ Contam Toxicol 88:515–520

Das SK, Mukherjee I (2014) Influence of microbial community on degradation of flubendiamide in two Indian soils. Environ Monit Assess 186:3213–3219

Das SK, Mukherjee I, Flubendiamide (2012a) Transport through packed soil columns. Bull Environ Contam Toxicol 88:229–233

Das SK, Mukherjee I, Das SK (2012b) Dissipation of flubendiamide in/on okra [Abelmoschus esculenta (L.) Moench] fruits. Bull Environ Contam Toxicol 88:381–384

Das SK, Avasthe RK, Gopi R (2014a) Vermiwash: use in organic agriculture for improved crop production. Pop Kheti 2:45–46

Das SK, Avasthe RK, Singh R, Babu S (2014b) Biochar as carbon negative in carbon credit under changing climate. Curr Sci 107:1090–1091

Das SK, Mukherjee I, Kumar A (2015) Effect of soil type and organic manure on adsorption–desorption of flubendiamide. Environ Monit Assess 187:403. https://doi.org/10.1007/s10661-015-4623-2

Gaskin JW, Steiner C, Harris K, Das KC, Bibens B (2008) Effect of low-temperature pyrolysis conditions on biochar for agricultural use. Trans Am Soc Agric Eng 51:2061–2069

Githinji L (2013) Effect of biochar application rate on soil physical and hydraulic properties of a sandy loam. Arch Agron Soil Sci 60:457–470

Gryze DS, Cullen M, Durschinger L, Lehmann J, Bluhm D, Six J (2010) Evaluation of opportunities for generating carbon offsets from soil sequestration of biochar. In: An issues paper commissioned by the Climate Action Reserve, final version. http://www.terraglobalcapital.com/press/Soil_Sequestration_Biochar_Issue_Paper1.pdf

Hammes K, Schmidt WI (2009) Changes of biochar in soil. In: Lehmann J, Joseph S (eds) Biochar for environmental management: science and technology. Earthscan, London, pp 169–182

Hass A, Gonzalez JM, Lima IM, Godwin HW, Halvorson JJ, Boyer DG (2012) Chicken manure biochar as liming and nutrient source for acid Appalachian soil. J Environ Qual 41:1096–1106

Jeffery S, Martijn BT, Cornelissen G, Kuyper TW, Lehmann J, Mommer L, Sohi SP, Van De Voorde TFJ, Wardle DA, van Groenigen JW (2015) The way forward in biochar research: targeting trade-offs between the potential wins. GCB Bioenergy, 7:1–13

Jefferym S, Verheijen FGA, van der Velde M, Bastos AC (2011) A quantitative review of the effects of biochar application to soils on crop productivity using meta-analysis. Agron Ecosyst Environ 144:175–187

Jones DL, Rousk J, Edwards-Jones G, DeLuca TH, Murphy DV (2012) Biochar-mediated changes in soil quality and plant growth in a three year field trial. Soil Biol Biochem 45:113–124

Keith A, Singh B, Singh BP (2011) Interactive priming of biochar and labile organic matter mineralization in a smectite-rich soil. Environ Sci Technol 45:9611–9618

Kimetu JM, Lehmann J (2010) Stability and stabilisation of biochar and green manure in soil with different organic carbon contents. Aust J Soil Res 48:577–585

Knowles OA, Robinson BH, Contangelo A, Clucas L (2011) Biochar for the mitigation of nitrate leaching from soil amended with biosolids. Sci Total Environ 409:3206–3210

Kookana RS (2010) The role of biochar in modifying the environmental fate, bioavailability, and efficacy of pesticides in soils: a review. Soil Res 48:627–637

Kour D, Rana KL, Yadav N, Yadav AN, Rastegari AA, Singh C, Negi P, Singh K, Saxena AK (2019) Technologies for biofuel production: current development, challenges, and future prospects. In: Rastegari AA et al (eds) Prospects of renewable bioprocessing in future energy systems, Biofuel and biorefinery technologies, vol 10. Springer, Cham, pp 1–50

Laird DA, Fleming P, Davis DD, Horton R, Wang BQ, Karlen DL (2010) Impact of biochar amendments on the quality of a typical midwestern agricultural soil. Geoderma 158:443–449

Lehmann J, Czimczik C, Laird D, Sohi S (2009) Stability of biochar in the soil. In: Lehmann J, Joseph S (eds) Biochar for environmental management. Earthscan, London, pp 183–206

Liu XY, Qu JJ, Li LQ, Zhang AF, Zheng JF, Zheng JW, Pan GX (2012) Can biochar amendment be an ecological engineering technology to depress N2O emission in rice paddies?—a cross site field experiment from south China. Ecol Eng 42:168–173

Major J, Steiner C, Downie A, Lehmann J (2009) Biochar effects on nutrient leaching. In: Lehmann J, Joseph S (eds) Biochar for environmental management. Science and Technology, Earthscan, London, pp 271–287

Mate CH, Mukherjee I, Das SK (2015) Persistence of spiromesifen in soil: influence of moisture, light, pH and organic amendment. Environ Monit Assess 187:7

Mohan D, Sarswat A, Ok YS, Pittman CU (2014) Organic and inorganic contaminants removal from water with biochar, a renewable, low cost and sustainable adsorbent – a critical review. Bioresour Technol 160:191–202

Nelissen V, Rütting T, Huygens D, Staelens J, Ruysschaert G, Boeckx P (2012) Maize biochars accelerate short-term soil nitrogen dynamics in a loamy sand soil. Soil Biol Biochem 55:20–27

Novak JM, Busscher WJ, Laird DL, Ahmedna M, Watts DW, Niandu MAS (2009) Impact of biochar amendment on fertility of a southeastern coastal plain soil. Soil Sci 174:105–112

Ogawa M, Okimori Y (2010) Pioneering works in biochar research, Japan. Aust Soil Res 48:489–500

Rajkovich S, Enders A, Hanley K, Hyland C, Zimmerman AR, Lehmann J (2012) Corn growth and nitrogen nutrition after additions of biochars with varying properties to a temperate soil. Biol Fertil Soils 48:271–284

Renner R (2007) Rethinking biochar. Environ Sci Technol 41:5932–5933

Rodríguez-Vila A, Covelo EF, Forján R, Asensio V (2014) Phytoremediating a copper mine soil with *Brassica juncea* L., compost and biochar. Environ Sci Pollut Res 21:11293–11304

Singh C, Tiwari S, Boudh S, Singh JS (2017a) Biochar application in management of paddy crop production and methane mitigation. In: Singh JS, Seneviratne G (eds) Agro-environmental sustainability: vol-2: managing environmental pollution. Springer, Cham, pp 123–146

Singh C, Tiwari S, Singh JS (2017b) Impact of rice husk biochar on nitrogen mineralization and methanotrophs community dynamics in paddy soil. Int J Pure Appl Biosci 5(5):428–435

Singh C, Tiwari S, Singh JS (2017c) Application of biochar in soil fertility and environmental management: a review. Bull Environ Pharmacol Life Sci 6(12):07–14

Singh C, Tiwari S, Gupta VK, Singh JS (2018) The effect of rice husk biochar on soil nutrient status, microbial biomass and paddy productivity of nutrient poor agriculture soils. Catena 171:485–493

Singh C, Tiwari S, Singh JS (2019) Biochar: a sustainable tool in soil 2 pollutant bioremediation. In: Bharagava RN, Saxena G (eds) Bioremediation of industrial waste for environmental safety. Springer, pp 475–494

Sohi S, Krull E, Lopez-Capel E, Bol R (2010) A review of biochar and its use and function in soil. Adv Agron 105:47–82

Steinbeiss S, Gleixner G, Antonietti M (2009) Effect of biochar amendment on soil carbon balance and soil microbial activity. Soil Biol Biochem 41:1301–1310
Tiwari S, Singh C, Singh JS (2018) Land use changes: a key ecological driver regulating methanotrophs abundance in upland soils. Energ Ecol Environ 3(6):355–371
Tiwari S, Singh C, Boudh S, Rai PK, Gupta VK, Singh JS (2019a) Land use change: a key ecological disturbance declines soil microbial biomass in dry tropical uplands. J Environ Manag 242:1–10
Tiwari S, Singh C, Singh JS (2019b) Wetlands: a major natural source responsible for methane emission. In: Upadhyay et al (eds) Restoration of wetland ecosystem: a trajectory towards a sustainable environment. Springer, Cham, pp 59–74

Chapter 10
Utilization of Agricultural Waste as Biochar for Soil Health

A. G. Rajalakshmi

Abstract The utilization of agricultural wastes are considered to be the important step in environmental protection, energy structure and agricultural development. The agricultural straw disposition of agricultural wastes not only results in environmental pollution, but also waste a lot of valuable biomass resources. Biochar the viable organic amendment product derived from organic sources and store carbon on a long term basis in the terrestrial ecosystem and also capable of reducing greenhouse gases (GHG) emission from soil to the atmosphere to combat climate change and sustain the soil health with sustainable crop production. The role of biochar in developing a sustainable agriculture production system is immense and so is its potential in mitigating climate change, which stands much beyond its uses in agriculture. The addition of biochar to soils resulted, on average, in increased above ground productivity, crop yield, soil microbial biomass, rhizobia nodulation, plant K tissue concentration, soil phosphorus (P), soil potassium (K), total soil nitrogen (N), and total soil carbon (C). The effects of biochar on multiple ecosystem functions and the central tendencies suggest that biochar holds promise in being a win-win-win solution to energy, carbon storage, and ecosystem function. However, biochar's impacts on a fourth component, the downstream non target environments, remain unknown and present a critical research gap.

Keywords Utilization · Biochar · Carbon sequestration · Soil health

10.1 Introduction

The global food system is estimated to contribute minimum one third anthropogenic emissions (Scialabba and Muller-Lindenlauf 2010). The rice and wheat system (RWS) is one of the widely practiced cropping systems in northern India. About 90–95% of the rice area is used under intensive rice wheat system (RWS) in Punjab (Gadde et al. 2009). Burning of straw emits emission of trace gases like CO_2, CH_4,

A. G. Rajalakshmi (✉)
Department of Biotechnology, SNMV College of Arts and Science, Coimbatore, India

J. S. Singh, C. Singh (eds.), *Biochar Applications in Agriculture and Environment Management*, https://doi.org/10.1007/978-3-030-40997-5_10

CO, N_2O, NOX, SO_2 and large amount of particulates which cause adverse impacts on human health. It is estimated that India annually emits 144,719 Mg of total particulate matter from open field burning of rice straw (Gadde et al. 2009).

Agriculture contributes between 10% and 25% of annual GHGs, both directly and indirectly, through land-use changes, land management, and production practices (Scialabba and Muller-Lindenlauf 2010). Agriculture is an important contributor to climate change, accounting directly for 10–12% of anthropogenic greenhouse gas (GHG) by land use change emissions (Hosonuma et al. 2012; IPCC 2014; Tubiello 2015). Agriculture influences global warming due to related direct and indirect GHG emissions from C and N dynamics. GHG emissions from soils a key topic in global change issues, climate research, and for agricultural and forestry management. SOC sequestration through improved crop and grassland management offers the possibility to sequester significant amounts of carbon in the soil, improving soil quality and productivity, and subsequently food security (Smith 2016).

Improved agriculture practices can reduce the amount of GHGs entering the atmosphere (Scialabba and Muller-Lindenlauf 2010; Smith et al. 2007), and carbon sequestration is considered a partial solution to short- and medium-term removal of atmospheric carbon (Hutchinson et al. 2007; Lal 2010; Morgan et al. 2010). Mitigation and adaptation differ in at least three ways including: (1) temporal and spatial scales at which the options are effective; (2) methods by which costs and benefits can be inventoried, estimated, and compared; and (3) stakeholders and governance drivers involved in their implementation (Klein et al. 2005; Tiwari et al. 2019a, b; Singh et al. 2019; Kour et al. 2019). Climate change adaptation for agriculture involves building resistance (the ability to resist the impact of a disturbance) and resilience (the ability to recover from disturbance) within agro-ecosystems, communities, and governance operations to prepare for climatic change and its impacts (Holt-Giménez 2002).

The Indian Ministry of New and Renewable Energy, estimated at ~500 million metric tons of biomass per year. The biomass residues used as animal feed, home thatching, and for domestic and industrial fuel, a large portion of unused crop residues are burned in the fields to clear the left-over straw and stubble after harvest, causing serious air pollution and producing CO_2 contributing to global warming. It also causes a huge loss of carbon feedstock which can be used to improve soil fertility (Fig. 10.1).

10.2 Agricultural Waste Management

Agricultural wastes production resulted in increased quantities of livestock waste, agricultural crop residues and agro-industrial by-products. It is estimated that about 998 million tonnes of agricultural waste is produced yearly (Agamuthu 2009) AWMS consist of six basic functions includes production, collection, storage, treatment, transfer, and utilization. Streets et al. (2003) reveal that 16% of total crop residues were burnt about 116 million tons of crop residues were burnt in India in

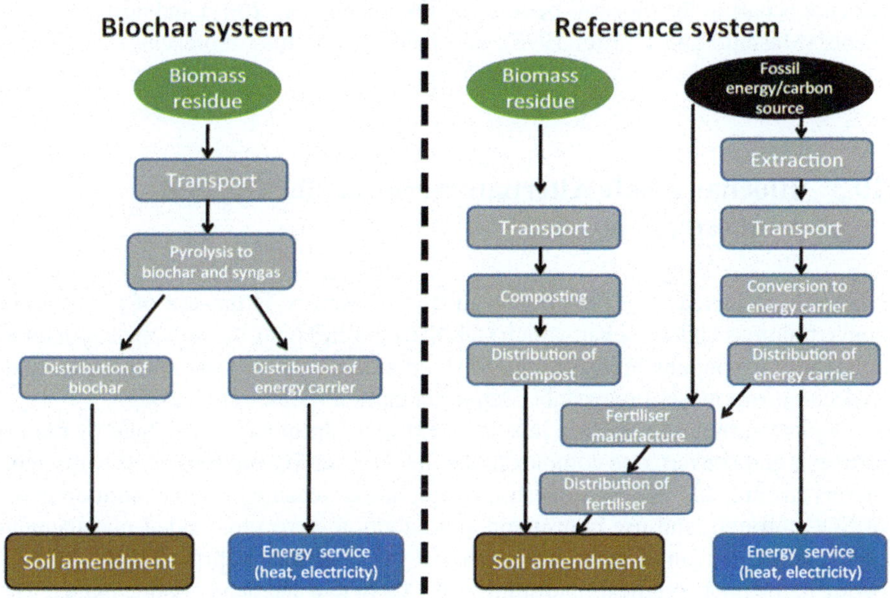

Fig. 10.1 Biochar as renewable energy source

2001, but with a strong regional variation (Venkataraman et al. 2006). The current availability of biomass in India (2010–2011) is estimated at about 500 million tons/year. Globally 78 ± 12 Gt C (this is equivalent to 29% of total CO2-C emission due to fossil fuel combustion of 270 ± 30 Gt.

Nguyen and Lehmann (2009) found that biochar formed during the conversion of undisturbed land to agricultural land by the burning of the natural vegetation, led to the formation of biochar and irrespective of its origin, the initial biochar content per unit soil mass decreased rapidly by 30% over a period of 30 years. Pyrolysis is incomplete combustion of biomass in oxygen limited condition, providing the slow cycling of organic carbon (Ameloot et al. 2013). The effective utilization of agricultural waste is a good option to convert these wastes in energy. Production of energy from biomass can provide farmers with new prospects and possibilities to diversify agricultural activities. Some of the crops may compete for land and other resources with traditional crops, while other crops may be grown on marginal lands or even ecologically degraded areas and thus have a positive effect on the environment.

Soil C sequestration implies increasing the concentration pools of SOC through land-use conversion and adoption of recommended management practices (RMPs) in agriculture. Application of manure and other organic amendments is another important SOC sequestration strategy (Anderson et al. 2011). Terrestrial ecosystems comprise a major C sink owing to the photosynthesis and storage of CO_2 in live and dead organic matter. Terrestrial C sequestration is often termed as a win-win strategy (Lal 2004) because of its numerous ancillary benefits. The quantity of carbon contained in soils is directly related to the diversity and health of soil life.

Biochar is part of the oldest C pool in soil (Pessenda et al. 2001) and deep-sea sediments (Masiello and Druffel 1998), and that black C may represent a significant global sink of C (Schmidt and Noack 2000)

10.3 Biochar a Safe Alternative Sources for Carbon Sequestration

Diminishing increased levels of CO_2 in the atmosphere is the use of pyrolysis to convert biomass into biochar, which stabilizes the carbon (C) that is then applied to soil. Biochar contains high concentrations of carbon that can be rather recalcitrant to decomposition, so it may stably sequester carbon (Glaser et al. 2002).

Organic carbon sequestered in soils is extracted from the atmosphere by photosynthesis and converted to complex molecules by bacteria and fungi in synergy with insects and animals. Soil C sequestration implies increasing the concentration pools of SOC through land-use conversion and adoption of recommended management practices (RMPs) in agriculture. Terrestrial ecosystems comprise a major C sink owing to the photosynthesis and storage of CO_2 in live and dead organic matter. The quantity of carbon contained in soils is directly related to the diversity and health of soil life. Formation of charcoal and use of biochar as a fertilizer is another option (Singh et al. 2017a, b, c, 2018; Tiwari et al. 2018; Fowles 2007) for carbon sequestration. Biochar is part of the oldest C pool in soil (Pessenda et al. 2001) and deep-sea sediments (Masiello and Druffel 1998), and that black C may represent a significant global sink of C (Schmidt and Noack 2000) (Fig. 1.2).

Biochar a new era with innovation and technological solution to reduce CO_2 emission and acts as a sequester almost 400 billion tonnes of carbon by 2100 and to lower atmospheric CO_2 concentrations by 37 parts per million (Tim Lenton 2009). Biochar needs two essential qualities to meet profitable agriculture: adoption of a carbon market and the market price for biochar must be low enough to make farmer friendly (Galinato et al. 2011). Apart from all the environmental stresses biochar exhibits a long mean residence times in soil, ranging from 1000 to 10,000 years, with 5000 years (Krull and Lyons 2009), this susceptible factor is mainly due to the complex chemical structure, aromatic nature, and graphitic C (Glaser et al. 2002). It is estimated that use of this method to "tie up" carbon has the potential to reduce current global carbon emissions by 10%.

10.4 Effect of Biochar on Soil Amendment

The char an energy source and as a soil amendment called biochar (Glaser et al. 2001) resist physical and microbial breakdown, allowing it to persist in soil due to the presence of crystalline morphology, the proportion of which may change with pyrolysis temperature (Cao and Harris 2010). The cations in the biochar after pyrolysis transformed into oxides, hydroxides, and carbonates (ash) acts as a liming agent when applied to soil. Application of biochar to soils contribute to carbon storage but at the same time act as fertilizers (Glaser et al. 2001; Marris 2006). It has been observed in several studies that biochar addition to soils improves soil fertility and thus increased crop yields on agricultural lands (Marris 2006; Chan et al. 2007). The possible improvements of soil's properties and fertility after biochar application (Fig. 1.3) due to the high surface area, amount of functional groups, and the content of liming. The well-developed pore structure enhances the capacity of water retention, shelter for soil's microorganisms, thus nutrient retention and cycling could be improved. The content of liming contained in biochar may increase soil's pH values.

10.5 Application of Biochar

The effect of biochar amendment on soil nutrient content, charcoal amendments have a positive effect on nutrient retention, particularly in highly weathered soils with low ion-retention capacities (Glaser et al. 2002). Biochar application elevates total C, organic C, total N, available P, and exchangeable cations like Ca, Mg, Na, and K increase, and Al decreases in soil (Chan et al. 2007; Major et al. 2010) the plant uptakes several of these nutrients after biochar application (Chan et al. 2007; Major et al. 2010) (Fig. 1.4).

Biochar is a low density material that reduces soil bulk density (Laird et al. 2010) and thereby increases water infiltration, root penetration, and soil aeration, increase soil aggregate stability (Glaser et al. 2002). Soil enriched with biochar improves soil fertility and to mitigate climate change by reducing emissions of greenhouse gases from cultivated soils (Yanai et al. 2007). Application of biochar in soil may provide a novel soil management practice because of its potential to improve soil fertility, enhance soil carbon, mitigate soil greenhouse gas emissions and enhance agricultural productivity (Fig. 10.2).

10.5.1 Biochar in Crop Production

Biochar improves plant growth and enhances crop yields, increasing food production and sustainability in areas with depleted soils, limited organic resources, insufficient water, access to agrochemical fertilizers. Beneficial effects of biochar with increased crop yield and improved soil quality (Glaser et al. 2002). Steiner et al. (2008) measured both higher soil N retention and an enhanced N cycling, biochar stimulation of crop yield mostly related to its higher stability in comparison to other organic amendments as well as the native soil organic matter (SOM) (Steiner et al. 2007). Biochar can capture high amounts of exchange cations (Lehmann et al.

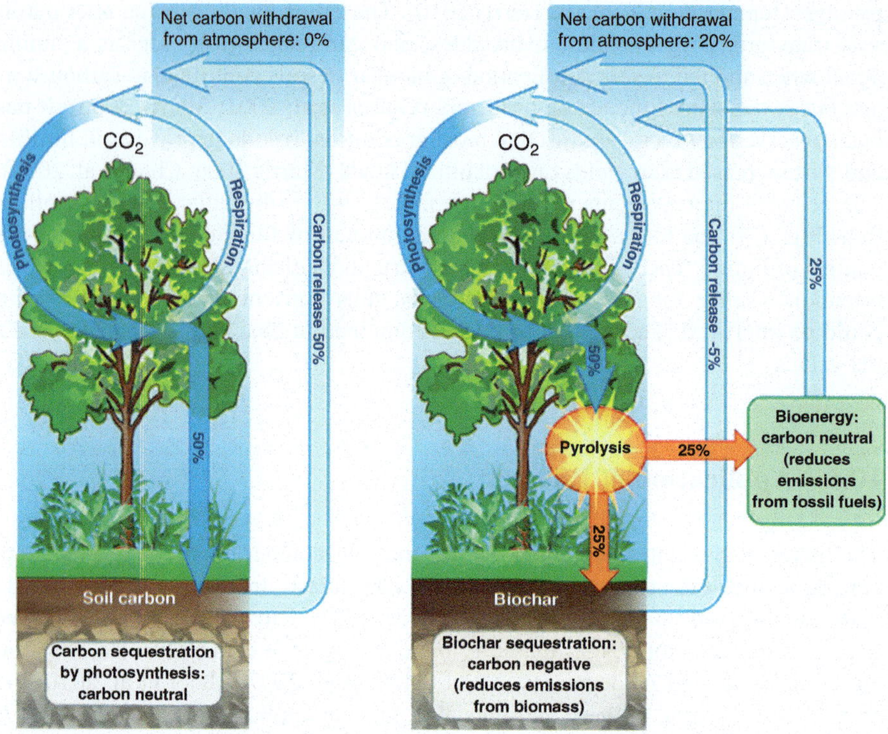

Fig. 10.2 Benefits of biochar application (Source Lehmann and Joseph 2009)

2003) because of its high porosity and surface/volume ratio and can improve plant nutrients uptake and P, Ca, K availability (Chan et al. 2007). In soils biochar is slowly oxidized, carboxylic groups are produced, cation exchange capacity and oxygen carbon ratio on the biochar surface increases (Brodowski et al. 2006), improving the capacity of biochar to retain nutrients in the long term Table 10.1.

Biochar application increases soil organic carbon levels (McHenry 2009) and improves soil structure (Glaser et al. 2002). Its application improves the soil's ability to retain moisture (Laird et al. 2010; Steiner et al. 2007), prevents nutrient leaching (Taghizadeh-Toosi et al. 2011; Spokas et al. 2011) and increases cation exchange capacity (Clough and Condron 2010; Yuan and Xu 2011). Application of biochar reduces aluminum toxicity (Van Zwieten et al. 2010) and bioavailability of heavy metals (Méndez et al. 2012), increases soil pH (Yuan and Xu 2011; Deal et al. 2012), supplies essential plant nutrients and decreases the need for chemical fertilizers (Bird et al. 2011). Biochar improves the biological condition of soils (Kwapinski et al. 2010), increases soil microbial biomass and supports beneficial organisms like earthworms. The conversion of biomass to biochar reduces greenhouse gas emissions (Wang et al. 2011) and helps in sequestering atmospheric carbon in to the soil (Bolan et al. 2012).

Table 10.1 Effect of biochar on the crop yields

Study	Results	References
Comparison of maize yields between disused charcoals production sites and adjacent fields Kotokosu watershed, Ghana	Grain yield 91% higher and biomass yield 44% on charcoal site than control	Oguntunde et al. (2004))
Soyabean on volcanic ash loam, Japan	0.5 mg ha^{-1} char increased yield 151%	Chidumayo (1994),
	5 mg ha^{-1} char decreased yield to 63%	
	15 mg ha^{-1} char decreased yield to 9%.	
Bauhinia trees on alfisol/ultisol	Charcoal increased biomass by 13% and height by 24%.	Kishimoto and Sugiura (1985)
Cowpea on xanthic ferrasal	67 mg ha^{-1} char increased biomass 150%	Glaser et al. (Glaser et al. 2002)
	135 mg ha^{-1} char increased biomass 200%.	
Pea, India	0.5 mg ha^{-1} char increased biomass 160%.	Iswaran et al. (1980)
Mung bean, India	0.5 mg ha^{-1} char increased biomass 122%.	Iswaran et al. (1980)

10.5.2 Impact of Biochar on Nitrogen Fixation

Nitrogen is a plant macronutrient essential to the survival of ecosystems. Yet, in most cases the amount of N available to plants is low (Robertson and Groffman 2007), which limits the gross primary productivity (GPP) of the site (Gundale et al. 2011).

Biochar is a promising fertilizer reduces N losses from the soil and alters the nitrogen (N) dynamics in soils (Robertson 2012). The anthropogenically induced global N cascade is resulting in enhanced fluxes of nitrous oxide (N_2O), ammonia (NH_3), and nitrate (NO_3) leaching as a consequence of the increasing intensification of agricultural systems (Galloway et al. 2008). Biocharcontain bioavailable N forms, its mineralization soil N and C pools on the soil in ecosystems. Biochar forms on immobilisation and mineralization determines the of relatively short duration or more long-term. Besides the release of N intrinsically embodied in the biochar (Wang et al. 2012; Schouten et al. 2012) there have been attempts to further enhance the delivery of N using biochar by adding nutrients to the biochar prior to soil incorporation.

Beneficial agricultural management tool, the most promising prospects for biochar, to date, are: (1) the reduction of NH3 volatilisation via adsorption processes (urine patches, animal housing filters, composting); (2) the development of slow release N fertilisers; and (3) the reduction of N_2O emissions using fresh biochar additions to soils. However, even these areas require further research since the use of biochar as a mitigation tool demands a deeper mechanistic understanding and at the same time an increase in our ability to predict net effects over time.

Microbial immobilization (Ippolito et al. 2012), biochar, especially when pyrolyzed at low temperatures, usually contains considerable amounts of labile carbon (Nelissen et al. 2012). This carbon can serve as a microbial substrate, resulting in microbial demand for inorganic N, which thereby immobilizes the N through biotic processes (Nelissen et al. 2012). Ammonia volatilization is another mechanism that accounts for the loss of fertilizer derived N.

Biochar a safe alternative reduce or eliminate the need for commercial fertilizers. Fertilizer in rainwater runoff can damage river systems and the surface application of commercial fertilizers can be eroded by wind and rainfall which may mix with water and leads to toxicity. Incorporation of biochar into soil, reducing carbon stocks could be replenished and long-term storage of carbon can be increased. According to a CSIRO report, biochar has the potential to remove 1 billion tons of carbon from the atmosphere per year. The interaction between biochar and other organic amendments in soil should now be the focus of future research. This is a simplistic low cost means of adding nutrients to soil and helping agriculture flourish. Environmental protection and human health will be the leading benefactors in large scale biochar production.

10.5.3 Biochar Interaction with Soil Rhizosphere

Agricultural intensification transfer carbon (C) to the atmosphere in the form of carbon dioxide (CO_2), thereby reducing ecosystem C pools. Agriculture contributes 10–12% of the total global anthropogenic greenhouse gas emissions. Diminishing increased levels of CO_2 in the atmosphere by pyrolysis to convert biomass into biochar, which stabilizes the carbon (C) that is then applied to soil. Biochar with high concentrations of carbon that can be rather recalcitrant to decomposition, it stably sequester carbon. The immediate beneficial effects of biochar additions for nutrient availability are largely due to higher potassium, phosphorus, and zinc availability, and to a lesser extent, calcium and copper. The presence of biochar in the soil can improve soil chemical (e.g. pH, CEC), and physical properties (e.g. soil water retention, hydraulic conductivity). Acting as a habitat and substrate for soil microorganisms, biochar added in the soil can increase microbial activities (Pietikäinen and Fritze 2000).

Biochar addition to soil increases in root colonization of AMF. Arbuscular mycorrhizal (AM) fungi are symbiotic soil organisms. AM fungi play role in vegetative succession of ecosystem, plant diversification and productivity, as well as restoration and re-establishment of degraded ecosystems. Arbuscular mycorrhizal fungi (AMF) increases plant nutrition. The increase in the availability of major plant nutrients due to application of biochar and mycorrizhae, the plants form mycorrhizal symbioses with specialized soil fungi (Fig. 1.5). The combination of biochar, mycorrhizal fungi approaches the goal of a viable soil environment for sustainable plant growth. It has often been observed that application of organic biochar amendments results in a higher level of C sequestration when compared to other management strategies including fertilizer application and conservation tillage. The opportunities for carbon sequestration and the reduction of greenhouse gas emissions have not been explored at all, but they are potentially significant (Fig. 10.3).

10.5.4 Influence of Biochar on Soil Biota

Biochar increase microbial biomass in soil across ecosystems, ranging from boreal forests (Wardle et al. 2008; Zackrisson et al. 1996) to Amazonian uplands (Liang et al. 2010; Steiner et al. 2007). The increase in microbial biomass may be caused by improved soil habitability or retention of microbes in the soil via adsorption to biochar (Thies and Rillig 2009). Biochar with organic fertilizers have been reported to significantly improve soil tilth, crop productivity, and the availability of nutrients to plants. Improved crop response as a result of biochar amendment can be attributed to its nutrient content including the neutralization of phytotoxic compounds in the soil, the promotion of mycorrhizal fungi, and the alteration of soil microbial populations and functions (Steiner et al. 2008). Biochar mediated microbial com-

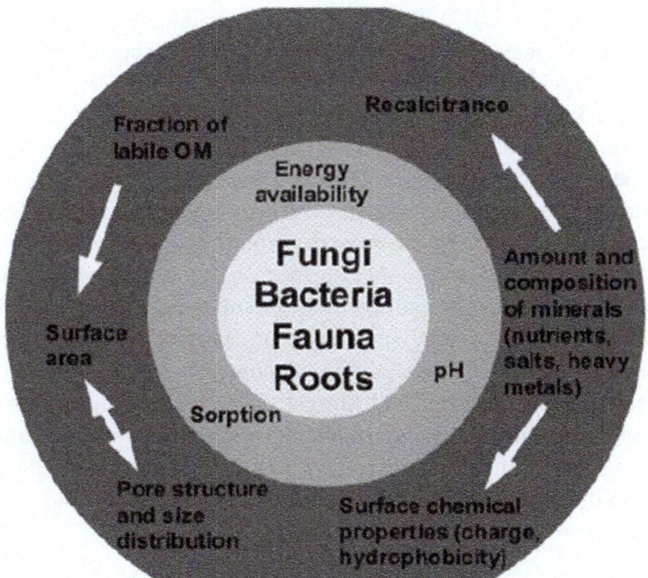

Fig. 10.3 Connection between primary biochar properties (outer circle), the soil process they may influence (intermediate circle) and the soil biota (inner circle), white arrows indicate the influence between biochar properties

munity compositional changes deserve further research, as these alterations can have essential implications for carbon sequestration and biogeochemical cycling. Biochar can promote a beneficial, self-sustaining soil biota, which also discourages plant-antagonistic organisms and pathogens. Soil biological activity, including the quantity, diversity, and activity level of soil microbes, affects soil productivity for crops. One study determined that biochar supported more microbial activity than pumice or activated-charcoal biochar, due in part to a higher water-holding capacity (Pietikäinen and Fritze 2000).

10.6 Outcome and Future Research Direction of Biochar

Soil is an essential component of the terrestrial ecosystem and has an important ecological function in biogeochemical cycling of resources needed for plant growth. An individual plant depends on soil for anchorage, water, oxygen and nutrients (Plaster 2009).

Contaminated soils and sediments are a significant worldwide environmental problem (The Norwegian Environment Agency 2017). Contaminated sites pose an environmental and human health hazard via ingestion and/or inhalation of contaminated dust or soil particles, consumption of crops produced on these sites as well as skin contact (Janus et al. 2015). Biochar as the sorbent for remediation of polluted soils is rapidly gaining popularity (Denyes et al. 2013). Sorbent amendment added to soils can alter the geochemistry of the soil, increase contaminant binding, reduce contaminant exposure risks to people and the environment as well as limit bioremediation (Cornelissen et al. 2005)

Biochar is a kind of insoluble, stable, highly aromatic and carbon-rich solid material produced by abandoned biomass under the condition of hypoxia and high temperature slow pyrolysis (usually The exchange adsorption of biochar surface is one of important reasons for the reduction of heavy metal activities. The bigger the number of cation exchange, the stronger the retention of heavy metals.

The larger surface area and higher surface energy are helpful for biocharstostrongly absorb the heavy metal pollutants and remove them from the soil (Fig. 1.7). The removal mechanism of heavy metals by biochars. Biochar is emerging tool to optimize the reduction of bioavailability of contaminants in the environment by making benefits to soil fertility and mitigating climate change (Sohi 2012). The inorganic contaminants in the environment (metals) derive from anthropogenic sources (Zhang et al. 2013).contains a fraction not carbonized, which could interact with soil contaminants and water the surface of the biochar could retain the contaminants (Uchimiya et al. 2010).

10.7 Conclusions

The biochar research has progressed considerably with important key findings on agronomic benefits, carbon sequestration, greenhouse gas emissions, soil fertility and health, removal of hazardous pollutants. Long-term field research focusing on an optimal combination of nutrient use, water use, carbon sequestration, avoided greenhouse gas emissions, and changes in soil quality, crop productivity and to promote the application of biochar as a soil amendment and also as a climate change abatement option, research, development and demonstration on biochar production and application is very vital. It is necessary to develop low-cost biochar kilns to make the technology affordable to small and marginal farmers. Efficient use of biomass by converting it as a useful source of soil amendment/nutrients is one way to manage soil health and fertility. The interaction between biochar and other organic amendments in soil should now be the focus of future research. This is a simplistic low cost means of adding nutrients to soil and helping agriculture flourish. Environmental protection and human health will be the leading benefactors in large scale biochar production.

Acknowledgement The author grateful to the authorities of Shri Nehru MahaVidyalaya college of Arts and Science college, Coimbatore, Tamil Nadu, India for the use of research facilities and encouragement.

References

Agamuthu P (2009) Challenges and opportunities in Agrowaste management: an Asian perspective. Inaugural meeting of First Regional 3R Forum in Asia 11–12 Nov., Tokyo, Japan

Ameloot N, Neve SD, Jegajeevagan K, Yildiz G, Buchan D, Funkuin YN, Prins W, Bouckaert L, Sleutel S (2013) Short-term CO2 and N2O emissions and microbial properties of biochar amended sandy loam soils. Soil Biol Biochem 57:401–410

Anderson CR, Condron LM, Clough TJ, Fiers M, Stewart A, Hill RA, Sherlock RR (2011) Biochar induced soil microbial community change: implications for biogeochemical cycling of carbon, nitrogen and phosphorus. Pedobiologia 54:309–320

Bird MI, Wurster CM, de Paula Silva PH, Bass AM, de Nys R (2011) Algal biochar – production and properties. Bioresour Technol 102(2):1886–1891

Bolan NS, Kunhikrishnan A, Choppala GK, Thangarajan R, Chung JW (2012) Stabilization of carbon in composts and biochars in relation to carbon sequestration and soil fertility. Sci Total Environ 424:264–270

Brodowski S, John B, Flessa H, Amelung W (2006) Aggregate-occluded black carbon in soil. Eur J Soil Sci 57:539–546

Cao X, Harris W (2010) Properties of dairy-manure-derived biochar pertinent to its potential use in remediation. Bioresour Technol 101(14):5222–5228

Chan KY, Van Zwieten L, Meszaros I, Downie A, Joseph S (2007) Agronomic values of green waste biochar as a soil amendment. Aust J Soil Res 45:629–634

Chidumayo EN (1994) Effects of wood carbonization on soil and initial development of seedlings in miombo woodland, Zambia. For Ecol Manag 70:353–357

Clough TJ, Condron LM (2010) Biochar and the nitrogen cycle: introduction. J Environ Qual 39(4):1218–1223
Cornelissen G, Gustafsson Ö, Bucheli TD, Jonker MTO, Koelmans AA, Van Noort PCM (2005) Extensive sorption of organic compounds to black carbon, coal, and kerogen in sediments and soils: mechanisms and consequences for distribution, bioaccumulation, and biodegradation. Environ Sci Technol 39:6881–6895
Deal C, Brewer CE, Brown RC, Okure MAE, Amoding A (2012) Comparison of kiln-derived and gasifier-derived biochars as soil amendments in the humid tropics. Biomass Bioenergy 37:161–168
Denyes MJ, Rutter A, Zeeb BA (2013) In situ application of activated carbon and biochar to PCB-contaminated soil and the effects of mixing regime. Environ Pollut 182:201–208
Fowles M (2007) Black carbon sequestration as an alternative to bioenergy. Biomass Bioenergy 31:426–432
Gadde B, Menke C, Wassmann R (2009) Rice straw as a renewable energy source in India, Thailand and the Philippines, overall potential and limitations for energy contribution and greenhouse gas mitigation. Biomass Bioenergy 33:1532–1546
Galinato SP, Yoder JK, Granatstein D (2011) The economic value of biochar in crop production and carbon sequestration. Energy Policy 39(10):6344–6350
Galloway JN, Townsend AR, Erisman JW, Bekunda M, Cai Z, Freney JR, Martinelli LA, Seitzinger SP, Sutton MA (2008) Transformation of the nitrogen cycle: recent trends, questions, and potential solutions. Science 320:889–892
Glaser B, Haumaier L, Guggenberger G, Zech W (2001) The "Terra Preta" phenomenon: a model for sustainable agriculture in the humid tropics. Naturwissenschaften 88:37–41
Glaser B, Lehmann J, Zech W (2002) Ameliorating physical and chemical properties of highly weathered soils in the tropics with charcoal -a review. Biol Fertil Soils 35(4):219–230
Gundale MJ, Deluca TH, Nordin A (2011) Bryophytes attenuate anthropogenic nitrogen inputs in boreal forests. Glob Chang Biol 17(8):2743–2753
Holt-Giménez E (2002) Measuring farmers' agroecological resistance after Hurricane Mitch in Nicaragua: a case study in participatory, sustainable land management impact monitoring. Agric Ecosyst Environ 93(1–3):87–105
Hosonuma N, Herold M, De Sy V, De Fries RS, Brockhaus M, Verchot L, Angelsen A, Romijn E (2012) An assessment of deforestation and forest degradation drivers in developing countries. Environ Res Lett 7(4):044009
Hutchinson JJ, Campbell CA, Desjardins RL (2007) Some perspectives on carbon sequestration in agriculture. Agric Forest Meteorol 142(2–4):288–302
IPCC (2014) Climate change 2014: mitigation of climate change contribution of working group III to the fifth assessment report of the intergovernmental panel on climate change. Cambridge University Press, Cambridge/New York
Ippolito JA, Novak JM, Busscher WJ, Ahmedna M, Rehrah D, Watts DW (2012) Switch grass biochar affects two aridisols. J Environ Qual 41:1123–1130
Iswaran V, Jauhri KS, Sen A (1980) Effect of charcoal, coal and peat on the yield of moong, soybean and pea. Soil Biol Biochem 12:191–192
Janus A, Elie A, Ene P, Heymans S, Deboffe C, Douay F, Waterlot C (2015) Elaboration, characteristics and advantages of biochars for the management of contaminated soils with a specific overview on Miscanthusbiochars. J Environ Manag 162:275–289
Kishimoto S, Sugiura G (1985) Charcoal as a soil conditioner. Int Achiev Future 5:12–23
Klein R, Schipper L, Dessai S (2005) Integrating mitigation and adaptation into climate and development policy: three research questions. Environ Sci Pol 8(6):579–588
Kour D, Rana KL, Yadav N, Yadav AN, Rastegari AA, Singh C, Negi P, Singh K, Saxena AK (2019) Technologies for biofuel production: current development, challenges, and future prospects. In: Rastegari AA et al (eds) Prospects of renewable bioprocessing in future energy systems, Biofuel and biorefinery technologies, vol 10. Springer, Cham, pp 1–50
Krull E, Lyons S (2009) What is biochar? ABC Science, viewed 26th April 2010

Kwapinski W, Byrne CMP, Kryachko E, Wolfram P, Adley C, Leahy JJ, Novotny EH, Hayes MHB (2010) Biochar from biomass and waste. Waste Biomass Valoriz 1:177–189
Laird DA, Fleming PD, Davis DD, Horton R, Wang B, Karlen DL (2010) Impact of biochar amendments on the quality of a typical Midwestern agricultural soil. Geoderma 158:443–449
Lal R (2004) Soil carbon sequestration impact on global climate change and good security. Science 304:1623–1627
Lal R (2010) Managing soils and ecosystems for mitigating anthropogenic carbon emissions and advancing global food security. BioScience 60:708–721
Lehmann J, Joseph S (2009) Biochar for environmental management: science and technology. Earthscan, London
Lehmann J, Kern DC, German LA, McCann J, Martins GC, Moreira A (2003) Soil fertility and production potential. Amazonian dark earths: origin, properties, management. Kluwer Academic Publishers, Dordrecht, pp 105–124
Liang B, Lehmann J, Sohi SP et al (2010) Black carbon affects the cycling of non-black carbon in soil. Org Geochem 41:206–213
Major J, Lehmann J, Rondon M, Goodale C (2010) Fate of soil-applied black carbon: downward migration, leaching and soil respiration. Glob Chang Biol 16:1366–1379
Marris E (2006) Putting the carbon back: black is the new green. Nature 442:624–626
Masiello CA, Druffel ERM (1998) Black carbon in deep-sea sediments. Science 280:1911–1913
McHenry MP (2009) Agricultural bio-char production, renewable energy generation and farm carbon sequestration in Western Australia: certainty, uncertainty and risk. Agric Ecosyst Environ 129(1–3):1–7
Méndez A, Gomez A, Paz-Ferreiro J, Gasco G (2012) Effects of sewage sludge biochar on plant metal availability after application to a Mediterranean soil. Chemosphere 89:1354–1359
Morgan JA, Follett RF, Allen LH, Del Grosso S, Derner JD, Dijkstra F, Schoeneberger MM (2010) Carbon sequestration in agricultural lands of the United States. J Soil Water Conserv 65(1):6A–13A
Nelissen V, Rütting T, Huygen D, Staelens J, Ruysschaerta G, Boeckx P (2012) Maize biochars accelerate short-term soil nitrogen dynamics in a loamy sand soil. Soil Biol Biochem 55:20–27
Nguyen BT, Lehmann J (2009) Black carbon decomposition under varying water regimes. Org Geochem 40:846–853
Oguntunde P, Fosu M, Ajayi A, Giesen N (2004) Effect of charcoal production on maize yield, chemical properties and texture of soil. Biol Fertil Soils 39:295–299
Pessenda LCR, Gouveia SEM, Aravena R (2001) Radiocarbon dating of total soil organic matter and humin fraction and its comparison with C-14 ages of fossil charcoal. Radiocarbon 43:595–601
Pietikäinen J, Fritze H (2000) Charcoal as a habitat for microbes and its effect on the microbial community of the underlying humus. Oikos 89:231–242
Plaster EJ (2009) Soil science and management, 5th edn. Delmar Cengage learning
Robertson SJ (2012) Biochar enhances seedling growth and alters root symbioses and properties of sub-boreal forest soils. Can J Soil Sci 92(2):329–340
Robertson GP, Groffman PM (2007) Nitrogen transformations. In: Paul EA (ed) Soil microbiology, ecology, and biochemistry. Elsevier Inc, p 532
Schmidt MWI, Noack AG (2000) Black carbon in soils and sediments: analysis, distribution, implications, and current challenges. Glob Biogeochem Cycles 14:777–793
Schouten S, Van Groenigen JW, Oenema O, Cayuela ML (2012) Bioenergy from cattle manure? Implications of anaerobic digestion and subsequent pyrolysis for carbon and nitrogen dynamics in soil. Glob Chang Biol Bioenerg 4:751–760
Scialabba N, Muller-Lindenlauf M (2010) Organic agriculture and climate change. Renew Agric Food Syst 25(2):158–169
Singh C, Tiwari S, Boudh S, Singh JS (2017a) Biochar application in management of paddy crop production and methane mitigation. In: Singh JS, Seneviratne G (eds) Agro-environmental sustainability: Vol-2: managing environmental pollution. Springer, Cham, pp 123–146

Singh C, Tiwari S, Singh JS (2017b) Impact of rice husk biochar on nitrogen mineralization and methanotrophs community dynamics in paddy soil. Int J Pure Appl Biosci 5(5):428–435

Singh C, Tiwari S, Singh JS (2017c) Application of biochar in soil fertility and environmental management: a review. Bull Environ Pharmacol Life Sci 6(12):07–14

Singh C, Tiwari S, Gupta VK, Singh JS (2018) The effect of rice husk biochar on soil nutrient status, microbial biomass and paddy productivity of nutrient poor agriculture soils. Catena 171:485–493

Singh C, Tiwari S, Singh JS (2019) Biochar: a sustainable tool in soil 2 pollutant bioremediation. In: Bharagava RN, Saxena G (eds) Bioremediation of industrial waste for environmental safety. Springer, Cham, pp 475–494

Smith P (2016) Soil carbon sequestration and biochar as negative emission technologies. Glob Chang Biol 22:1315–1324

Smith P, Martino D, Cai Z, Gwary D, Janzen H, Kumar P, McCarl B, Sirotenko OP (2007) Agriculture: mitigation, contribution of working group III to the fourth assessment report of the intergovernmental panel on climate change. Cambridge University Press, Cambridge/New York, pp 497–540

Sohi SP (2012) Carbon storage with benefits. Science 338:1034–1035

Spokas KA, Novak JM, Venterea RT (2011) Biochar's role as an alternative N-fertilizer: ammonia capture NH3. Plant Soil 350:35–42

Steiner C, Teixeira WG, Lehmann J, Nehls T, de Macedo JLV, Blum WEH, Zech W (2007) Long term effects of manure, charcoal and mineral fertilization on crop production and fertility on a highly weathered central Amazonian upland soil. Plant Soil 291(1–2):275–290

Steiner C, Glaser B, Teixeira WG, Lehmann J, Blum WEH, Zech W (2008) Nitrogen retention and plant uptake on a highly weathered central Amazonian Ferralsolamended with compost and charcoal. J Plant Nutr Soil Sci 171(6):893–899

Streets D, Yarber K, Woo J, Carmichael G (2003) Biomass burning in Asia: annual and seasonal estimates and atmospheric emissions. Glob Biogeochem Cycles 17(4):1099

Taghizadeh-Toosi A, Clough TJ, Sherlock RR, Condron LM (2011) Biochar adsorbed ammonia is bioavailable. Plant Soil 350:57–69

The Norwegian Environment Agency (2017) Environment.no. At http://www.environment.no/topics/hazardous-chemicals/contaminated-soil. Accessed Sept 2017

Thies JE, Rillig MC (2009) Characteristics of biochar: biological properties. In: Lehmann J, Joseph S (eds) Biochar for environmental management: science and technology. Earthscan, London

Tim Lenton (2009). http://climatechangepsychology.blogspot.com/2009/03/chris-goodalljohanneslehmanntim.htm

Tiwari S, Singh C, Singh JS (2018) Land use changes: a key ecological driver regulating methanotrophs abundance in upland soils. Energ Ecol Environ 3(6):355–371

Tiwari S, Singh C, Boudh S, Rai PK, Gupta VK, Singh JS (2019a) Land use change: a key ecological disturbance declines soil microbial biomass in dry tropical uplands. J Environ Manag 242:1–10

Tiwari S, Singh C, Singh JS (2019b) Wetlands: a major natural source responsible for methane emission. In: Upadhyay et al (eds) Restoration of wetland ecosystem: a trajectory towards a sustainable environment. Springer, Cham, pp 59–74

Tubiello FN (2015) The contribution of agriculture forestry and other land use activities to global warming, 1990–2012. Glob Chang Biol 21:2655–2660

Uchimiya M, Wartelle LH, Lima IM, Klasson KT (2010) Sorption of deisopropylatrazine on broiler litter biochars. J Agric Food Chem 58:12350–12356

Van Zwieten L, Kimber S, Morris S, Chan KY, Downie A, Rust J, Joseph S, Cowie A (2010) Effects of biochar from slow pyrolysis of paper mill waste on agronomic performance and soil fertility. Plant Soil 327:235–246

Venkataraman C, Habib G, Kadamba D, Shrivastava M, Leon J, Crouzille B, Boucher O, Streets D (2006) Emissions from open biomass burning in India: integrating the inventory approach with high-resolution moderate resolution imaging Spectroradiometer (MODIS) active-fire and land cover data. Glob Biogeochem Cycles 20(2):GB2013

Wang J, Zhang M, Xiong Z, Liu P, Pan G (2011) Effects of biochar addition on N2O and CO2 emissions from two paddy soils. Biol Fertil Soils 47:887–896

Wang T, Camps Arbestain M, Hedley M, Bishop P (2012) Chemical and bioassay characterisation of nitrogen availability in biochar produced from dairy manure and biosolids. Org Geochem 51:45–54

Wardle DA, Nilsson MC, Zackrisson O (2008) Fire-derived charcoal causes loss of forest humus. Science 320:629–630

Yanai Y, Toyota K, Okazaki M (2007) Effect of charcoal addition on N_2O emissions from soil resulting from rewetting air-dried soil in short-term laboratory experiments. Soil Sci Plant Nutr 53:181–188

Yuan JH, Xu RK (2011) The amelioration effects of low temperature biochar generated from nine crop residues on an acidic Ultisol. Soil Use Manag 27:110–115

Zackrisson O, Nilsson MC, Wardle DA (1996) Key ecological function of charcoal from wildfire in the boreal forest. Oikos 77:10–19

Zhang X, Wang H, He L, Lu K, Sarmah A, Li J, Bolan NS, Pei J, Huang H (2013) Using biochar for remediation of soils contaminated with heavy 44 metals and organic pollutants. Environ Sci Pollut Res 20:8472–8483

Chapter 11
Biochar: A New Environmental Paradigm in Management of Agricultural Soils and Mitigation of GHG Emission

Palakshi Borah, Nijara Baruah, Lina Gogoi, Bikram Borkotoki, Nirmali Gogoi, and Rupam Kataki

Abstract Biochar, a co-product of the pyrolytic conversion of biomass and bio-wastes to biofuel is a carbon rich recalcitrant material. It has received much attention in the recent times for its prospective application in various fields viz. as a soil amendment for improving the physical, chemical, and biological qualities of agricultural soils, as an adsorbent for removal of various organic and inorganic contaminants in water, for removal of pesticides residues in soil, for correcting soil acidity, as a precursor for chemical synthesis, for industrial applications such as supercapacitor application, as a support material for fuel cells, for enhancement in biogas generation to name a few. In addition to all these, biochar's green-house gas mitigation potential, and C-sequestration potential are two most significant attributes that has made biochar a suitable component for SDGs. Further, these applications have made biochar as one of the most researched topics in recent times. The ease of biochar production is also another advantage which can be beneficial for farmers even with a marginal land holding. In this chapter, an attempt has been made to discuss the role of biochar in management of agricultural soils, as well as its vast environmental application possibilities.

Keywords Biochar · Soil amendment · Environmental management

P. Borah · N. Baruah · N. Gogoi (✉)
Plant Physiology and Biochemistry Laboratory, Department of Environmental Science, Tezpur University, Tezpur, Assam, India

L. Gogoi · R. Kataki
Department of Energy, Tezpur University, Tezpur, Assam, India

B. Borkotoki
Biswanath College of Agriculture, Assam Agricultural University, Biswanath Chariali, Assam, India

J. S. Singh, C. Singh (eds.), *Biochar Applications in Agriculture and Environment Management*, https://doi.org/10.1007/978-3-030-40997-5_11

11.1 Introduction: Biochar as a Soil Amendment

Biochar is a charred carbon-enriched material intended to be used as a soil amendment to sequester carbon and enhance soil quality. Sustainable biochar is produced from waste biomass using modern thermochemical technologies. Addition of sustainable biochar to soil has many environmental and agricultural benefits, including waste reduction, energy production, carbon sequestration, water resource protection, and soil improvement. When used as a soil amendment, biochar has been reported to boost soil fertility and improve soil quality by raising soil pH, increasing moisture holding capacity, attracting more beneficial fungi and microbes, improving cation exchange capacity (CEC), and retaining nutrients in soil (Lehmann et al. 2006; Lehmann 2007a). Another major benefit associated with the use of biochar as a soil amendment is its ability to sequester carbon from the atmosphere-biosphere pool and transfer it to soil (Winsley 2007; Gaunt and Lehmann 2008; Laird 2008). Biochar usually has a greater sorption ability than natural soil organic matter due to its greater surface area, negative surface charge, and charge density (Liang et al. 2006). Biochar can not only efficiently remove many cationic chemicals including a variety of metal ions, but also absorb anionic nutrients such as phosphate ions, though the removal mechanism for this process is not fully understood (Lehmann 2007a). Thus, the addition of biochar to soil offers a potential environmental benefit by preventing the loss of nutrients and thereby protecting water resources. Biochar is considered much more effective than other organic matter in retaining and making nutrients available to plants. Its surface area and complex pore structure are hospitable to bacteria and fungi that plants need to absorb nutrients from the soil. Moreover, biochar is a more stable nutrient source than compost and manure (Chan et al. 2007).

11.1.1 Agronomic and Environmental Benefits of Biochar

Biochars can provide agronomic and environmental benefits in soils through increased cation exchange capacity, reduced nutrient leaching, enhanced water holding capacity, reduced soil acidity and stimulation of microbial activity (Kookana et al. 2011; Lehmann and Joseph 2015).

11.1.1.1 Crop Disease Management

A positive influence of biochar on reducing plant diseases such as rust in wheat and mildew in other crops was first reported some 170 years ago (Allen 1847) and drew attention in the last decade where several pathosystems were studied by different groups worldwide (Elad et al. 2010; Elmer and Pignatello 2011; Jaiswal et al. 2014; Copley et al. 2015; Jaiswal et al. 2015). Pathosystems included both foliar pathogens and soil borne pathogens (Elad et al. 2011; Graber et al. 2014a).Biochar application can enhance crop response to disease (Elad et al. 2011), and this enhancement can be attributed to an increase in soil pH (Novak et al. 2009), nutrient retention (Chan et al. 2007; Steiner et al. 2007), cation exchange capacity in soil (Steiner et al. 2007), transformations and turnover of P and S (Lehmann and Joseph 2009), and neutralization of phytotoxic compounds in soil (Wardle et al. 1998).

Biochar can reduce fungal foliar diseases caused by *Botrytis cinerea* and *Oidiopsis sicula* in tomato (*Solanum lycopersicum* L.) and pepper (*Capsicum annuum* L.) (Elad et al. 2010). Biochar induced defense responses of strawberry are functionally similar to induced systemic resistance (Harel et al. 2012). Moreover, biochar can reduce soil borne diseases caused by bacteria and fungi (Jaiswal et al. 2014). Incidence of bacterial wilt (*R. solanacearum*) in tomato was reduced due to biochar application derived from municipal biowaste (Nerome et al. 2005). Biochar induced plant disease suppression were attributed to several mechanisms (Hoitink and Fahy 1986; Lehmann et al. 2011; Noble and Coventry 2005) such as chemical components of biochar that directly inhibit growth of pathogens and the porous structure of biochar provide microbial habitats beneficial for bacterial abundance. Biochar promotes plant growth by providing nutrients and improving nutrient solubilization and uptake. The sorption property of biochar may change the mobility and activity of pathogens or modify signaling between pathogens and plants (Lehmann et al. 2011).

Adding biochar to soil and soilless media was found to suppress plant diseases caused by both foliar and soilborne pathogens (Elad et al. 2011; Frenkel et al. 2017; Graber et al. 2014b; Jaiswal et al. 2014). Biochar-elicited suppression of foliar fungal diseases is related to activation of plant defense system, given that biochar is spatially distant from the site of pathogen attack. Mechanisms responsible for biochar-related attenuation of soil borne diseases can be much more diverse. This is because the biochar and pathogens both reside in the soil, and can have direct and indirect interactions with each other (Graber et al. 2014b). Ways in which biochar could influence the progress of diseases caused by soil borne pathogens includes (1) changes in nutrient supply and availability (Elmer and Pignatello 2011); (2) alterations in soil physiochemical characteristics (Rogovska et al. 2017); (3) induction of systematic plant defenses (De Tender et al. 2016; Zwart and Kim 2012); (4) alteration of soil microbial abundance in terms of taxonomic, functional diversity and activity (De Tender et al. 2016; Jaiswal et al. 2017, 2018); (5) modification of pathogen growth, survival, virulence and activity (Akhter et al. 2016; Copley et al. 2015; Jaiswal et al. 2015, 2017, 2018) and (6) adsorption and inactivation of pathogenic enzymes and/or toxins. Ad-sorption of toxic metabolites by 3% biochar significantly reduced the severity of the disease-like symptoms caused by the toxic metabolites as compared to no-biochar control toxic metabolites treatments.

Biochar application at a rate of 3% and 5% by weight under tomato and pepper cultivation documented significant reduction in leaf symptoms caused by two common fungal pathogens i.e. powdery mildew and grey mould. Whole plant peppers after 60 days had 59% of powdery mildew infection in plots with biochar and only 17% infection with 5% biochar application. Grey mould in tomato after 59 days was significantly reduced from 66% infection in untreated plants to 2% under 3% biochar application. Biochar addition at a rate of 5% in the pepper crop reduced leaf infection from 18% (no biochar) to 6%. The reduced level of residual tars present in biochar induced resistance to the diseases and pest (Elad et al. 2010). Jaiswal et al. (2018) documented that biochars obtained from eucalyptus wood and pepper plant wastes can significantly adsorbed and deactivated enzyme exudates of pathogenic fungi *Fusarium oxysporum.*

Bonanomi et al. (2015) reviewed and summarized the data from 13 pathosystems that tested the effect of biochar on plant disease. In their analysis, 85% of the studies showed a positive influence of biochar in reducing plant disease severity, 12% had no effect, and only 3% showed that biochar additions were conducive to plant disease. However, their analysis did not consider the dose of the as a crucial factor on plant susceptibility/resistance to a disease.

11.1.1.2 Abiotic Stress Management

Abiotic stress such as, drought, high soil salinity, heat, cold, oxidative stress and heavy metal toxicity is the common adverse environmental conditions that affect and limit crop productivity worldwide (Singh 2013, 2014, 2015, 2016; Singh and Boudh 2016; Kumar et al. 2017; Kumar and Singh 2017; Tiwari and Singh 2017). The abiotic stress conditions that most adversely affect crop yield are associated with water deficiency ion imbalance and temperature extremes (Gupta et al. 2014). Biochar is known to have a number of positive effects on plant ecophysiology. However, limited research has been carried out to date on the effects and mechanisms of biochar on plant ecophysiology under abiotic stresses. A series of experiments on rice seedlings treated with different concentrations of biochar leacheates (between 0 and 10% by weight) under cold stress (10 °C) was conducted by Yuan et al. 2017. Quantitative real-time PCR (qRT-PCR) and cold-resistant physiological indicator analysis at low temperatures revealed that the cold tolerance of rice seedlings increased after treatment with high concentrations of biochar leacheates (between 3% and 10% by weight). Results also show that the organic molecules in biochar leacheates enhance the cold resistance of plants when other interference factors are excluded. The positive influence of biochar on plant cold tolerance is because of surface organic molecules and their interaction with stress-related proteins (Yuan et al. 2017). As a direct source of plant soil nutrients; presence of biochar impact root growth, and plant performance (Prendergast-Miller et al. 2013). Thomas et al. 2013 reported that biochar mitigates negative effects on two herbaceous plant species via salt sorption and application is known to preserve rice pollen under high-temperature stress (Fahad et al. 2015). Biochar addition enhance drought tolerance of quinoa crop with improve the growth and higher leaf nitrogen content (Kammann et al. 2011).

The beneficial effects of biochar under limited water conditions have been widely reported (Akhtar et al. 2015b; Paneque et al. 2016; Ramzani et al. 2017; Rogovska et al. 2014). Biochar as soil amendment improved growth and biomass of plants under drought-stress. Use of biochar exhibited the highest vegetative growth and seed production of field-grown sunflower under non-irrigated conditions (Paneque et al. 2016). Enhanced tomato fruit quality, growth and yield was reported under deficit irrigation due to biochar application (Agbna et al. 2017). Similarly, use of biochar supports the growth of winter rapeseed under drought conditions (Bamminger et al. 2016). Likewise, Basso et al. (2013) found that application of hardwood biochar significantly increase soil water holding capacity and might be the reason of enhanced available water capacity (AWC – available water between field capacity and permanent wilting point) for crops. Tomato seedlings were protected from wilting due to improved soil moisture content with higher (30% v/v) rates of biochar as soil amendment in sandy soils (Mulcahy et al. 2013). Studies have shown that biochar may minimize water stress in plants when applied with microorganisms (Liu et al. 2017b; Nadeem et al. 2017). Egamberdieva et al. (2017) reported inoculation of biochar with *Bradyrhizobium sp.* enhance the growth, biomass, phosphorus and nitrogen uptake, and nodulation in lupin (*Lupinus angustifolius* L.) seedlings under drought stress as compared to the only microbial inoculation. Nadeem et al. (2017) reported inoculation of biochar with *Pseudomonas fluorescens* reduced the harmful impact of drought stress on cucumber (*Cucumis sativus L.*). Significant improvements were observed in chlorophyll and relative water contents, as well as a reduction in leaf electrolyte leakage demonstrating the effectiveness of this approach. In another study, Liu et al. (2017b) reported that inoculation of birch wood biochar with *Rhizophagus irregularis* under limited root zone water decreased water use efficiency, leaf area, nitrogen and phosphorous in potato and did not adversely impacted the root biomass and soil pH as compared to control. However, under limited irrigation soil amendment with wood derived biochar (30 mg.ha^{-1}) had no significant effect on soil biota groups such as protozoa, bacteria, fungi, nematodes and arthropods (Pressler et al. 2017). Application of biochar with arbuscular mycorrhizal (AM) fungi and other beneficial microbes (Vimal et al. 2018; Singh 2019; Vimal and Singh 2019) can enhance drought and salt tolerance of the host plant by physiological mechanisms in nutrient adsorption and biochemical mechanisms, e.g. hormones, osmotic adjustment and antioxidant systems. However, application of BC to the agricultural soil with AM fungi stimulated the growth of extra-radical hyphae in soil and increased mycorrhizal colonization of roots. As the water potential of the soil was the same with and without biochar amendment, it is unlikely that the observed effects on plant growth were related to possible benefits from the water holding capacity of the biochar (Mickan et al. 2016).

Biochar not only improves crop productivity under normal conditions but also improves crop yield under adverse conditions such as salinity and drought (Thomas et al. 2013; Haider et al. 2014). For example, biochar enhanced the permanent wilting point (Abel et al. 2013; Cornelissen et al. 2013a), while the quantity of water retained at field capacity improved to a larger extent compared to the water held at permanent wilting point, i.e., increased plant available water. Therefore, the increase in WHC of biochar amended soils can be used as an indicator of the overall rise in plant available water (Liu et al. 2015). Because of its porous nature, biochar can improve your soil's water retention and water holding capacity. This can be attributed to the micropores present in biochar, where a larger volume of pores correlates to better water retention and better water holding capacity. Biochar with a fine particle size can also improve these characteristics by packing with the soil to create tight pores that will hold the water against gravity (https://char-grow.com/biochar-impact-nutrient-water-retention).

In another study, biochar addition to a fertile sandy clay loam soil in a boreal climate relieved the temporary water deficit leading improvement in harvestable yield (Tammeorg et al. 2014). Haider et al. (2014) quoted that biochar induced plant growth in a poor sandy soil is due to better soil-plant water relations as observed in terms of improved relative water content and leaf osmotic potential) and photosynthesis (due to lowered stomatal resistance and increased electron transport rate of photosystem II) under both well-watered and drought conditions. Biochar application at higher rates can mitigate adverse effects of salt stress for plant growth (Kim et al. 2016; Akhtar et al. 2015a). For instance, topdressing with biochar at 50 t ha^{-1} mitigated salt-induced mortality in *Abutilon theophrasti* and extended the survival rate of *Prunella vulgaris*. Plants of *A.theophrasti* receiving both biochar and salts had growth rates similar to plants devoid of salt addition (Thomas et al. 2013). Recently, Akhtar et al. (2015a) reported enhanced tuber productivity of potato crop in salt-affected soils under application of biochar due to enhanced Na+ absorption and mainteinance of higher K+ content in xylem. The authors further observed positive residual effects of biochar application in lowered Na+ uptake in the following wheat crop under salinity stress (Akhtar et al. 2015b). Therefore, biochar has the potential to mitigate salinity-induced reductions in mineral uptake, and may be a novel technique to alleviate the effects of salinization in arable and salt contaminated soils (Thomas et al. 2013; Kim et al. 2016).

11.1.1.3 Crop Productivity

Soil organic carbon (SOC) is known to play an important role in maintaining soil fertility and crop productivity (Díaz-Zorita et al. 2002; Lal 2004; Pan et al. 2009). Enhancing SOC stocks in croplands with good management practices has the significant contribution to climate change mitigation in agriculture (Smith et al. 2007b, 2008a). Direct incorporation of crop residues as well as organic manure to soils has been traditionally performed to maintain soil resilience and carbon (C) stocks. However, the residence time of these C sources in soil is relatively short because of mineralization, perhaps less than 30 years (Lehmann et al. 2006). Moreover, such an incorporation of fresh organic matter would potentially lead to an increase in the production of methane (CH_4) in rice fields (Yan et al. 2005; Shang et al. 2011). In contrast, C from biochar could be stabilized in soil for long periods, potentially hundreds of years (Lehmann et al. 2006; Kleber 2010; Schmidt et al. 2011; Woolf and Lehmann 2012). Furthermore, biochar soil amendment (BSA) has been shown to effectively reduce nitrogen (N) fertilizer-induced nitrous oxide (N_2O) emissions from agricultural soils (Yanai et al. 2007; Liu et al. 2012; Zhang et al. 2010) with no or minimal increase in CO_2 and CH_4 emissions (Spokas and Reicosky 2009; Karhu et al. 2011; Zhang et al. 2012a). Thus, biochar, produced via pyrolysis of biomass, has been recommended as an option to enhance SOC sequestration and mitigate greenhouse gas (GHG) emissions with the co-benefits of improving soil productivity and ecosystem functioning in world agriculture (Lehmann et al. 2006; Sohi 2012; Sohi et al. 2010; Woolf et al. 2010).

Many earlier studies on biochar focused on the potential of biochar from bio wastes to mitigate GHG emissions in agriculture (Lehmann 2007a; Spokas and Reicosky 2009; van Zwieten et al. 2009; Knoblauch et al. 2011; Singh et al. 2010a; Sohi et al. 2010; Taghizadeh-Toosi et al. 2011; Vaccari et al. 2011; Liu et al. 2012). Sohi (2012) addressed co benefits of biochar for soil and environmental quality, plant nutrition, and health as well as ecosystem functioning. The significant and persistent increase in crop productivity with BSA suggests a major benefit for agricultural production besides its role in mitigating GHG emission. BSA could provide a practical option to meet the challenge of food security in a changing climate.

Crop productivity responses to BSA also varied with crop type. Generally, greater positive responses were found in experiments with legumes, vegetables and grasses. The average increase in crop productivity was 30.3, 28.6, and 13.9% respectively for legume crops, vegetables, and grasses and 8.4, 11.3, and 6.6% respectively for maize, wheat, and rice. Yield increases with BSA were greater than biomass increases for maize. Whereas, the reverse was true for wheat. This indicates the differential influence of biochar on crop productivity.

Biochars used in the reported experiments were derived from almost 20 different types of biomass and were grouped into six general types of crop residues, wood, manure, sludge, municipal waste, and mixtures of wood and sludge. Wood and crop residue biochars documented an average (12.1 and 2.6% respectively) increase of constant crop productivity while manure biochar showed generally greater (29%) productivity with variable responses across the experiments. However, biochar from municipal waste significantly decreased crop productivity by 12.8% on average. Crop productivity response was also dependent on the pyrolyzing temperature during biochar production. Greater increase in crop productivity were seen with biochar produced at temperatures of >350 °C from wood, >550 °C for crop residues and 350– 550 °C for manure biochar. Meanwhile, crop productivity responses were generally negative (−7.9% on average) with non-alkaline (pH <7.0) biochars though generally positive with alkaline biochar (pH >7.0). Finally, crop productivity changes with BSA were not shown to be proportional to biochar application rate up to 20–40 t ha^{-1} although the increase in crop productivity diminished at biochar application rate of >40 t ha^{-1} (Liu et al. 2013)

The response of crop productivity was shown to vary with biochar type, pyrolysis temperature and the feedstock used. It had been well established that both the physicochemical properties and nutrient contents of biochars are affected by the feedstock type (Spokas and Reicosky 2009; Qin et al. 2012). While biochar from wood and crop residues exerted consistent positive yield increase, the greatest mean increase was observed with manure biochar. Manure biochars have been generally considered very significant for improving soil fertility by promoting soil structure development (Joseph et al. 2010) in addition to their large amounts of plant available nutrients (Hass et al. 2012). The negative effects with municipal waste biochar observed by Rajkovich et al. (2012) reported a great decline in crop productivity by 80% under application of food waste biochar at a higher (91 t ha^{-1}) rate. Presence of higher sodium (ten times) food waste biochar compared to wood and straw biochar increased soil salinity and inhibited plant growth. Crop productivity was significantly increased with biochar produced at higher pyrolyzing temperatures; presumably as a result of the liming effect as biochar pH generally increases with increasing temperature for pyrolysis (Rajkovich et al. 2012). However, there was an interaction of feedstock and pyrolysis temperature on crop productivity to BSA. Biochars produced at both low and high pyrolysis temperatures generally contained very limited N. Pyrolyzing at temperature more than 450 °C would result in losses of N in manure biochar. Higher nutrient contents and crop yields were found with the application of manure biochar pyrolyzed at temperatures of <500 °C compared to more recalcitrant biochar produced at even higher temperatures (Chan et al. 2008).

Along with improved soil health, increased crop yield is generally reported with application of biochar to soils. However, many of the published experiments are highly variable and dependent on many factors, mainly the initial soil properties and biochar characteristics. Positive crop and biomass yield was found for biochars produced from wood, paper pulp, wood chips and poultry litter. Liu et al. (2012) reviewed published data from 59 pot experiments and 57 field experiments from 21 countries and found increased crop productivity by 11% on average Benefits at field application was noted at a rate below 30 tons/ha. They reported that increases in crop productivity varied with crop type with greater increases for legume crops (30%), vegetables (29%), and grasses (14%) compared to cereal crops corn (8%), wheat (11%), and rice (7%). Biederman and Harpole (2013) analyzed the results of 371 independent studies. This meta-analysis showed that the addition of biochar to soils resulted in increased aboveground productivity, crop yield, soil microbial biomass, rhizobia nodulation, plant tissue content of K, soil phosphorus (P), soil potassium (K), total soil nitrogen (N), and total soil carbon (C) compared with control conditions. The yield gains were attributed to the combined effect of increased nutrient availability (P and N) and improved soil chemical conditions. However, there exists the concern of heavy metal contamination from biochars produced from sewage sludge. The inconsistency of sewage sludge might contain differing amounts of toxic metals which limit the land application due to the possibility of food chain contamination. Several studies have indicated the strong potential of biochar application for improving crop yields, particularly on nutrient-poor soils (Van Zwieten et al. 2010a; Zhang et al. 2012a) (Table 11.1).

Biochar application may substantially improve soil fertility and crop productivity (Lehmann and Joseph 2015). For instance, biochar application (68 t ha^{-1}) increased rice (*Oryza sativa* L.) and cowpea (*Vigna unguiculata* (L.) Walp) biomass by 20 and 50% respectively. Increased grain yields in durum wheat (*Triticum durum* L.) by up to 30%, was observed due to biochar application but there was no effect was noted on grain N content (Vaccari et al. 2011). Oguntunde et al. (2004) recorded increases of 91 and 44% in grain and biomass yield, respectively, in maize (Zea

Table 11.1 Influence of biochar application on crop yields based on the literatures

Crops	Bio char feed stocks	Type of soil	Doses	Yield response	References
Amaranthus	Water hyacinth, domestic organic waste	Calcareous Fluvisols	10 t ha^{-1}	17–64% increase in yield	Piash et al. (2019)
Lettuce	Fecal matter	Silty loam and sandy loam	0,10,20,30 t ha^{-1}	Increased crop yield	Woldetsadik et al. (2017)
Maize	Corncob	Alfisols	2% w/w	Increased crop yield	Mensah and Frimpong (2018)
Cotton	Hardwood	Fine, kaolinitic, thermic Rhodic Kandiudults	0, 22.4, 44.8, 89.6, and 134.4 Mg ha^{-1}	No difference in yield	Sorensen and Lamb (2016)
Corn	Hardwood	Fine, kaolinitic, thermic Rhodic Kandiudults	0, 22.4, 44.8, 89.6, and 134.4 Mg ha^{-1}	No difference in yield	Sorensen and Lamb (2016)
Peanut	Hardwood	Fine, kaolinitic, thermic Rhodic Kandiudults	0, 22.4, 44.8, 89.6, and 134.4 Mg ha^{-1}	No difference in yield	Sorensen and Lamb (2016)
Cotton	Corn straw	Inceptisol	0, 5, 10, and 20 t ha^{-1}	Increased yields by 8.1–17.1%, 9.6–13.5%, and 8.1–18.6% in 2013, 2014, and 2015, respectively	Tian et al. (2018)
Maize	Acacia wood	Clay	50 + 50 Mg ha^{-1}	Seasonal yield increase was average around 1.2 Mg ha^{-1}	Katterer et al. (2019)
Soybean	Acacia wood	Clay	50 + 50 Mg ha^{-1}	Seasonal yield increase was average around 0.4 Mg ha^{-1}	Katterer et al. (2019)
Corn	Pine chips	Ultisols (loamy sand)	30,000 kg ha^{-1}	No significant difference in yield	Novak et al. (2019)

mays L.) on charcoal-amended soils when compared with adjacent field soils in Ghana. Likewise, almost double maize yield in degraded soils was obtained from application of Eucalyptus-derived biochar in in Kenya (Kimetu et al. 2008). Improvement of rice grain yield (upland) in soils with lower P availability was found with addition of biochar in Loas. However, at sites with low native N supply, biochar application reduced the leaf chlorophyll contents suggesting that biochar may reduce grain yield in N-deficient soils if additional N is not applied (Asai et al. 2009; Nelson et al. 2011). The effect on crop yields particularly in nutrient-rich soils remains uncertain. Several other studies have revealed only small improvements or even reductions in grain yield with biochar application in nutrient-rich soils (Deenik et al. 2010; Gaskin et al. 2010; Van Zwieten et al. 2010a). For instance, Gaskin et al. (2010) noted a linear decrease in grain yield with increasing rates of biochar application. Meta-analysis on biochar application and crop productivity (either yield or aboveground biomass) by Jeffery et al. (2011) documented an overall small (~10%) but significant improvement in grain yield from biochar application, and identified a liming effect and increase in soil WHC as principal reasons for biochar-induced yield gain (Jeffery et al. 2011). Among biochar feedstocks, poultry litter was the best (28%), while biosolids had a negative effect (−28%) on crop productivity (Jeffery et al. 2011). In another study conducted for 3 years by Feng et al. (2014) reported that annual yield of either summer maize or winter wheat was not enhanced significantly due to biochar application; however, cumulative yield over the first 4 growing seasons were significantly higher. Spokas et al. (2012) analysed 44 published articles on biochar and found that about half of them claimed biochars improved crop yield while the others had no or negative effect on crop yield. Biochar-induced increases in specific surface area, CEC, soil porosity (Thies and Rillig 2009), WHC, nutrient retention (Glaser et al. 2002; Lehmann and Rondon 2006; Yamato et al. 2006), and liming effect (Rondon et al. 2007; Liu et al. 2013) are mainly responsible for improved crop productivity. For example, biochar obtained from crop biomass ashes can provide a P source similar to that of commercial P and K fertilizer (Schiemenz and Eichler-Loebermann 2010; Luo et al. 2014) or may improve the supply of Ca and Mg (Major et al. 2010).

Biochar amendment has a synergistic effect with fertilizers in improving crop yield; for example, maize yield increased with biochar and fertilizer application more than fertilizer alone in acidic soil in Indonesia (Yamato et al. 2006). In another study, Steiner et al. (2007) harvested 4–12 times more rice and sorghum (*Sorghum bicolor* L.) yield by application of charcoal (11.25 t ha^{-1}) with compost and/or fertilizer than by using fertilizer alone. Similar results on biochar induced doubling of rice and sorghum grain yield was reported while applied with NPK fertilizers (Christoph et al. 2007). Mau and Utami (2014) also recorded increase in maize yield due to increased P availability and uptake under combined application of biochar and inoculation of AM fungal spores; however, biochar amendment alone did not improve maize growth or P uptake. In a field study conducted on a boreal sandy clay loam, biochar as soil amendment (10 t ha^{-1}) improved grain numbers in wheat (dry year) probably by alleviating the water deficit (Tammeorg et al. 2014). In crux, biochar application has the potential to improve crop productivity on a variety of soils under normal and less than optimal environmental conditions if prepared and used wisely.

11.1.2 Soil Health Management

Soil health is the capacity of a soil to sustain biological productivity, environmental quality and promote plant and animal health through self-regulation, stability, resilience and lack of stress symptoms within ecosystem boundaries. Although, another terminology 'Soil Quality' has often been used simultaneously, they involve two different concepts. What constitutes a high-quality soil may depend on the intended use or the role of soil management system. For example, a good quality soil for engineering construction many not suited for agricultural production (Brady and Well 2012).

Let's have an example of Soil Health of *Terra Preta* (= Dark in Portuguese) soil of Amazon basin. In general, the highly weathered Oxisols and Ultisols of Amazon basin are dominated by iron and aluminium oxides clays. Due to high soil acidity and low Cation Exchange Capacity (CEC) these soils possess very low level of soil fertility, have little capacity to sustain nutrients and therefore poor in health. Conversely, soil scientists exploring this area were mystified and surprised when they found around twenty hectares of dark coloured, high organic matter containing, fertile healthy soils along the Amazon river and some of its tributaries. When they conducted ^{14}C isotopic study they found that most of the carbon of these soils were accumulated several thousand years ago. Now it is believed that these patches of highly fertile healthy soils were created by the ancient dwellers of that vicinity that lived in miniature agricultural settlements carved out of the Amazon rainforest. These dwellers farmed this soil regularly for many years in such a way that they enhanced the soil health rather than degraded by agricultural use. Even today some of the *Terra Preta* soils are dug and sold in local markets for their high fertility value. Soil analysis revealed that *Terra Preta* soils are rich in calcium and phosphorus than that of the surrounding soils of Amazon basin because of the amendments with human excrement and bones of animals eaten by the ancient inhabitants. However, the unique aspect of these soils is that much of the carbon in them is present as Charcoal. The complex aromatic structure makes the charcoal recalcitrant because of its resistance to microbial degradation results in very high and stable accumulation of organic carbon as well as high nutrient availability for plant growth. Again, the small bits of charcoal found in Terra Preta soils are very porous in nature that greatly enhances water holding capacity of these soils and capacity to retain nutrients in the form of dissolve organic compounds. Thus soil scientists studying the effect of this unique anthropogenic activity on soil health reported that adding charcoal or biochar to soils may significantly enliven soil health and make the soil defiant to degradation in agricultural use. It is also reported that the conversion of biomass carbon to biochar leads to sequestration of about 50% of the initial carbon compared to the low amounts retained after burning (3%) and biological decomposition (less than 10–20% after 5–10 years) (Lehmann et al. 2006).

Soil Health with application of biochar may be governed by a number of physicochemical and biological attributes and processes and expressed by different quantitative and qualitative measures of these attributes as also by outcomes that are

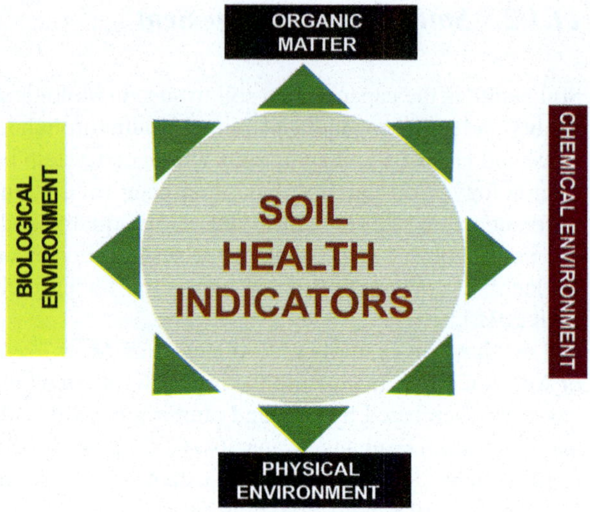

Fig. 11.1 Indicators of soil health

governed by the soil such as productivity, nutrient and water use efficiencies and quality of produce. Gaunt and Lehmann, (Gaunt and Lehmann 2008) pointed out that the application of biocahr may improve soil health by altering its physical, chemical and biological environment principally by improving soil organic carbon status. Therefore, apart from soil physical, chemical and biological environment organic matter must be kept as distinct indicator of soil health (Fig. 11.1).

11.1.2.1 Soil Physical, Chemical and Biological Health

Hypothetically, four mechanisms have been proposed to elucidate how application of biochar might help in improvement of soil physico-chemical and biological health. These are: (1) direct modification of soil chemistry through inherent chemical composition of biochar. (2) provides chemically active surfaces that alters the dynamics of available nutrients or otherwise catalyze important soil reactions and (3) modifies physical character of the soil in a way that benefits root growth and/or nutrient and water retention and acquisition and (4) modifies soil biological health through priming effect.

The first mechanism may result in a momentary shift in crop productivity in positive direction; the extent and duration of which will be governed by the natural phenomenon of biochar weathering and the upshots of crop uptake. This could happen where the biochar has considerable mineral nutrients content, or equally enhance in CEC in due course of time as the weathering progress. The benefits provided by the second and third mechanisms depend upon recalcitrant nature of biochar. It depends upon the half life of biochar carbon that principally varies depending on feed stock and temperature at which biochars are produced and may also accordingly be finite, even though over a much longer period of time. This would include the impact of pours biochar on water retention or lowering bulk density and increasing the total porosity of soil. The fourth mechanism of priming effect depicts about raise in soil organic matter decomposition rate after addition of fresh organic matter in soil which is often supposed to as a result from escalation of microbial activity on account of higher availability of energy released from decay of fresh organic matter (Zimmerman et al. 2011). However, both positive and negative priming have also been reported by earlier workers (Wardle et al. 2008; Kuzyakov et al. 2000; Kuzyakov 2010; Zimmerman et al. 2011). Possible causes of positive priming might be the positive co-metabolic effect of labile part of organic matter on growth of microorganisms and provision of habitat for microbes that protects them from predation and simultaneously supports microbial growth through co-locating labile organic matter on surfaces of the biochar. Other mechanisms include altering soil reaction, availability of nutrients and/or water holding capacity, which have a say in positive priming (Fontaine et al. 2003). Probable reasons of negative priming are adsorption of organic matter through encapsulation and absorptive protection where encapsulation takes place within biochar pores that exclude biota and their extracellular enzymes from access to the organic matter and absorptive protection onto external bio-char surfaces. Biochar induced stabilization or protection of comparatively labile organic matter in soil within organo-mineral fractions and a transient shift of microbial communities to exploit relatively more labile C in biochar, a phenomenon known as preferential substrate utilization, may also contribute to negative priming and predominant in low C soil receiving nutrient application (Fontaine et al. 2003).

The magnitude and relative importance of first three mechanisms in a particular setting will evolve over time as the slow process of chemical and physical modification results in a gradually increasing concentration of smaller, partially oxidized particles.

11.1.2.2 Soil Fertility

Soil fertility is the capacity of the soil to supply nutrients to plants in adequate amounts and in suitable proportions to produce crop of economic value and to maintain soil health. The addition of biochar to agricultural soil is receiving considerable interest because of its positive impact on soil fertility (Quayle 2010). At local scale, increase in soil organic carbon levels due to addition of biochar shape agro-ecosystem function and influence soil fertility by altering soil physical, chemical and biological properties (Milne et al. 2007). The ability of soil to retain nutrients can be increased using biochar (Sohi et al. 2010). Biochar application reduces leaching loss of soil nutrients, enhances plant nutrient availability, reduces toxicity of aluminium to plant roots and micro-biota in acid soils and bio-availability of heavy metals like Pb, Cd etc. (Lehmann et al. 2006; Rondon et al. 2005; Yanai et al. 2007; Mukherjee and Lal 2014). It reduces soil acidity (Zwieten et al. 2010), sequesters recalcitrant carbon in soils and thus improves soil fertility and mitigate climate change (Fowles 2007; Glaser et al. 2002; Laird 2008; Lehmann 2007a, b; Lehmann et al. 2006; Marris 2006; Sohi et al. 2010). This help to uphold the growth of microbes specially the bioactivity of beneficial soil microorganisms (Marris 2006), improves soil organic matter and consequently plant growth (Sanchez et al. 2009; Glaser et al. 2002), soil porosity. Reduces bulk density and improved water holding capacity of soil (Rasa et al. 2018). Biochar has the potential to boost up conventional agricultural productivity and augment the capacity of the farmers to play a part in carbon markets ahead of the routine approach by directly applying carbon into the soil (McHenry 2009). The combined application of biochar along with inorganic fertilizer has the potential to increase crop productivity, therefore providing additional income, and reducing quantity of inorganic fertilizer use and importation (De Gryze et al. 2010; Quayle 2010).

11.2 Biochar in Environmental Management

In recent time research interest on use of biochars produced from different feedstock as environmental management is increasing prominently. Earlier research reports on positive impact of biochar application on seedling growth (Retan 1915) and soil chemistry (Tryon 1948) are available. According to Lehmann (2009) and Schmidt et al. (2002) biochar contains higher percentage of recalcitrant organic C which is more stable (hundreds to thousands of years) in soil than any other commonly used amendments and also enhances the availability of nutrients and maintain soil quality beyond a fertilizer effect. Biochar can be used for environmental management through improving productivity, reduce pollution, climate change mitigation, waste management and energy production. In this section we will discuss the use of biochar in environmental management from the fact of its ability to soil carbon sequestration, mitigation of greenhouse gas (GHGs) emission and soil and water pollution.

11.2.1 Soil Carbon Sequestration

Soil carbon sequestration is a process of long term or permanent (100 years) storage of CO_2 from atmosphere to soil (Stockmann et al. 2013; Shin et al. 2019). Biochar is a carbon rich product derived from biomass burning in anaerobic environment, it contains high amount of recalcitrant organic carbon and least prone to chemical and microbial degradation and remain in soil for hundreds to thousands of years increasing soil carbon storage capacity (Schmidt et al. 2002; Roberts et al. 2010; Wang et al. 2014a; Lehmann 2007b), can improve soil physicochemical property, fertility and crop productivity. Highly porous, high cation exchange capacity, larger surface area and adsorption ability are some significant properties of biochar (Luo et al. 2016). As soil amendment; biochar is a good carbon sequester and is suggested as an effective countermeasure for increasing GHGs emission to atmosphere (Lal

1999; Mukherjee et al. 2014; Deng et al. 2017). An increase (0.4% per year in global scale) of agricultural soil organic carbon due to biochar addition can compensate global emission of GHGs by anthropogenic sources (Minasny et al. 2017). Downie et al. (2011) reported significantly elevated soil carbon stocks, compared to the adjacent soil in a 650 and 1,609 years old historic charcoal added soil in ancient Australian Aboriginal oven mounds. Biochar can be implemented in global scale to mitigate climate change by potentially sequester up to 12% of anthropogenic GHGs emissions (Woolf et al. 2010) in an ecologically sustainable system. Biomass like straw, when converted to biochar through gasification process contain aromatic carbon compounds having high stability and potential for carbon sequestration than the original feedstock in amended soil (Hansen et al. 2015, 2016; Wiedner et al. 2013). Along with reduced soil organic carbon decomposition, biochar can also adsorb significant amount of soil- dissolved carbon (Lu et al. 2014). Hailegnaw et al. (2019) also reported reduction of nitrate and dissolved organic carbon in soil amended with wood chip biochar. In a study by Béghin-Tanneau et al. (2019) documented significant ability of anaerobically digested exogenous organic matter (EOM) to sequester carbon in soil compared to undigested EOM. It was due to higher stability and negative priming effect of digested-EOM that reduced native soil organic matter (SOM) respiration compared to low stability and positive priming effect of undigested-EOM that enhanced native SOM respiration. Béghin-Tanneau et al. (2019) also noted a reduction of CO_2 emission by 27% along with carbon sequestration compared to maize silage amendment in soil. Huang et al. (2018) found reduction of CO_2 flux from a rape-maize cropping system with increased net C sequestration without reducing crop yield and net primary productivity under sole biochar treatment compared to straw, straw with straw decay bacterium, mixed straw and biochar treatment. This may be due to lower labile organic carbon (LOC), especially microbial biomass carbon fraction in biochar treated soil. While increase CO_2 emission from crop straw added soil is because of availability of higher C substrate for microorganism causing lower carbon sequestration (Dendooven et al. (2012). Chemical property of biochar, such as lower hydrogen to carbon (H/C) and oxygen to carbon (O/C) ration makes it highly stable for microbial degradation (Schimmelpfennig and Glaser 2012). Similarly, Hansen et al. (2015) also found higher microbial degradation of straw carbon compared to straw gasification biochar resulting in 80% of added straw carbon respiring as CO_2 compared to 3% of the biochar added after 110 days of incubation. This indicates the potentiality of the biochar in soil carbon sequestration. Although, some researcher documented negative response of biochar amendment to crop yield such as lettuce and ryegrass (Marks et al. 2014). While biochar pellet blended with biochar and pig manure compost (4:6 ratio) application was found effective for carbon sequestration in rice cultivation without decreasing crop yield (Shin et al. 2019). Thammasom et al. (2016) noted an increase of soil carbon sequestration (1.87 to 13.37 tons ha^{-1}) with application of wood biochar, while a reduction (0.92 to 2.56 tons ha^{-1}) of the same was noted under rice straw application. Kimetu and Lehmann (2010) documented reduced loss of soil CO_2–C by 27% under biochar application contrarily *Tithionia diversifolia* green manure increased soil CO_2 C loss by 22%, while biochar also increased intra aggregate C per respired C by 6.8 times relative to the *Tithionia diversifolia* green manure additions, indicating more efficient stabilisation capacity of biochar. Charcoal produce from sustainable system using biomass on application to soil can remove carbon from short term photosynthesis cycle to long term reservoir. Thus the energy generated through biochar production can be certified as carbon negative and can act as revenue source from both sale and tradable carbon credits by virtue of increase forest cover and reduce greenhouse gases emission (Mathews 2008). Pyrolysis temperature plays a crucial role on aromaticity of the produced biochar (Yip et al. 2010), which in turn affect the recalcitrance property in soil and thus the carbon sequestration potential. As an option to carbon sequestration, reduce or delay nutrient leaching like nitrate is beneficial for both environment and plants (Liu et al. 2017a; Ghorbani et al. 2019). Holding nutrients for long and improving soil aggregation, biochar application can consequently mitigate CO_2 emission (Xu et al. 2011a). Biochar production method greatly affect the carbon sequestration capability. In a study Santín et al. (2017) found lower carbon sequestration potential of wildfire charcoal produced at high temperature than the most slow-pyrolysed biochars. Their findings challenge the common opinion "natural charcoal and biochar are well suited as proxies for each other".

11.2.2 Greenhouse Gas Emission

Mitigation of greenhouse gas emission is an area of growing importance and concern due to global warming and increasing rate of GHG emission globally. CO_2, CH_4 and N_2O are three main greenhouse gases responsible for 90% of anthropogenic climate warming (IPCC 2013). IPCC (2013) also reported an increase of global average surface temperature by 0.85 °C during 1880–2012 based on multiple independently produce datasets and suggest an increase of 0.3–4.8 °C temperature by the end of this century. Environmental management through greenhouse gas mitigation include reduction and avoidance of emission along with removal of GHGs existing in the atmosphere (Smith et al. 2007a). Choosing biochar as soil amendment is an approach to mitigate climate change by reducing greenhouse gas emissions from soil. Application of biochar in soil has direct and indirect influence on soil physico-chemical properties, soil microbial diversity, abundance and function of soil microbial diversity that in turn affect production and emission of greenhouse gas. However, the potentiality of biochar in reducing GHG emission is controversial. Soil GHG significantly decreased or remain unchanged in some studied (Case et al. 2014; Quin et al. 2015; Liu et al. 2016; Scheer et al. 2011) while increased in others (Wang et al. 2012; He et al. 2017; Song et al. 2016). Some of the variables that influence the GHGs emission from soil environment under biochar treatments are study duration, soil texture and pH, feedstock used for biochar preparation, pH of the produced biochar and application rate and vegetation (Sohi et al. 2010; Woolf and Lehmann 2012; Hilscher and Knicker 2011; Lorenz and Lal 2014). Jones et al. (2011b), Luo et al. (2017) and Liang et al. (2010) documented the effect of biochar application on soil bulk density, pH, water holding capacity, cation exchange capacity, carbon and nitrogen dynamics and plant productivity, which have a significant effect on soil CO_2 and N_2O emissions. According to He et al. (2017); application of fertilizer and experimental condition influence CO_2 emission; they noted significant increase in CO_2 emission when biochar was applied in unfertilized soil, while it decreased when applied in fertilized soil in laboratory condition, but did not find any significant effect of biochar in field condition. Sun et al. (2014) found that biochar application in the at a rate of 30 t ha^{-1} reduced (31.5%) CO_2 emission from a pine forest soil. While in a field experiment in paddy soil amended with wheat straw biochar enhanced CO_2 emissions (12%), but N_2O emissions was reduced (41.8%) (Zhang et al. 2012b). However, studies of Wang et al. (2014b), Malghani et al. (2013) and Zhou et al. (2017) revealed no significant effect of biochar on CO_2 emission. Elevated CO_2 emission was noted in a temperate forest soil under application of sugar maple biochar at a rate of 5, 10 and 20 t ha^{-1} (Mitchell et al. (2015). Hawthorne et al. (2017) also reported significantly greater CO_2 fluxes from application of 10% biochar compared to 1% biochar in a Douglas-fir forest soil. The enhancement of CO_2 emission might be the addition of labile C from biochar and increased belowground net primary productivity (BNPP) (Zimmerman et al. 2011; Yoo and Kang 2012; Mukherjee and Lal 2013). While the cause behind suppressed CO_2 emission might be the absorption of soil CO_2 molecules by the large biochar surfaces and reduced enzymatic activity of microbes (Case et al. 2014; Liang et al. 2010; Liu et al. 2009).

Widespread use of synthetic nitrogen (N) fertilizer is the primary cause of agricultural soil emission of N_2O (Smith et al. 2008b). Rondon et al. (2005) first reported reduction in N_2O emissions in a greenhouse experiment after biochar amendment in soil. They recorded reduction of N_2O emissions by 50% under soybean cropping and by 80% for grass growing in a low-fertile oxisol. In a meta-analysis of biochar effect on N_2O emissions both in long and short term studies, Cayuela et al. (2015) found that soil N_2O emissions was reduced by 54 ± 3% at lab scale and 28 ± 16% at the field scale. Sun et al. (2014) also noted a significant decreased of cumulative N_2O emissions (25.5%) in a pine forest, when biochar was incorporated to the soil at 30 t ha^{-1}. Bass et al. (2016) found an interaction of cropping system with biochar on N_2O emission, they noted a decrease of N_2O emission under papaya cultivation while no effect was noted under banana cultivation. Fidel et al. (2019) noted a suppressive effect of biochar on N_2O emission in a continuous corn cropping system while did not found any effect on CO_2 emission. They also suggested that both soil moisture and temperature play role in CO_2 and N_2O emission. Contrastingly enhanced emission of soil N_2O was recorded by Hawthorne et al. (2017) in a forest soil when 10% biochar was applied but did not find any significant effect under 1% of biochar application. Cayuela et al. (2013) also found direct correlation between N_2O emission and biochar application rate. While in a study in temperate hardwood forest, no significant effect was noted for 5 t ha^{-1} biochar application on soil N_2O emission (Sackett et al. 2015). The primary mechanism of reduction of soil N_2O emission by biochar application might be the increased oxygen in soil due to soil aeration, which will inhibit denitrification of soil by microorganisms, that mostly occurs in low oxygen condition (Bateman and Baggs 2005; Taghizadeh-Toosi et al. 2011; Van Zwieten et al. 2010b; Hale et al. 2012). Another reason is absorption of inorganic nitrogen pool (NH^{4+}, NO^{3-} etc) by the biochar (Cornelissen et al. 2013b), which in turn will decrease nitrogen availability for nitrifiers and denitrifiers, reducing N_2O emission (Singh et al. 2010a; Clough et al. 2013). While rises in N_2O emissions may be due to increased soil water content influenced by biochar addition, that helps denitrification, or due to release of biochar embodied-N (Lorenz and Lal 2014). Soil pH is another important factor which is influenced due to biochar addition. An optimum rage of pH is preferable for reducing N_2O emission from agricultural soil because denitrifies have a wider pH optimum in the range of pH 4–8, while for the nitrifiers, the optimum range of pH is slightly acidic to slightly alkaline (Mørkved et al. 2007; Liu et al. 2010).

Biochar is also used to reduce the emission of soil methane (CH_4). Chicken manure biochar (10%, w/w) was found to significantly increase CH_4 uptake in forest soils (Yu et al. (2013). Xiao (2016) reported significantly higher efficiency of biochar in CH_4 uptake regardless of the application rate in a chestnut plantation in china. While, Sackett et al. (2015) reported no significant difference in CH_4 flux in biochar-treated and control soils in a temperate hardwood forest. Hawthorne et al. (2017) observed contrasting results, where significant reduction in soil CH_4 uptake under biochar application (1 or 10% w/w) was noted. This increased uptake of CH_4 in soil might be due to increase soil pH under biochar addition, which in turn facilitate the growth of methanotrophs (Jeffery et al. 2016; Anders et al. 2013) along with biochar induced decrease in soil bulk density and porosity favours aerobic methanotrophs and CH_4 oxidation and uptake by soil microbes (Brassard et al. 2016; Feng et al. 2012; Karhu et al. 2011; Van Zwieten et al. 2009). Enhanced CH_4 emission might be ascribed to the chemicals inhibitory effect of biochar on soil methanotrophs (Spokas 2013). Thus, the efficacy of biochar for GHGs mitigation is largely uncertain due to various factors involved in the reduction and enhancement of soil GHG emissions.

11.2.3 Soil and Water Pollution

Anthropogenic contaminants caused by rapid urbanization and industrialization are triggering degradation of water and soil quality in ecological environment. Soil and water are two basic needs for survival of lives. Soil serves as the main medium for plant growth, support human and animals, sustain plant and animal productivity, improve the quality of water and air (Zhou and Song 2004; Zhang et al. 2012c). Soil pollution can be remediated through physical, chemical and biological methods (Mendez and Maier 2008), but physical and chemical methods are not suitable for large scale management of arable soil, due to higher cost and the disadvantages of complexity and secondary pollution (Houben et al. 2013). Though biological remediation approaches are cheap and feasible, but its efficiency on improvement of soil quality is not constant because of its susceptibility to environment (Arthur et al.

2005) Biochar seized attention as a promising multi-beneficial remediating agent to stabilize soil contaminants, such as organic molecules, heavy metals, pesticides, herbicide etc. (Kong et al. 2014; Cheng and Lehmann 2009; Yuan et al. 2018; O'Connor et al. 2018). It acts as a soil conditioner via enhancing cation exchange capacity (CEC), pH and water holding capacity. Inorganic nutrients such as potassium, phosphorous, calcium, silica, boron and molybdenum are added to soil, making them bioavailable for plants as biochar is rich in inorganic nutrients derived from the feedstock (Page-Dumroese et al. 2015; Liu et al. 2014; Xiao et al. 2018; Xu et al. 2013). Cao and Harris (2010) confirm increased availability of nutrients (P, Mg and Ca) with increase of pyrolysis temperature. Moreover, addition of biochar to soil may effectively reduce eutrophication of nearby water bodies, and also underground waters pollution due to reduced leaching of nitrogen and phosphorus (Laird et al. 2010; Kookana et al. 2011). Soil contaminated with both organic and inorganic pollutants can be remediated with addition of biochar by reducing toxin bioavailability by both organic and inorganic pollutants (Ajayi and Horn 2017; Yao et al. 2012). Among the inorganic pollutants; heavy metals are non-biodegradable and persist in soil for very long (Sun et al. 2008). Thus lowering bioavailability of heavy metal is crucial to remediate contaminated soil. Biochar have negatively charged surfaces and functional groups that can strongly attract (electrostatic adsorption, ion exchange) metal ions having small ionic radii and high charges, or can stabilized metal via complexation or precipitation due to high soil pH introduced by biochar (Kong et al. 2014; Kumar et al. 2018; Mukherjee et al. 2011; Lu et al. 2012; Li et al. 2017). Biochar can transfer soluble metals forms to insoluble one by binding it to organic matter, oxides, carbonates and can fixed in soil (Xiao et al. 2018). It can also stabilize heavy metals through reduction. Choppala et al. (2015) reported efficiency of chicken manure biochar and black carbon for reducing Cr(VI) (extremely toxic and highly mobile) to Cr(III) (generally nontoxic) and subsequent immobilization in soils. Pyrolysis temperature of biochar plays an important role in removal efficiency of metal by biochar. Wang et al. (2018) reported better removal efficiency of higher temperature pyrolyzed biochar for Hg than biochar pyrolized at lower temperature. Contrastingly, Cao et al. (2009) reported dairy manure biochar pyrolyzed at 200 °C have better potential to remove Pb from soil than the same biochar pyrolyzed at 350 °C. Skjemstad et al. (2002) and Cheng et al. (2006) documented effectiveness of bamboo biochar to adsorb Cu, Ni, Hg and Cr from both water and soil, and Cd only in contaminated soil. While cotton stalk biochar can reduce Cd bioavailability in polluted soil through adsorption or co-precipitation (Zhou et al. (2008). Salt affected soil can also be remediated to a larger extend with addition of biochar, that reduce salt stress and enhanced plant growth and improve soil nutrients which in turn will counteract the adverse effect of Na (Kim et al. 2016; Wakeel 2013). Cao et al. (2011) and Jones et al. (2011a) reported reduction (66–97%) of pesticides such as atrazine, simazine, carbaryl and ethion, when biochar was applied to soil. Zhelezova et al. (2017) reported absorption of two herbicides, glyphosate (N-(phosphonomethyl)-glycine) and diuron (3-(3,4-dichlorophenyl)-1,1-dimethylurea) by biochar. They also noted a decrease of adsorption by aged biochar comparison with freshly prepared biochar. Similarly,

Martin et al. (2012) also found reduction (47%) of sorption by aged biochars for herbicide diuron. However, Trigo et al. (2014) found that, in some cases, biochar could serve for at least 2 years as effective sorbent of herbicides (indaziflam and fluoroethyldiaminotriazin). Polycyclic aromatic hydrocarbons (PAHs) and polychlorinated biphenyls (PCBs) in water and soils can also be removed by incorporation of biochar (Wang et al. 2013a; Beesley et al. 2010). Apart from persistent organic pollutant (POPs) such as PAHs, PCBs, PCDD (polychlorinated dibenzo-p-dioxins) and DFs (dibenzofurans), some emerging organic pollutants such as phthalate acid esters (dibutyl phthalate and di(2-ethylhexyl) phthalate), pharmaceutical and personal care products (PPCPs, trimethoprim and triclosan), naturally released estrogenic steroid hormone and its metabolites (estradiol and estrone) are becoming threat to the soil quality (WHO 2010; Petrović et al. 2001).

Biochar has been reported to be very effective in the uptake of a variety of organic chemicals including fungicides, pesticides, PAHs, and emerging contaminants such as steroid hormones (Beesley et al. 2010; Kookana et al. 2011; Song et al. 2012; Sarmah et al. 2010). Qin et al. (2013) also found significantly higher removal efficiency of contaminants with rice straw biochar on petroleum-contaminated soil than that of the unrestored soils. Molecular diameter of contaminants determines the strength of biochar sorption. Small molecules can penetrate to the micro and mesopores of biochar, while larger molecules tend to adsorb on the biochar surface, that may block pores (Nguyen et al. 2007).

Thus, biochar can improve the physicochemical properties of degraded land and immobilize both organic and inorganic pollutants, based on feedstocks, production methods, application rates, soil types and age of biochars (Obia et al. 2016). Despite the immense benefits of biochar, it can also be harmful if it contains PAHs, chlorinated hydrocarbon, dioxin and heavy metals derived from carbonization temperature and feedstock chosen (Chagger et al. 1998; Brown et al. 2006; Singh et al. 2010b). Therefore, to remediate polluted soil by applying biochar emphasis should be given in selection of proper feedstock and pyrolysis condition.

Removal of contaminants from water is most commonly done by chemical precipitation employing hydroxide, sulfide, phosphate and carbonate (Sharma and Bhattacharya 2017). But it creates problem when the sludge produced during chemical precipitation need to dispose. Biochar is a low cost sorbent for contaminants and pathogens, can absorb hydrocarbons, dyes, phenolics, pesticides, PAHs, antibiotics, inorganic metal ions. Potentiality of biochar as water purifier has been studied by many researchers (Klasson et al. 2013; Tong et al. 2011). Biochars obtained from straw and bamboo were reported to remove dyes from wastewater (Xu et al. 2011b; Yang et al. 2014), that were stable to light, oxidizing agents and aerobic digestion during conventional waste treatment. Xu et al. 2011b also documented efficiency of biochar derived from canola straw, peanut straw, soybean straw, and rice hulls to remove methyl violet from water.

11.3 Factors Influencing the Efficacy of Biochar

A range of process conditions like the feedstock composition, temperature and heating rate during pyrolysis can be optimized to obtain diverse amounts and properties of biochars. The physiochemical properties of biochars contribute to their function as a tool for environmental management (Lehmann and Joseph 2009).

Feedstock materials and temperature duringpyrolysis mostly influence the nutrient content in biochar. Screening Electron Micrograph (SEM) images of biochar material shows its resemblance with the composition of feedstock materials. Loss of nutrient during production is affected by the pyrolytic temperature. The concentration of nutrients like nitrogen reduces with increasing rate of pyrolytic temperature while, the availability of phosphorus increases. This nutrient content finally affects pH and electrical conductivity of the produced bichar (Chan and Xu 2009; Singh et al. 2010b). Maximum biochar yield obtained from low operational temperatures and low heating rate (Kwapinski et al. 2010). With increasing operational temperature, biochar yield decreases but the concentration of carbon increases. Biochar produced at high temperature have a high surface and also highly aromatic in nature that results it chemically recalcitrant (Keiluweit et al. 2010; Chen et al. 2011). Therefore, biochar produced at low temperature are considered to be more reactive in soil which have a less condensed carbon structure contributing soil fertility (Singh and Cowie 2008; Steinbeiss et al. 2009).

Higher reactivity of the biochar surfaces with soil particles is partly attributed to the presence of a range of reactive functional groups. Surface area of biochar increases with increasing HTT (High Heating Temperature) until it reaches the temperature at which deformation occurs, resulting subsequent decrease in surface area. The fundamental physical changes that occur in biochar are all temperature dependent. Heating rate and pressure affect the physical mass transfer of volatiles evolving at the given temperature from the reacting particles. Lua et al. (2004) reported that with increasing pyrolytic temperature from 250 °C to 500 °C, the Brunauer–Emmett–Teller (BET) surface area of biochar also found to increase.

11.4 Constrains of Biochar Application

Research related to biochar and its application has developed with time and important key findings were found related to agriculture, forestry, and global environment. More research in this field is required as the benefits vary from soil to soil and various other parameters like feedstocks, production of biochar, etc. Vaccari et al. (2015) reported that the effect of biochar on agricultural productivity also depended on plant species. Therefore, a synergetic effort is required to understand the limitations in biochar applications and the problems related to its applications. In a study carried out by Anyanwu et al. (2018), it was reported that aged biochar has a negative effect on the growth of both earthwarm and fungi in soil ecosystem. Studies also found that the aged biochar also led to reduction in underground root biomass of *Oryza sativa* and *Solanum lycopersicum*. Biochars also found to reduce the soil thermal diffusivity. In addition, an increase in weed growth was found with higher rate of biochar application. Khorram et al. (2018) found 200% increase in weed growth with an application rate of 15 t ha^{-1} of biochar. A delay in flowering of plant was also reported in some studies with addition of biochar.

In some cases, biochar also act as a soil contaminant due to the presence of some chemical compounds that may be form during the conversion processes. Studies are needed to see the presence of heavy metals and the plant-available organic compounds on the biochar surface. According to some researchers, these compounds may act as a fungicide or bactericide on the other hand; some others reported that they can serve as carbon source for some microbes (Painter 2001; Ogawa 1994). Also to understand the long term effect of biochar, a long term field study should be done on different soil types using biochars. Lack of long term studies on biochar application limits the actual scenario where various natural parameters are active. Moreover, the cost related to the feedstock preparation, biochar production, transportation and application is full of uncertainties that need to be clear.

There is also lack of a uniform system for the classification and governance regarding the commercialization of biochars for land and other applications. This will be helpful to the consumers in using biochar for various applications. The environmental agencies of different locations can play an active role regarding this issue. No standard biochar rate for application is available for specific type of soil regarding specific result. There is also lack of a decision support system for choosing a particular type and rate of biochar to fulfill a particular need. Mechanisms related to biochar-soil interaction are very complex and multiples assumptions have been made. In recent years, biochar is attracting a huge attention in the research field. The research outcome should be updated and make available for the benefit of people to apply on practical field.

11.5 Nano-biochar and Its Prospects

Biochar is gaining a huge attention in recent time from scientists, policymakers, farmers, and investors due to its properties that directly or indirectly helps humankind. Bulk biochar mostly applied for agronomic and environmental purposes. Recent studies found that generation of nano biochar (N-BC) from the physical degradation of bulk biochar (B-BC). Nano biochar is characterized by having a size smaller than 100 nm than the bulk biochar (Wang et al. 2013b; Chen et al. 2017). Due to its size, nano biochar have an excellent mobility both in soils and water and can act as a carrier particle for natural solutes and contaminants (Ahmad et al. 2014; Lian and Xing 2017). With increasing the application of biochar in soil, the degree of formation of nano biochar will increase. However, the knowledge on the formation of nano biochar particle is much limited. Some mechanisms that results in the formation of nano biochar are pore collapse and matrix fracturing during production of biochar and also weathering process in the environment. The carbon matrixes that can be easily fragmentized are readily mineralized through various chemical and microbial processes (Lin et al. 2012; Warnock et al. 2007). Degradation and conversion of nano biochar from the bulk biochar can be against longevity of biochar within soil systems. Due to its size particle and mobility, the toxic effects of nano biochar is considered more than the bulk biochar (Wang et al. 2016). Exposure of nano biochar may also trigger risks to organisms in waters and soils. It is considered that hetero aggregation formation can prevent the vertical transport of these nano biochar in soil ecosystem and therefore weaken its negative effect.

References

Abel S, Peters A, Trinks S, Schonsky H, Facklam M, Wessolek G (2013) Impact of biochar and hydrochar addition on water retention and water repellency of sandy soil. Geoderma 202:183–191

Agbna GHD, Dongli S, Zhipeng L, Elshaikh NA, Guangcheng S, Timm LC (2017) Effects of deficit irrigation and biochar addition on the growth, yield, and quality of tomato. Sci Hortic 222:3–10

Ahmad M, Rajapaksha AU, Lim JE, Zhang M, Bolan N, Mohan D, Vithanage M, Lee SS, Ok YS (2014) Biochar as a sorbent for contaminant management in soil and water: a review. Chemosphere 99:19–33

Ajayi AE, Horn R (2017) Biochar-induced changes in soil resilience: effects of soil texture and biochar dosage. Pedosphere 27:236–247

Akhtar SS, Andersen MN, Liu F (2015a) Biochar mitigates salinity stress in potato. J Agron Crop Sci. https://doi.org/10.1111/jac.12132

Akhtar SS, Andersen MN, Liu F (2015b) Residual effects of biochar on improving growth, physiology and yield of wheat under salt stress. Agric Water Manage 158:61–68

Akhter A, Hage-Ahmed K, Soja G, Steinkellner S (2016) Potential of Fusarium wilt inducing chlamydospores, in vitro behaviour in root exudates and physiology of tomato in biochar and compost amended soil. Plant Soil 406:425–440

Allen RL (1847) A brief compend of American agriculture, 2nd edn. C. M. Saxton, New York. https://doi.org/10.5962/bhl.title.20151

Anders E, Watzinger A, Rempt F et al (2013) Biochar affects the structure rather than the total biomass of microbial communities in temperate soils. Agric Food Sci 22:404–423

Anyanwu IN, Alo MN, Onyekwere AM, Crosse JD, Nworie O, Chamba EB (2018) Influence of biochar aged in acidic soil on ecosystem engineers and two tropical agricultural plants. Ecotoxicol Environ Safe 153:116–126

Arthur EL, Rice PJ, Rice PJ, Anderson TA, Baladi SM, Henderson KLD et al (2005) Phytoremediation—an overview. Crit Rev Plant Sci 24:109–122

Asai H, Samson BK, Stephan HM, Songyikhangsuthor K, Homma K, Kiyono Y, Inoue Y, Shiraiwa T, Horie T (2009) Biochar amendment techniques for upland rice production in northern Laos. Soil physical properties, leaf SPAD and grain yield. Field Crops Res 111:81–84

Bamminger C, Poll C, Sixt C, Högy P, Wüst D, Kandeler E, Marhan S (2016) Short-term response of soil microorganisms to biochar addition in a temperate agroecosystem under soil warming. Agric Ecosyst Environ 233:3–10

Bass AM, Bird MI, Kay G, Muirhead B (2016) Soil properties, greenhouse gas emissions and crop yield under compost, biochar and co-composted biochar in two tropical agronomic systems. Sci Total Environ 550:459–470

Basso AS, Miguez FE, Laird DA, Horton R, Westgate M (2013) Assessing potential of biochar for increasing water-holding capacity of sandy soils. Gcb Bioenergy 5(2):3–10

Bateman EJ, Baggs EM (2005) Contributions of nitrification and denitrification to N_2O emissions from soils at different water-filled pore space. Biol Fertil Soils 41:379–388

Beesley L, Moreno-Jiménez E, Gomez-Eyles JL (2010) Effects of biochar and greenwaste compost amendments on mobility, bioavailability and toxicity of inorganic and organic contaminants in a multi-element polluted soil. Environ Pollut 158(6):2282–2287

Béghin-Tanneau R, Guérin F, Guiresse M, Kleiber D, Scheiner JD (2019) Carbon sequestration in soil amended with anaerobic digested matter. Soil Till Res 192:87–94

Biederman LA, Harpole WS (2013) Biochar and its effects on plant productivity and nutrient cycling: a meta-analysis. Glob Chang Biol Bioener 5:202–214

Bonanomi G, Ippolito F, Scala F (2015) A "black" future for plant pathology? Biochar as a new soil amendment for controlling plant diseases. J Plant Pathol 97:223–234

Brady NC, Well RR (2012) The nature and properties of soils, 14th edn. Dorling Kindersley (India) Pvt. Ltd., licensees of Pearson Education in South Asia

Brassard P, Godbout S, Raghavan V (2016) Soil biochar amendment as a climate change mitigation tool: key parameters and mechanisms involved. J Environ Manag 181:484–497

Brown RA, Kercher AK, Nguyen TH, Nagle DC, Ball WP (2006) Production and characterization of synthetic wood chars for use as surrogates for natural sorbents. Org Geochem 37:321–333

Cao X, Harris W (2010) Properties of dairy-manure-derived biochar pertinent to its potential use in remediation. Bioresour Technol 101(14):5222–5228

Cao XD, Ma LN, Gao B, Harris W (2009) Dairy-manure derived biochar effectively sorbs lead and atrazine. Environ Sci Technol 43:3285–3291

Cao X, Ma L, Liang Y, Gao B, Harris W (2011) Simultaneous immobilization of lead and atrazine in contaminated soils using dairy-manure biochar. Environ Sci Technol 45(11):4884–4889

Case SD, McNamara NP, Reay DS, Whitaker J (2014) Can biochar reduce soil greenhouse gas emissions from a Miscanthus bioenergy crop? Gcb Bioenergy 6(1):76–89

Cayuela ML, Sanchez-Monedero M, Roig A, Hanley K, Enders A, Lehmann J (2013) Biochar and denitrification in soils: when, how much and why does biochar reduce N2O emissions? Sci Rep 3:1732

Cayuela ML, Jeffery S, van Zwieten L (2015) The molar H: Corg ratio of biochar is a key factor in mitigating N2O emissions from soil. Agric Ecosyst Environ 202:135–138

Chagger HK, Kendall A, McDonald A, Pourkashanian M, Williams A (1998) Formation of dioxins and other semi-volatile compounds in biomass combustion. Appl Energy 60:101–114

Chan KY, Xu K (2009) Biochar: nutrient properties and their enhancement. In: Lehmann J, Joseph S (eds) Biochar for environmental management: science and technology. Earthscan Publishers Ltd, London

Chan KY, Van Zwieten L, Meszaros I, Downie A, Joseph S (2007) Agronomic values of greenwaste biochar as a soil amendment. Soil Res 45(8):629–634

Chan KY, van Zwieten L, Meszaros I, Downie A, Joseph S (2008) Using poultry litter biochars as soil amendments. Aust J Soil Res 46:437–444

Chen X, Chen G, Chen L, Chen Y, Lehmann J, McBride MB, Hay AG (2011) Adsorption of copper and zinc by biochars produced from pyrolysis of hardwood and corn straw in aqueous solution. Bioresour Technol 102:8877–8884

Chen M, Wang D, Yang F, Xu X, Xu N, Cao X (2017) Transport and retention of biochar nanoparticles in a paddy soil under environmentally-relevant solution chemistry conditions. Environ Pollut 230:540–549

Cheng CH, Lehmann J (2009) Ageing of black carbon along a temperature gradient. Chemosphere 75(8):1021–1027

Cheng CH, Lehmann J, Thies JE, Burton SD, Engelhard MH (2006) Oxidation of black carbon by biotic and abiotic processes. Organ Geochem 37:1477–1488

Choppala G, Bolan N, Kunhikrishnan A et al (2015) Concomitant reduction and immobilization of chromium in relation to its bioavailability in soils. Environ Sci Pollut Res 22(12):8969–8978

Christoph GT, Wenceslau L, Johannes N, Thomas M, de Vasconcelos JL, Winfried EHB, Zech W (2007) Long-term effects of manure, charcoal and mineral fertilization on crop production and fertility on a highly weathered central Amazonian upland soil. Plant Soil 291:275–290

Clough TJ, Condron LM, Kammann C, Müller C (2013) A review of biochar and soil nitrogen dynamics. Agronomy 3:275–293

Copley TR, Aliferis KA, Jabaji S (2015) Maple bark biochar affects Rhizoctonia solani metabolism and increases damping-off severity. Phytopathology 105:1334–1346

Cornelissen G, Martinsen V, Shitumbanuma V, Alling V, Breedveld GD, Rutherford DW, Sparrevik M, Hale SE, Obia A, Mulder J (2013a) Biochar effect on maize yield and soil characteristics in five conservation farming sites in Zambia. Agron J 3:256–274

Cornelissen G, Rutherford DW, Arp HPH, Dörsch P, Kelly CN, Rostad CE (2013b) Sorption of pure N_2O to biochars and other organic and inorganic materials under anhydrous conditions. Environ Sci Technol 47:7704–7712

De Gryze S, Cullen M, Durschinger L, Lehmann J, Bluhm D, Six J (2010) Evaluation of opportunities for generating carbon offsets from soil sequestration of biochar. In: An issue paper commissioned by climate action reserve, final version

De Tender CA, Debode J, Vandecasteele B, D'Hose T, Cremelie P, Haegeman A, Ruttink T, Dawyndt P, Maes M (2016) Biological, physicochemical and plant health responses in lettuce and strawberry in soil or peat amended with biochar. Appl Soil Ecol 107:1–12

Deenik JL, McClellan T, Uehara G, Antal MJ, Campbell S (2010) Charcoal volatile matter content influences plant growth and soil nitrogen transformations. Soil Sci Soc Am J 74:1259–1270

Dendooven L, Patino-Zúniga L, Verhulst N, Luna-Guido M, Marsch R, Govaerts B (2012) Global warming potential of agricultural systems with contrasting tillage and residue management in the central highlands of Mexico. Agric Ecosyst Environ 152:50–58

Deng W, Van Zwieten L, Lin Z, Liu X, Sarmah AK, Wang H (2017) Sugarcane bagasse biochars impact respiration and greenhouse gas emissions from a latosol. J Soils Sediments 17:632–640

Díaz-Zorita M, Duarte GA, Grove JH (2002) A review of no-till systems and soil management for sustainable crop production in the sub-humid and semiarid Pampas of Argentina. Soil Tillage Res 65:1–18

Downie AE, van Zwieten L, Smernik RJ, Morris S, Munroe PR (2011) Terra preta Australis: reassessing the carbon storage capacity of temperate soils. Agric Ecosyst Environ 140:137–147

Egamberdieva D, Reckling M, Wirth S (2017) Biochar-based Bradyrhizobium inoculum improves growth of lupin (*Lupinus angustifolius* L.) under drought stress. Eur J Soil Biol 78:3–10

Elad Y, David DR, Meller Harel Y, Borenshtein M, Ben Kalifa H, Silber A, Graber ER (2010) Induction of systemic resistance in plants by biochar, a soil-applied carbon sequestering agent. Phytopathology 100:913–921

Elad Y, Cytryn E, Harel YM, Lew B, Graber ER (2011) The biochar effect: plant resistance to biotic stresses. Phytopathol Mediterr 50(3):335–349

Elmer WH, Pignatello JJ (2011) Effect of biochar amendments on mycorrhizal associations and Fusarium crown and root rot of asparagus in replant soils. Plant Disease 95:960–966

Fahad S, Hussain S, Saud S, Tanveer M, Bajwa AA, Hassan S et al (2015) A biochar application protects rice pollen from high-temperature stress. Plant Physiol Biochem 96:281–287

Feng Y, Xu Y, Yu Y, Xie Z, Lin X (2012) Mechanisms of biochar decreasing methane emission from Chinese paddy soils. Soil Biol Biochem 46:80–88

Feng LIANG, Gui-tong LI, Qi-mei LIN, Xiao-rong ZHAO (2014) Crop Yield and Soil Properties in the First 3 Years After Biochar Application to a Calcareous Soil. J Integr Agric 13(3):525–532

Fidel RB, Laird DA, Parkin TB (2019) Effect of biochar on soil greenhouse gas emissions at the laboratory and field scales. Soil Syst 3(1):8

Fontaine S, Mariotti A, Abbadie L (2003) The priming effect of organic matter: a question of microbial competition? Soil Biol Biochem 35(6):837–843

Fowles M (2007) Black carbon sequestration as an alternative to bioenergy. Biomass Bioener 31:426–432

Frenkel O, Jaiswal AK, Elad Y, Lew B, Kammann C, Graber ER (2017) The effect of biochar on plant diseases: what should we learn while designing biochar substrates? J Environ Eng Landsc Manag 25:105–113

Gaskin JW, Speir RA, Harris K, Das KC, Lee RD, Morris LA, Fisher DS (2010) Effect of peanut hull and pine chip biochar on soil nutrients, corn nutrient status, and yield. Agron J 102:623–633

Gaunt J, Lehmann J (2008) Energy balance and emissions associated with biochar sequestration and pyrolysis bioenergy production. Environ Sci Technol 42:4152–4158

Ghorbani M, Asadi H, Abrishamkesh S (2019) Effects of rice husk biochar on selected soil properties and nitrate leaching in loamy sand and clay soil. Int Soil Water Conserv Res 7:258–263

Glaser B, Lehmann J, Zech W (2002) Ameliorating physical and chemical properties of highly weathered soils in the tropics with charcoal—a review. Biol Fertil Soils 35:219–230

Graber E, Tsechansky L, Lew B, Cohen E (2014a) Reducing capacity of water extracts of biochars and their solubilization of soil Mn and Fe. Eur J Soil Sci 65:162–172

Graber ER, Frenkel O, Jaiswal AK, Elad Y (2014b) How may biochar influence severity of diseases caused by soil borne pathogens? Carbon Manage 5:169–183

Gupta VK, Schmoll M, Estrella AH, Upadhyay RS, Druzhinina I, Tuohy MG (2014) Biotechnol Biol Trichoder:543–549

Haider G, Koyro HW, Azam F, Steffens D, Müller C, Kammann C (2014) Biochar but not humic acid product amendment affected maize yields via improving plant-soil moisture relations. Plant Soil 395:141–157

Hailegnaw NS, Mercl F, Pračke K, Száková J, Tlustoš P (2019) High temperature-produced biochar can be efficient in nitrate loss prevention and carbon sequestration. Geoderma 338:48–55

Hale SE, Lehmann J, Rutherford D et al (2012) Quantifying the total and bioavailable polycyclic aromatic hydrocarbons and dioxins in biochars. Environ Sci Technol 46:2830–2838

Hansen V, Müller-Stöver D, Ahrenfeldt J, Holm JK, Henriksen UB, Hauggaard-Nielsen H (2015) Gasification biochar as a valuable by-product for carbon sequestration and soil amendment. Biomass Bioenergy 72:300–308

Hansen V, Müller-Stöver D, Munkholm LJ, Peltre C, Hauggaard-Nielsen H, Jensen LS (2016) The effect of straw and wood gasification biochar on carbon sequestration, selected soil fertility indicators and functional groups in soil: an incubation study. Geoderma 269:99–107

Harel YM, Elad Y, Rav-David D et al (2012) Biochar mediates systemic response of strawberry to foliar fungal pathogens. Plant and Soil 357(1):245–257

Hass A, Gonzalez JM, Lima IM, Godwin HW, Halvorson JJ, Boyer DG (2012) Chicken manure biochar as liming and nutrient source for acid Appalachian soil. J Environ Qual 41:1096–1106

Hawthorne I, Johnson MS, Jassal RS, Black TA, Grant NJ, Smukler SM (2017) Application of biochar and nitrogen influences fluxes of CO_2, CH_4 and N_2O in a forest soil. J Environ Manage 192:203–214

He Y, Zhou X, Jiang L, Li M, Du Z, Zhou G, Shao J, Wang X, Xu Z, Hosseini Bai S, Wallace H (2017) Effects of biochar application on soil greenhouse gas fluxes: a meta analysis. Gcb Bioenergy 9(4):743–755

Hilscher A, Knicker H (2011) Carbon and nitrogen degradation on molecular scale of grass-derived pyrogenic organic material during 28 months of incubation in soil. Soil Biol Biochem 43:261–270

Hoitink HA, Fahy PC (1986) Basis for the control of soil borne plant pathogens with composts. Ann Rev Phytopathol 24(1):93–114

Houben D, Evrard L, Sonnet P (2013) Beneficial effects of biochar application to contaminated soils on the bioavailability of Cd, Pb and Zn and the biomass production of rapeseed (Brassica napus L.). Biomass Bioener 57:196–204. https://char-grow.com/biochar-impact-nutrient-water-retention

Huang R, Tian D, Liu J, Lv S, He X, Gao M (2018) Responses of soil carbon pool and soil aggregates associated organic carbon to straw and straw-derived biochar addition in a dryland cropping mesocosm system. Agric Ecosyst Environ 265:576–586

IPCC (2013) Summary for policy markers. In: Stocker et al (eds) Climate change 2013: In: The physical science basis. Contribution of working group I to the fifth assessment report of the intergovernmental panel on climate change. Cambridge University Press, Cambridge/New York, pp 3–29

Jaiswal AK, Elad Y, Graber ER, Frenkel O (2014) *Rhizoctonia solani* suppression and plant growth promotion in cucumber as affected by biochar pyrolysis temperature, feedstock and concentration. Soil Biol Biochem 69:110–118

Jaiswal AK, Frenkel O, Elad Y, Lew B, Graber ER (2015) Non-monotonic influence of biochar dose on bean seedling growth and susceptibility to Rhizoctonia solani: the "Shifted Rmax-Effect". Plant Soil 395:125–140

Jaiswal AK, Elad Y, Paudel I, Graber ER, Cytryn E, Frenkel O (2017) Linking the belowground microbial composition, diversity and activity to soilborne disease suppression and growth promotion of tomato amended with biochar. Sci Rep 7. https://doi.org/10.1038/srep44382

Jaiswal AK, Elad Y, Cytryn E, Graber ER, Frenkel O (2018) Activating biochar by manipulating the bacterial and fungal microbiome through pre-conditioning. New Phytol. https://doi.org/10.1111/nph.15042

Jeffery S, Verheijen FGA, van der Velde M, Bastos AC (2011) A quantitative review of the effects of biochar application to soils on crop productivity using meta-analysis. Agric Ecosyst Environ 144:175–187

Jeffery S, Verheijen FG, Kammann C, Abalos D (2016) Biochar effects on methane emissions from soils: a meta-analysis. Soil Biol Biochem 101:251–258

Jones DL, Edwards-Jones G, Murphy DV (2011a) Biochar mediated alterations in herbicide breakdown and leaching in soil. Soil Biol Biochem 43(4):804–813

Jones DL, Murphy DV, Khalid M, Ahmad W, Edwards-Jones G, Deluca TH (2011b) Short-term biochar-induced increase in soil CO_2 release is both biotically and abiotically mediated. Soil Biol Biochem 43:1723–1731

Joseph SD, Camps-Arbestain M, Lin Y, Munroe P, Chia CH, Hook J, van Zwieten L, Kimber S, Cowie A, Singh BP, Lehmann J, Foidl N, Smernik RJ, Amonette JE (2010) An investigation into the reactions of biochar in soil. Aust J Soil Res 48:501–515

Kammann CI, Linsel S, Gößling JW, Koyro HW (2011) Influence of biochar on drought tolerance of Chenopodium quinoa, willd and on soil-plant relations. Plant Soil 345:195–210. https://doi.org/10.1007/s11104-011-0771-5

Karhu K, Mattila T, Bergström I, Regina K (2011) Biochar addition to agricultural soil increased CH4 uptake and water holding capacity–Results from a short-term pilot field study. Agric Ecosyst Environ 140(1–2):309–313

Katterer T, Ronbroeck D, Andren O et al (2019) Bio char addition persistently increased soil fertility and yields in Maize-soybean rotation over 10 years in sub-humid regions of Kenya. Field Crops Res 235:18–26

Keiluweit M, Nico PS, Johnson MG, Kleber M (2010) Dynamic molecular structure of plant biomass-derived black carbon (biochar). Environ Sci Technol 44:1247–1253

Khorram SM, Fatemi A, Khan MA, Kiefer R, Jafarnia S (2018) Potential risk of weed outbreak by increasing biochar's application rates in slow-growth legume, lentil (*Lens culinaris medik.*). J Sci Food Agric 98(6):2080–2088

Kim HS, Kim KR, Yang JE, Ok YS, Owens G, Nehls T, Wessolek G, Kim KH (2016) Effect of biochar on reclaimed tidal land soil properties and maize (*Zea mays* L.) response. Chemosphere 142:153–159

Kimetu JM, Lehmann J (2010) Stability and stabilisation of biochar and green manure in soil with different organic carbon contents. Aus J Soil Res 48(7):577–585

Kimetu JM, Lehmann J, Ngoze SO, Mugendi DN, Kinyangi JM, Riha S, Verchot L, Recha JW, Pell AN (2008) Reversibility of soil productivity decline with organic matter of differing quality along a degradation gradient. Ecosystems 11:726–739

Klasson KT, Ledbetter CA, Uchimiya M, Lima IM (2013) Activated biochar removes 100% dibromochloropropane from field well water. Environ Chem Lett 11(3):271–275

Kleber M (2010) What is recalcitrant soil organic carbon? Environ Chem 7:320–332
Knoblauch C, Maarifat AA, Pfeiffer EM, Haefele SM (2011) Degradability of black carbon and its impact on trace gas fluxes and carbon turnover in paddy soils. Soil Biol Biochem 43:1768–1778
Kong LL, Liu WT, Zhou QX (2014) Biochar: an effective amendment for remediating contaminated soil. In: Reviews of environmental contamination and toxicology. Springer, Cham, pp 83–99
Kookana RS, Sarmah AK, Van Zwieten L, Krull E, Singh B (2011) Chapter three—biochar application to soil: agronomic and environmental benefits and unintended consequences. In: Donald LS (ed) Advances in agronomy. Academic, New York, pp 103–143
Kumar A, Singh JS (2017) Cyanoremediation: a green-clean tool for decontamination of synthetic pesticides from agro-and aquatic ecosystems. In: Singh JS, Seneviratne G (eds) Agro-environmental sustainability: Vol (2) managing environmental pollution. Springer, Cham, pp 59–83
Kumar A, Kaushal S, Saraf S, Singh JS (2017) Cyanobacterial biotechnology: an opportunity for sustainable industrial production. Clim Change Environ Sustain 5(1):97–110
Kumar A, Joseph S, Tsechansky L, Privat K, Schreiter IJ, Schüth C et al (2018) Biochar aging in contaminated soil promotes Zn immobilization due to changes in biochar surface structural and chemical properties. Sci Total Environ 626:953–961
Kuzyakov Y (2010) Priming effects: interactions between living and dead organic matter. Soil Biol Biochem 42:1363–1371
Kuzyakov Y, Friedel JK, Stahr K (2000) Review of mechanisms and quantification of priming effects. Soil Biol Biochem 32:1485–1498
Kwapinski W, Byrne CMP, Kryachko E, Wolfram P, Adley C, Leahy JJ, Novotny E, Hayes MHB (2010) Biochar from biomass and waste. Waste Biomass Valor 1:177–189
Laird DA (2008) The charcoal vision: a win-win-win scenario for simultaneously producing bioenergy, permanently sequestering carbon, while improving soil and water quality. Agron J 100:178–181
Laird D, Fleming P, Wang B, Horton R, Karlen D (2010) Biochar impact on nutrient leaching from a Midwestern agricultural soil. Geoderma 158(3–4):436–442
Lal R (1999) World soils and greenhouse effect. IGBP Global Change Newsl 37:4–5
Lal R (2004) Soil carbon Sequestration impacts on global climate change and food security. Science 304:1623–1627
Lehmann J (2007a) Bio-energy in the black. Front Ecol Environ 5:381–387
Lehmann J (2007b) A handful of carbon. Nature 447:143–144
Lehmann J (2009) Terra preta Nova–where to from here? In: Amazonian dark Earths: Wim Sombroek's vision. Springer, Dordrecht, pp 473–486
Lehmann J, Joseph S (2009) Biochar for environmental management: an introduction. In: Lehmann J, Joseph S (eds) Biochar for environmental management: science and technology. Earthscan Publishers Ltd., London
Lehmann J, Joseph S (2015) Biochar for environmental management: an introduction. In: Lehmann J, Joseph S (eds) Biochar for environmental management: science, technology and implementation. Earthscan/Routledge, London, pp 1–1214
Lehmann J, Rondon M (2006) Bio-char soil management on highly weathered soils in the humid tropics. In: Uphoff N, Ball AS, Herren H, Husson O, Laing M, Palm C, Pretty J, Sanchez P, Sanginga N, Thies J (eds) Biological approaches to sustainable soil systems. Taylor and Francis, Boca Raton, pp 517–530
Lehmann J, Gaunt J, Rondon M (2006) Bio-char sequestration in terrestrial ecosystems—a review. Mitig Adapt Strateg Glob Change 11:403–427
Lehmann J, Rillig MC, Thies J, Masiello CA, Hockaday WC, Crowley D (2011) Biochar effects on soil biota—a review. Soil Biol Biochem 43(9):1812–1836
Li HB, Xiao L, Evandro B et al (2017) Mechanisms of metal sorption by biochars: biochar characteristics and modifications. Chemosphere 178:466–478

Lian F, Xing B (2017) Black carbon (biochar) in water/soil environments: molecular structure, sorption, stability, and potential Risk. Environ Sci Technol 51:13517–13532
Liang B, Lehmann J, Solomon D, Kinyangi J, Grossman J, O'Neill B, Skjemstad JO, Thies J, Luizão FJ, Petersen J, Neves EG (2006) Black carbon increases cation exchange capacity in soils. Soil Sci Soc Am J 70:1719–1730
Liang BQ, Lehmann J, Sohi SP, Thies JE, O'Neill B, Trujillo L, Gaunt J, Solomon D, Grossman J, Neves EG, Luizãoc FJ (2010) Black carbon affects the cycling of non-black carbon in soil. Org Geochem 41:206–213
Lin Y, Munroe P, Joseph S, Kimber S, VanZwieten L (2012) Nanoscale organo-mineral reactions of biochars in ferrosol: an investigation using microscopy. Plant Soil 357:369–380
Liu QS, Liu Y, Show KY, Tay JH (2009) Toxicity effect of phenol on aerobic granules. Environ Technol 30:69–74
Liu B, Mørkved PT, Frostegård Å, Bakken LR (2010) Denitrification gene pools, transcription and kinetics of NO, N_2O and N_2 production as affected by soil pH. FEMS Microbiol Ecol 72(3):407–417
Liu XY, Qu JJ, Li LQ, Zhang AF, Zheng JF, Zheng JW, Pan GX (2012) Can biochar be an ecological engineering technology to depress N2O emission in rice paddies?—a cross site field experiment from South China. Ecol Eng 42:168–173
Liu XY, Zhang AF, Ji CY, Joseph S, Bian RJ, Li LQ, Pan GX, PazFerreiro J (2013) Biochar's effect on crop productivity and the dependence on experimental conditions—a meta-analysis of literature data. Plant Soil 373:583–594
Liu XY, Li LQ, Bian RJ, Chen D, Qu JJ, Wanjiru Kibue G, Pan GX, Zhang XH, Zheng JW, Zheng JF (2014) Effect of biochar amendment on soil-silicon availability and rice uptake. J Plant Nutr Soil Sci 177:91–96
Liu C, Wang H, Tang X, Guan Z, Reid BJ, Rajapaksha AU, Ok YS, Sun H (2015) Biochar increased water holding capacity but accelerated organic carbon leaching from a sloping farmland soil in China. Environ Sci Pollut Res 23:995–1006
Liu S, Zhang Y, Zong Y, Hu Z, Wu S, Zhou J, Jin Y, Zou J (2016) Response of soil carbon dioxide fluxes, soil organic carbon and microbial biomass carbon to biochar amendment: a meta-analysis. Gcb Bioenergy 8(2):392–406
Liu Z, He T, Cao T, Yang T, Meng J, Chen W (2017a) Effects of biochar application on nitrogen leaching, ammonia volatilization and nitrogen use efficiency in two distinct soils. J Soil Sci Plant Nutrit 17(2):515–528
Liu C, Liu F, Ravnskov S, Rubæk GH, Sun Z, Andersen MN (2017b) Impact of wood biochar and its interactions with mycorrhizal fungi, phosphorus fertilization and irrigation strategies on potato growth. J Agron Crop Sci 203(2):03–10
Lorenz K, Lal R (2014) Biochar application to soil for climate change mitigation by soil organic carbon sequestration. J Soil Sci Plant Nutrit Soil Sci 177:651–670
Lu H, Zhang W, Yang Yet al. (2012) Relative distribution of Pb^{2+} sorption mechanisms by sludge-derived biochar. Water Res 46(3):854–862
Lu W, Ding W, Zhang J, Li Y, Luo J, Bolan N, Xie Z (2014) Biochar suppressed the decomposition of organic carbon in a cultivated sandy loam soil: a negative priming effect. Soil Biol Biochem 76:12–21
Lua AC, Yang T, Guo J (2004) Effects of pyrolysis conditions on the properties of activated carbons prepared from pistachio-nut shells. J Anal Appl Pyrol 72:279–287
Luo Y, Jiao Y, Zhao X, Li G, Zhao L, Meng H (2014) Improvement to maize growth caused by biochars derived from six feedstock's prepared at three different temperatures. J Integ Agric 13:533–540
Luo Y, Yu ZY, Zhang KL, Xu JM, Brookes PC (2016) The properties and functions of biochars in forest ecosystems. J Soils Sed 16:2005–2020
Luo Y, Zang HD, Yu ZY, Chen ZY, Gunina A, Kuzyakov Y, Xu JM, Zhang KL, Brookes PC (2017) Priming effects in biochar enriched soils using a three-source-partitioning approach: ^{14}C labelling and ^{13}C natural abundance. Soil Biol Biochem 106:28–35

Major J, Lehmann J, Rondon M, Goodale C (2010) Fate of soil-applied black carbon: downward migration, leaching and soil respiration. Glob Change Biol 16:1366–1379
Malghani S, Gleixner G, Trumbore SE (2013) Chars produced by slow pyrolysis and hydrothermal carbonization vary in carbon sequestration potential and greenhouse gases emissions. Soil Biol Biochem 62:137–146
Marks EA, Alcañiz JM, Domene X (2014) Unintended effects of biochars on short-term plant growth in a calcareous soil. Plant Soil 385(1–2):87–105
Marris E (2006) Putting the carbon back: black is the new green. Nature 442(7103):624–626
Martin SM, Kookana RS, Van Zwieten L, Krull E (2012) Marked changes in herbicide sorption-desorption upon ageing of biochars in soil. J Hazard Mater 231:70–78
Mathews JA (2008) Carbon-negative biofuels. Energy Policy 36:940–945
Mau AE, Utami SR (2014) Effects of biochar amendment and arbuscular mycorrhizal fungi inoculation on availability of soil phosphorus and growth of maize. J Degr Min Lands Manage 1:69–74
McHenry MP (2009) Agricultural biochar production, renewable energy generation and farm carbon sequestration in Western Australia: Certainty, uncertainty and risk. Agric Ecosyst Environ 129:1–7
Mendez MO, Maier RM (2008) Phytostabilization of mine tailings in arid and semiarid environments—an emerging remediation technology. Environ Health Perspect 116:278–283
Mensah AK, Frimpong KA (2018) Bio char and/or compost applications improve soil properties, growth, and yield of maize grown in acidic rainforest and coastal savannah soils in Ghana. Interna J Agron 2018:8 pages. https://doi.org/10.1155/2018/6837404
Mickan BS, Abbott LK, Stefanova K, Solaiman ZM (2016) Interactions between biochar and mycorrhizal fungi in a water-stressed agricultural soil. Mycorrhiza 26(6):3–10
Milne E, Powlson DS, Cerri CE (2007) Soil carbon stocks at regional scales (preface). Agric Ecosyst Environ 122:1–2
Minasny B, Malone BP, McBratney AB, Angers DA, Arrouays D, Chambers A, Chaplot V, Chen ZS, Cheng K, Das BS, Field DJ (2017) Soil carbon 4 per mille. Geoderma 292:59–86
Mitchell PJ, Simpson AJ, Soong R, Simpson MJ (2015) Shifts in microbial community and water-extractable organic matter composition with biochar amendment in a temperate forest soil. Soil Biol Biochem 81:244–254
Mørkved PT, Dörsch P, Bakken LR (2007) The N_2O product ratio of nitrification and its dependence on long-term changes in soil pH. Soil Biol Biochem 39:2048–2057
Mukherjee A, Lal R (2013) Biochar impacts on soil physical properties and greenhouse gas emissions. Agronomy 3:313–339
Mukherjee A, Lal R (2014) The biochar dilemma. Soil Res 52:217–230
Mukherjee A, Zimmerman AR, Harris W (2011) Surface chemistry variations among a series of laboratory-produced biochars. Geoderma 163(3–4):247–255
Mukherjee A, Lal R, Zimmerman AR (2014) Effects of biochar and other amendments on the physical properties and greenhouse gas emissions of an artificially degraded soil. Sci Total Environ 487:26–36
Mulcahy DN, Mulcahy DL, Dietz D (2013) Biochar soil amendment increases tomato seedling resistance to drought in sandy soils. J Arid Environ 88:3–10
Nadeem SM, Imran M, Naveed M, Khan MY, Ahmad M, Zahir ZA, Crowley DE (2017) Synergistic use of biochar, compost and plant growth promoting rhizobacteria for enhancing cucumber growth under water deficit conditions. J Sci Food Agric 97(15):5139–5145
Nelson NO, Agudelo SC, Yuan W, Gan J (2011) Nitrogen and phosphorus availability in biochar-amended soils. Soil Sci 176:218–226
Nerome M, Toyota K, Islam T et al (2005) Suppression of bacterial wilt of tomato by incorporation of municipal biowaste charcoal into soil. Soil Microorganisms 59:9–14
Nguyen TH, Cho H-H, Poster DL, Ball WP (2007) Evidence for a Pore-Filling mechanism in the adsorption of aromatic hydrocarbons to a natural wood char. Environ Sci Technol 41(4):1212–1217

Noble R, Coventry E (2005) Suppression of soil-borne plant diseases with composts: a review. Biocontrol Sci Technol 15(1):3–20
Novak JM, Busscher WJ, Laird DL, Ahmedna M, Watts DW, Niandou MAS (2009) Impact of biochar amendment on fertility of a south eastern coastal plain soil. Soil Sci 174(2):105–112
Novak JM, Sigua GC, Ducey TF et al (2019) Designer Bio char impact on corn grain yields, biomass production, and fertility properties of a highly weathered ultisol. Environments 6(6):64
O'Connor D, Peng T, Zhang J et al (2018) Biochar application for the remediation of heavy metal polluted land: a review of in situ field trials. Sci Total Environ 619:815–826
Obia A, Mulder J, Martinsen V, Cornelissen G, Børresen T (2016) In situ effects of biochar on aggregation, water retention and porosity in light-textured tropical soils. Soil Till Res 155:35–44
Ogawa M (1994) Symbiosis of people and nature in the tropics. Farm Japan 28:10–34
Oguntunde PG, Fosu M, Ajayi AE, van de Giesen N (2004) Effects of charcoal production on maize yield, chemical properties and texture of soil. Biol Fertil Soils 39:295–299
Page-Dumroese DS, Coleman M, Thomas SC (2015) Opportunities and uses of biochar on forest sites in North America. In: Bruckman VJ, Varol EA, Uzun BB, Liu J (eds) Biochar: a regional supply chain approach in view of mitigating climate change. Cambridge University Press, Cambridge
Painter TJ (2001) Carbohydrate polymers in food preservation: an integrated view of the Maillard reaction with special reference to discoveries of preserved foods in *Sphagnum* dominated peat bogs. Carbohydr Polym 36:335–347
Pan GX, Smith P, Pan WN (2009) The role of soil organic matter in maintaining the productivity and yield stability of cereals in China. Agric Ecosyst Environ 129:344–348
Paneque M, José M, Franco-Navarro JD, Colmenero-Flores JM, Knicker H (2016) Effect of biochar amendment on morphology, productivity and water relations of sunflower plants under non-irrigation conditions. Catena 147:3–10
Petrović M, Eljarrat E, López de Alda MJ, Barceló D (2001) Analysis and environmental levels of endocrine-disrupting compounds in freshwater sediments. TrAC Trend Anal Chem 20:637–648
Piash MI, Hossain F, Parveen Z (2019) Effect of bio char and fertilizer application on the growth and nutrient accumulation of rice and vegetable in two contrast soils. Acta Sci Agric 3(2):74–83
Prendergast-Miller MT, Duvall M, Sohi SP (2013) Biochar-root interactions are mediated by biochar nutrient content and impacts on soil nutrient availability. Eur J Soil Sci 65:173–185
Pressler Y, Foster EJ, Moore JC, Cotrufo MF (2017) Coupled biochar amendment and limited irrigation strategies do not affect a degraded soil food web in a maize agroecosystem, compared to the native grassland. GCB Bioenergy 9(8):1344–1355
Qin HZ, Liu YY, Li LQ, Pan GX, Zhang XH, Zheng JW (2012) Adsorption of cadmium in solution by biochar from household biowaste. J Ecol Rural Environ 28:181–186
Qin G, Gong D, Fan MY (2013) Bioremediation of petroleum-contaminated soil by biostimulation amended with biochar. Int Biodeterior Biodegrad 85:150–155
Quayle WC (2010) Biocahr potential for soil improvement and soil fertility. In: IREC farmers Newsletter; Large Area No. 182
Quin P, Joseph S, Husson O, Donne S, Mitchell D, Munroe P, Phelan D, Cowie A, Van Zwieten L (2015) Lowering N_2O emissions from soils using eucalypt biochar: the importance of redox reactions. Sci Rep 5:16773
Rajkovich S, Enders A, Hanley K, Hyland C, Zimmerman AR, Lehmann J (2012) Corn growth and nitrogen nutrition after additions of biochars with varying properties to a temperate soil. Biol Fertil Soils 48:271–284
Ramzani PMA, Shan L, Anjum S, Ronggui H, Iqbal M, Virk ZA, Kausar S (2017) Improved quinoa growth, physiological response, and seed nutritional quality in three soils having different stresses by the application of acidified biochar and compost. Plant Physiol Biochem 116:3–10
Rasa K, Heikkinen J, Hannula M, Arstila K, Kulju S, Hyväluoma J (2018) How and why does willow biochar increase a clay soil water retention capacity? Biomass Bioener 119:346–353
Retan GA (1915) Charcoal as a means of solving some nursery problems. J Forest 13(1):25–30

Roberts KG, Gloy BA, Joseph S, Scott NR, Lehmann J (2010) Life cycle assessment of biochar systems: estimating the energetic, economic, and climate change potential. Environ Sci Technol 44:827–833

Rogovska N, Laird DA, Rathke SJ, Karlen DL (2014) Biochar impact on Midwestern Mollisols and maize nutrient availability. Geoderma 230:3–10

Rogovska N, Laird D, Leandro L, Aller D (2017) Biochar effect on severity of soybean root disease caused by *Fusarium virguliforme*. Plant Soil 413:111–126

Rondon M, Ramirez JA, Lehmann J (2005) Greenhouse gas emissions decrease with charcoal additions to tropical soils. In: Proceedings of the 3rd USDA symposium on greenhouse gases and carbon sequestration, baltimore, USA, vol 208

Rondon M, Lehmann J, Ramirez J, Hurtado M (2007) Biological nitrogen fixation by common beans (*Phaseolus vulgaris* L.) increases with bio-char additions. Biol Fert Soils 43:699–708

Sackett TE, Basiliko N, Noyce GL, Winsborough C, Schurman J, Ikeda C, Thomas SC (2015) Soil and greenhouse gas responses to biochar additions in a temperate hardwood forest. GCB Bioenergy 7:1062–1074

Sanchez ME, Lindao E, Margaleff D, Martinez O, Moran A (2009) Pyrolysis of agricultural residues from rape and sunflower: production and characterization of biofuels and biochar soil management. J Anal Appl Pyrolysis 85:142–144

Santín C, Doerr SH, Merino A, Bucheli TD, Bryant R, Ascough P, Gao X, Masiello CA (2017) Carbon sequestration potential and physicochemical properties differ between wildfire charcoals and slow-pyrolysis biochars. Sci Rep 7(1):11233

Sarmah AK, Srinivasan P, Smernik RJ, Manley-Harris M, Antal MJ Jr, Downie A, Van Zwieten L (2010) Retention capacity of biocharamended New Zealand dairy farm soil for an estrogenic steroid hormone and its primary metabolite. Aust J Soil Res 48:648–658

Scheer C, Grace PR, Rowlings DW, Kimber S, Van Zwieten L (2011) Effect of biochar amendment on the soil-atmosphere exchange of greenhouse gases from an intensive subtropical pasture in northern New South Wales, Australia. Plant Soil 345(1–2):47–58

Schiemenz K, Eichler-Loebermann B (2010) Biomass ashes and their phosphorus fertilizing effect on different crops. Nutr Cycl Agroecosyst 87:471–482

Schimmelpfennig S, Glaser B (2012) One step forward toward characterization: some important material properties to distinguish biochars. J Environ Qual 41(4):1001–1013

Schmidt MWI, Skjemstad JO, Jager C (2002) Carbon isotope geochemistry and nano-morphology of soil black carbon: black chernozemic soils in central Europe originate from ancient biomass burning. Glob Biogeochem Cyc 16:1123

Schmidt MWI, Torn MS, Abiven S, Dittmar T, Guggenberger G, Janssens IA, Kleber M, Kogel-Knabner I, Lehmann J, Manning DAC, Nannipieri P, Rasse DP, Weiner S, Trumbore SE (2011) Persistence of soil organic matter as an ecosystem property. Nature 478:49–56

Shang QY, Yang XX, Gao CM, Wu PP, Liu JJ, Xu YC, Shen QR, Zou JW, Guo SW (2011) Net annual global warming potential and greenhouse gas intensity in Chinese double rice cropping systems: a 3-year field measurement in long-term fertilizer experiments. Glob Chang Biol 17:2196–2210

Sharma S, Bhattacharya A (2017) Drinking water contamination and treatment techniques. Appl Water Sci 7(3):1043–1067

Shin J, Jang E, Park S, Ravindran B, Chang SW (2019) Agro-environmental impacts, carbon sequestration and profit analysis of blended biochar pellet application in the paddy soil-water system. J Environ Manage 244:92–98

Singh JS (2013) Anticipated effects of climate change on methanotrophic methane oxidation. Clim Change Environ Sustain 1(1):20–24

Singh JS (2014) Cyanobacteria: a vital bio-agent in eco-restoration of degraded lands and sustainable agriculture. Clim Change Environ Sustain 2:133–137

Singh JS (2015) Biodiversity: current perspective. Clim Change Environ Sustain 3(1):71–72

Singh JS (2016) Microbes play major roles in ecosystem services. Clim Change Environ Sustain 3:163–167

Singh JS (ed) (2019) New and future developments in microbial biotechnology and bioengineering: microbes in soil, crop and environmental sustainability. Elsevier, San Diego

Singh JS, Boudh S (2016) Climate change resilient crops to sustain Indian agriculture. Clim Change Environ Sustain 5:97–110

Singh BP, Cowie AL (2008) A novel approach, using ^{13}C natural abundance, for measuring decomposition of biochars in soil. In: Currie LD, Yates LJ (eds) Carbon and nutrient management in agriculture, fertilizer and lime research centre workshop proceedings. Massey University, Palmerston North

Singh BP, Hatton BJ, Singh B, Cowie AL, Kathuria A (2010a) Influence of biochars on nitrous oxide emission and nitrogen leaching from two contrasting soils. J Environ Qual 39:1224–1235

Singh B, Singh BP, Cowie AL (2010b) Characterisation and evaluation of biochars for their application as a soil amendment. Aust J Soil Res 48:516–525

Skjemstad JO, Reicosky DC, McGowan JA, Wilts AR (2002) Charcoal carbon in U.S. agricultural soils. Soil Sci Soc Am J 66:1249–1255

Smith P, Martino D, Cai Z, Gwary D, Janzen H, Kumar P, McCarl B, Ogle S, O'Mara F, Rice C, Scholes B (2007a) Greenhouse gas mitigation in agriculture. Philos Trans Royal Soc B Biol Sci 363(1492):789–813

Smith P, Martino Z, Cai ZC, Gwary D, Janzen H, Kumar P, McCarl B, Ogle S, O'Mara F, Rice C, Scholes B, Sirotenko O (2007b) Agriculture. In: Metz B, Davidson OR, Dave R, Meyer LA (eds) Climate change 2007: mitigation. Contribution of working group III to the fourth assessment report of the intergovernmental panel on climate change. Cambridge University Press, Cambridge, pp 498–540

Smith P, Marino D, Cai ZC, Gwary D, Janzen H, Kumar P (2008a) Greenhouse gas mitigation in agriculture. Philos Trans R Soc B 363:789–813

Smith P, Martino D, Cai Z, Gwary D, Janzen H, Kumar P, McCarl B, Ogle S, O'Mara F, Rice C, Scholes B, Sirotenko O, Howden M, McAllister T, Pan G, Romanenkov V, Schneider U, Towprayoon S, Wattenbach M, Smith J (2008b) Greenhouse gas mitigation in agriculture. Philos Trans R Soc Lond B Biol Sci 363:789–813

Sohi S (2012) Carbon storage with benefits. Science 338:1034–1035

Sohi SP, Krull E, Lopez-Capel E, Bol R (2010) A review of biochar and its use and function in soil. Adv Agron 105:47–82

Song Y, Wang F, Bian YR, Kengara FO, Jia MY, Xie ZB, Jiang X (2012) Bioavailability assessment of hexachlorobenzene in soil as affected by wheat straw biochar. J Hazard Mater 217:391–397

Song X, Pan G, Zhang C, Zhang L, Wang H (2016) Effects of biochar application on fluxes of three biogenic greenhouse gases: a meta-analysis. Ecosyst Health Sustain 2(2):e01202

Sorensen RB, Lamb MC (2016) Crop yield response to increasing biochar rates. J Crop Improve 30(6):703–712

Spokas KA (2013) Impact of biochar field aging on laboratory greenhouse gas production potentials. GCB Bioenergy 5:165–176

Spokas KA, Reicosky DC (2009) Impact of sixteen different biochars on soil greenhouse gas production. Ann Environ Sci 3:179–193

Spokas KA, Cantrell KB, Novak JM, Archer DA, Ippolito JA, Collins HP, Boateng AA, Lima IM, Lamb MC, McAloon AJ, Lentz RD, Nichols KA (2012) Biochar: a synthesis of its agronomic impact beyond carbon sequestration. J Environ Qual 41:973–989

Steinbeiss S, Gleixner G, Antonietti M (2009) Effect of biochar amendment on soil carbon balance and soil microbial activity. Soil Biol Biochem 41:1301–1310

Steiner C, Teixeira WG, Lehmann J, Nehls T, Macedo J, Blum WH, Zech W (2007) Long term effect of manure, charcoal and mineral fertilization on crop production and fertility on a highly weathered Central Amazonian upland soil. Plant Soil 291:275–290

Stockmann U, Adams MA et al (2013) The knowns, known unknowns and unknowns of sequestration of soil organic carbon. Agric Ecosyst Environ 164:80–99

Sun Y, Zhou Q, Diao C (2008) Effects of cadmium and arsenic on growth and metal accumulation of Cd-hyperaccumulator *Solanum nigrum* L. Bioresour Technol 99(5):1103–1110

Sun LY, Li L, Chen ZZ, Wang JY, Xiong ZQ (2014) Combined effects of nitrogen deposition and biochar application on emissions of N_2O, CO_2 and NH_3 from agricultural and forest soils. Soil Sci Plant Nutr 60:254–265

Taghizadeh-Toosi A, Clough TJ, Condron LM, Sherlock RR, Anderson CR, Craigie RA (2011) Biochar incorporation into pasture soil suppresses in situ nitrous oxide emissions from ruminant urine patches. J Environ Qual 40:468–476

Tammeorg P, Parviainen T, Nuutinen V, Simojoki A, Vaara E, Helenius J (2014) Effects of biochar on earthworms in arable soil: avoidance test and field trial in boreal loamy sand. Agric Ecosyst Environ 191:150–157

Thammasom N, Vityakon P, Lawongsa P, Saenjan P (2016) Biochar and rice straw have different effects on soil productivity, greenhouse gas emission and carbon sequestration in Northeast Thailand paddy soil. Agric Nat Res 50:192–198

Thies J, Rillig MC (2009) Characteristics of biochar: biological properties. In: Lehmann J, Joseph S (eds) Biocharfor environmental management: science and technology. Earthscan, London, pp 85–105

Thomas SC, Frye S, Gale N, Garmon M, Launchbury R, Machado N, Melamed S, Murray J, Petroff A, Winsborough C (2013) Biochar mitigates negative effects of salt additions on two herbaceous plant species. J Environ Manage 129:62–68

Tian X, Li C, Zhang M, Wan Y, Xie Z, Chen B et al (2018) Biochar derived from corn straw affected availability and distribution of soil nutrients and cotton yield. PLoS ONE 13(1):e0189924

Tiwari AP, Singh JS (2017) Plant growth promoting rhizospheric *Pseudomonas aeruginosa* strain inhibits seed germination in *Triticum aestivum* (L) and *Zea mays* (L). Microbiol Res 8(7233):73–79

Tong XJ, Li JY, Yuan JH, Xu RK (2011) Adsorption of Cu(II) by biochars generated from three crop straws. Chem Eng J 172(2–3):828–834

Trigo C, Spokas KA, Cox L, Koskinen WC (2014) Influence of soil biochar aging on sorption of the herbicides MCPA, nicosulfuron, terbuthylazine, indaziflam, and fluoroethyldiaminotriazine. J Agric Food Chem 62(45):10855–10860

Tryon EH (1948) Effect of charcoal on certain physical, chemical, and biological properties of forest soils. Ecolo Mono 18(1):81–115

Vaccari FP, Baronti S, Lugato E, Genesio L, Castaldi S, Fornasier F, Miglietta F (2011) Biochar as a strategy to sequester carbon and increase yield in durum wheat. Eur J Agron 34:231–238

Vaccari F, Maienza A, Miglietta F, Baronti S, Di Lonardo S, Giagnoni L, Lagomarsino A, Pozzi A, Pusceddu E, Ranieri R (2015) Biochar stimulates plant growth but not fruit yield of processing tomato in a fertile soil. Agric Ecosyst Environ 207:163–170

Van Zwieten L, Singh B, Joseph S, Kimber S, Cowie A, Chan KY (2009) Biochar and emissions of non-CO2 greenhouse gases from soil. In: Lehmann J, Joseph S (eds) Biochar for environmental management—science and technology. Earthscan, London, pp 227–249

Van Zwieten L, Kimber S, Morris S, Chan KY, Downie A, Rust J, Joseph S, Cowie A (2010a) Effects of biochar from slow pyrolysis of papermill waste on agronomic performance and soil fertility. Plant Soil 327:235–224

Van Zwieten L, Kimber S, Morris S, Downie A, Berger E, Rust J, Scheer C (2010b) Influence of biochars on flux of N_2O and CO_2 from Ferrosol. Soil Res 48:555–568

Vimal SR, Singh JS (2019) Salt tolerant PGPR and FYM application in saline soil paddy agriculture sustainability. Clim Change Environ Sustain 7(1):23–33

Vimal SR, Gupta J, Singh JS (2018) Effect of salt tolerant *Bacillus* sp. and *Pseudomonas* sp. on wheat (*Triticum aestivum* L.) growth under soil salinity: a comparative study. Microbiol Res 9(1):1–14

Wakeel A (2013) Potassium–sodium interactions in soil and plant under saline-sodic conditions. J Plant Nutr Soil Sci 176:344–354

Wang JY, Pan XJ, Liu YL, Zhang XL, Xiong ZQ (2012) Effects of biochar amendment in two soils on greenhouse gas emissions and crop production. Plant and Soil 360(1–2):287–298

Wang Y, Wang Y-J, Wang L, Fang G-D, Cang L, Herath HMSK, Zhou D-M (2013a) Reducing the bioavailability of PCBs in soil to plant by biochars assessed with triolein-embedded cellulose acetate membrane technique. Environ Pollut 174:250–256

Wang D, Zhang W, Hao X, Zhou D (2013b) Transport of biochar particles in saturated granular media: effects of pyrolysis temperature and particle size. Environ Sci Technol 47:821–828

Wang C, Tu Q, Dong D, Strong PJ, Wang H, Sun B, Wu W (2014a) Spectroscopic evidence for biochar amendment promoting humic acid synthesis and intensifying humification during composting. J Hazard Mater 280:409–416

Wang ZL, Li YF, Chang SX, Zhang JJ, Jiang PK, Zhou GM, Shen ZM (2014b) Contrasting effects of bamboo leaf and its biochar on soil CO_2 efflux and labile organic carbon in an intensively managed Chinese chestnut plantation. Biol Fertil Soils 50:1109–1119

Wang Z, Xu L, Zhao J, Wang X, White JC, Xing B (2016) CuO nanoparticle interaction with arabidopsis thaliana: toxicity, parent-progeny transfer, and gene expression. Environ Sci Technol 50:6008–6016

Wang M, Zhu Y, Cheng L, Andserson B, Zhao X, Wang D et al (2018) Review on utilization of biochar for metal-contaminated soil and sediment remediation. J Environ Sci (China) 63:156–173

Wardle D, Zackrisson O, Nilsson M (1998) The charcoal effect in Boreal forests: mechanisms and ecological consequences. Oecologia 115(3):419–426

Wardle DA, Nilsson MC, Zackrisson O (2008) Fire-derived charcoal causes loss of forest humus. Science 320:629–629

Warnock DD, Lehmann J, Kuyper TW, Rillig MC (2007) Mycorrhizal responses to biochar in soil – concepts and mechanisms. Plant Soil 300:9–20

Wiedner K, Rumpel C, Steiner C, Pozzi A, Maas R, Glaser B (2013) Chemical evaluation of chars produced by thermochemical conversion (gasification, pyrolysis and hydrothermal carbonization) of agro-industrial biomass on a commercial scale. Biomass Bioenergy 59:264–278

Winsley P (2007) Biochar and bioenergy production for climate change mitigation. NZ Sci Rev 64:5–10

Woldetsadik D, Drecshel P et al (2017) Effect of bio char derived from faecal matter on yield and nutrient content of Lettuce (Lactuca sativa) in two contrasting soils. Environ Syst Res 6(2). https://doi.org/10.1186/s40068-017-0082-9

Woolf D, Lehmann J (2012) Modelling the long-term response to positive and negative priming of soil organic carbon by black carbon. Biogeochemistry 111:83–95

Woolf D, Amonette JE, Street-Perrott FA, Lehmann J, Joseph S (2010) Sustainable biochar to mitigate global climate change. Nat Commun 1:56

World Health Organisation (2010) Persistent organic pollutants: impact on child health. WHO Document Production Services, Geneva

Xiao YH (2016) Effects of different application rates of biochar on the soil greenhouse gas emission in Chinese chestnut stands. Master thesis, Zhejiang A and F University, Hangzhou, Zhejiang (in Chinese)

Xiao X, Chen B, Chen Z, Zhu L, Schnoor JL (2018) Insight into multiple and multilevel structures of biochars and their potential environmental applications: a critical review. Environ Sci Technol 52:5027–5047

Xu M, Lou Y, Sun X, Wang W, Baniyamuddin M, Zhao K (2011a) Soil organic carbon active fractions as early indicators for total carbon change under straw incorporation. Biol Fertil Soils 47(7):745

Xu RK, Xiao SC, Yuan JH, Zhao AZ (2011b) Adsorption of methyl violet from aqueous solutions by the biochars derived from crop residues. Bioresour Technol 102(22):10293–10298

Xu G, Wei LL, Sun JN, Shao HB, Chang SX (2013) What is more important for enhancing nutrient bioavailability with biochar application into a sandy soil: direct or indirect mechanism. Ecol Eng 52:119–124

Yamato M, Okimori Y, Wibowo IF, Anshori S, Ogawa M (2006) Effects of the application of charred bark of Acacia mangium on the yield of maize, cowpea and peanut and soil chemical properties in south Sumatra, Indonesia. Soil Sci Plant Nutr 52:489–495

Yan XY, Yagi K, Akiyama H, Akimoto H (2005) Statistical analysis of the major variables controlling methane emission from rice fields. Glob Chang Biol 11:1131–1141

Yanai Y, Toyota K, Okazaki M (2007) Effect of charcoal addition on N2O emissions from soil resulting from rewetting air-dried soil in short-term laboratory experiments. Soil Sci Plant Nutr 53:181–188

Yang Y, Lin X, Wei B, Zhao Y, Wang J (2014) Evaluation of adsorption potential of bamboo biochar for metal-complex dye: equilibrium, kinetics and artificial neural network modeling. Int J Environ Sci Technol 11(4):1093–1100

Yao Y, Gao B, Chen H, Jiang L, Inyang M, Zimmerman AR et al (2012) Adsorption of sulfamethoxazole on biochar and its impact on reclaimed water irrigation. J Hazard Mater 209–210:408–413

Yip K, Xu M, Li CZ, Jiang SP, Wu H (2010) Biochar as a fuel: 3. Mechanistic understanding on biochar thermal annealing at mild temperatures and its effect on biochar reactivity. Energy Fuels 25(1):406–414

Yoo G, Kang H (2012) Effects of biochar addition on greenhouse gas emissions and microbial responses in a short-term laboratory experiment. J Environ Qual 41:1193–1202

Yu LQ, Tang J, Zhang RD, Wu QH, Gong MM (2013) Effects of biochar application on soil methane emission at different soil moisture levels. Biol Fertil Soils 49:119–128

Yuan J, Meng J, Liang X, Yang E, Xu Y, Wenfu C (2017) Organic molecules from biochar leacheates have a positive effect on rice seedling cold tolerance. Front Plant Sci 8:1624

Yuan P, Wang J, Pan Y, Shen B, Wu C (2018) Review of biochar for the management of contaminated soil: preparation, application and prospect. Sci Total Environ 659:473–490

Zhang AF, Cui LQ, Pan GX, Li LQ, Hussain Q, Zhang XH, Zheng JW, Crowley D (2010) Effect of biochar amendment on yield and methane and nitrous oxide emissions from a rice paddy from Tai Lake plain, China. Agric Ecosyst Environ 139:469–475

Zhang AF, Bian RJ, Pan GX, Cui LQ, Hussain Q, Li LQ, Zheng JW, Zheng JF, Zhang XH, Han XJ, Yu XY (2012a) Effect of biochar amendment on soil quality, crop yield and greenhouse gas emission in a Chinese rice paddy: a field study of 2 consecutive rice growing cycles. Field Crop Res 127:153–160

Zhang A, Liu Y, Pan G, Hussain Q, Li L, Zheng J, Zhang X (2012b) Effect of biochar amendment on maize yield and greenhouse gas emissions from a soil organic carbon poor calcareous loamy soil from Central China Plain. Plant Soil 351:263–275

Zhang M, Gao B, Yao Y, Xue Y, Inyang M (2012c) Synthesis of porous MgO-biochar nanocomposites for removal of phosphate and nitrate from aqueous solutions. Chem Eng J 210:26–32

Zhelezova A, Cederlund H, Stenström J (2017) Effect of biochar amendment and ageing on adsorption and degradation of two herbicides. Water Air Soil Pollut 228(6):216

Zhou Q, Song Y (2004) Contaminated soil remediation: principles and methods. Science Press, Beijing

Zhou JB, Deng CJ, Chen JL, Zhang QS (2008) Remediation effects of cotton stalk carbon on cadmium(Cd) contaminated soil. Ecol Environ 17:1857–1860

Zhou GY, Zhou XH, Zhang T, Du ZG, He YH, Wang XH, Shao JJ, Cao Y, Xue SG, Wang HL, Xu CY (2017) Biochar increased soil respiration in temperate forests but had no effects in subtropical forests. Forest Ecol Manage 405:339–349

Zimmerman AR, Gao B, Ahn MY (2011) Positive and negative carbon mineralization priming effects among a variety of biochar-amended soils. Soil Biol Biochem 43:1169–1179

Zwart DC, Kim SH (2012) Biochar amendment increases resistance to stem lesions caused by Phytophthora spp. in tree seedlings. Hort Sci 47:1736–1740

Zwieten VL, Kimber S, Morris S, Chan KY, Downie A, Rust J, Joseph S, Cowie A (2010) Effects of biochar from slow pyrolysis of paper mill waste on agronomic performance and soil fertility. Plant Soil 327:235–246

Chapter 12
Multifarious Benefits of Biochar Application in Different Soil Types

Umesh Pankaj

Abstract The extensive use of chemical fertilizers in agriculture have long term deleterious impact such as leading salinity, decline fertility of soil with fast growth of agricultural production and it is predicted that the fertilizer use to continue increase in the coming years. With current scenario, there has been keen interest on biochar, produced from various crop residues with multiple environmental applications such as soil amelioration, pollutants removal and carbon sequestration. Biochar has several unique properties like high alkaline pH, fixed carbon content, stability against decay, water holding capacity and cation exchange capacity, which makes it an efficient, cost-effective and environmentally-friendly material. Many study showed the effectiveness of biochar amendments in soil i.e. nutrient status improvement, increases soil porosity, soil pH, soil moisture-holding capacity and boost the growth of beneficial plant growth promoting microbial community.

Keywords Biochar · Microbial abundance · Soil physio-chemical property · Nutrient improvement

12.1 Introduction

Over the last 30 years, the huge amount of chemical fertilizers use in agriculture resulted fast growth of agricultural production, and it is predicted that the fertilizer use to continue increase in the coming decades. In country like India and China, the excessive use of chemical fertilizers is a common practice to achieve high crop yield. Though, continuous use of chemical fertilizers for the intensive cropping not only enhance soil nutrients but also decrease the soil organic carbon (C) and other negative effects on soils such as leading acidification, deplete soil structure and soil

U. Pankaj (✉)
Division of Biotechnology, CSIR-Institute of Himalayan Bioresource Technology (IHBT), Palampur, Himachal Pradesh, India

J. S. Singh, C. Singh (eds.), *Biochar Applications in Agriculture and Environment Management*, https://doi.org/10.1007/978-3-030-40997-5_12

productivity (Ge et al. 2008). Rapid industrial development and increasing adaption of agro-chemical based crop production practices since green revolution have increased the persistent organic adulterations in the food chain. The uses of agro-chemical in soil are considerable costly and also produce a substantial amount of chemical residues. The Chinese Ministry of Agriculture has immediate stopped the over use of chemical fertilizer and proposed the plan "zero increment in chemical fertilizer until 2020" to reduce the consumption of chemical fertilizer. Therefore, the high effectiveness of fertilizer and alternate of chemical fertilizer is demanding (Singh et al. 2017a, b, 2019a, b; Vimal et al. 2018; Kumar et al. 2018; Singh 2019; Singh and Singh 2019; Vimal and Singh 2019). So, we should need to find the replacement or substitute of chemical fertilizer, which would be a cost effective, sustainable and wide range of applicability.

The actual use and management of agricultural residues, paddy or wheat straw, green manure and beneficial microbes have become a key focus of sustainable agriculture in recent years (Singh 2013, 2014, 2015, 2016; Singh and Boudh 2016; Kumar et al. 2017; Kumar and Singh 2017; Tiwari and Singh 2017). Biochar is considered as a stable form of organic carbon which improves the soil properties and also sequestrates carbon. Biochar considered as a promising solution with various valuable properties (Joseph et al. 2010; Uras et al. 2012). Biochar can be formed from a numerous of agricultural biomass comprising straw, woody leftovers, animal manure, and other waste products. Its use can make available resourceful path for agricultural waste utilization. Due to its unique structure and composition, application of biochar can potentially enhanced the carbon sequestration, improve soil health, and lead to sustainable management of organic waste (Lehmann and Joseph 2009). Biochar can also improve the soil cation exchange capacity (CEC) (Zwieten et al. 2010), nutrient absorption (which prevents subsequent nutrient runoff), water holding capacity (Laird et al. 2010; Schulz and Glaser 2012; Zhang et al. 2013), and excessive soil acidification (Karami et al. 2011).

Biochar is a promising carbonaceous material and substitute to the activated carbon to remove various organic pollutants such as agrochemicals, polycyclic aromatic hydrocarbons (PAHs), volatile organic compounds (VOCs), polychlorinated biphenyls (PCBs), aromatic dyes and antibiotics (Beesley et al. 2010; Teixidó et al. 2011; Xu et al. 2012; Zheng et al. 2010), and also a series of inorganic contaminants (e.g., heavy metals, ammonia, nitrate, phosphate, sulfide etc.) from aqueous, gaseous and/or solid phases (Ahmad et al. 2014; Jung et al. 2015; Oliveira et al. 2017). Biochar application to the soil gives many beneficial effects (Fig. 12.1) such as increase microbial respiration, crop yield, improve soil health, water holding capacity etc. (Marjenah 1994; Yamato et al. 2006).

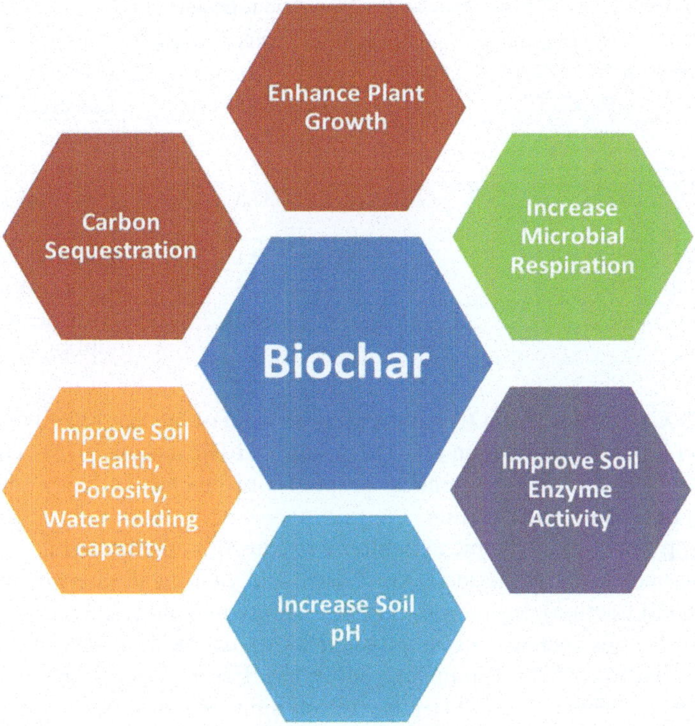

Fig. 12.1 Multi-benefits of biochar application into poor physico-chemical property soil

12.2 Biochar Production from Organic Residue

Generally, biochar was produced in a muffle furnace that was equipped with a digital temperature regulator (detection accuracy <5 °C). Biochar was obtained from slow and/or fast pyrolysis of organic residue (manure, organic waste, bioenergy crops, crop residues) at around 400–600 °C for 8–12 h in oxygen-free or low-oxygen environment. After pyrolysis of the biomass, an average the production yield was approximately 25–50% of the original biomass C remains in the biochar (Lehmann 2007). Under the pyrolysis process most of the Ca, Mg, K, P, and plant micronutrients, and about half of the N and S in the biomass feedstock are separated into the biochar fraction.

Major thermochemical technologies for biochar production include slow and fast pyrolysis, gasification, torrefaction, and hydrothermal carbonization (Kambo and Dutta 2015). Biochar yield greatly depends on adaption of pyrolysis type. Slow pyrolysis performed at longer residence time and at a moderate temperature

Table 12.1 Properties of different biochar from the various organic residues

Biochar type	pH	Density (g cm^{-3})	Ash content (%)	Fixed carbon (%)	References
Rice straw	9.0	0.13	23.0	51.8	Li-li et al. (2017)
Bamboo	8.6	0.56	11.9	69.0	Li-li et al. (2017)
Swine manure	8.4	–	32.5	17.7	Cantrell et al. (2012)
Mulberry wood	10.2	–	7.5	37.5	Zama et al. (2017)
Maize straw	9.8	0.40		59.1	Luo et al. (2017)
Peanut shell	7.0	–	7.0	32.5	Zama et al. (2017)
Oak biochar	10.2	–	3.49	68.2	Teutscherova et al. (2018)

(350–550 °C) in absence of O_2 results in higher yield of biochar (30%) than the fast pyrolysis (12%) or gasification (10%) (Inyang and Dickenson 2015). The various factors affecting the physicochemical properties of biochar during production are discussed in the following section.

During pyrolysis the organic agricultural residue (e.g. lignin, cellulose, hemicellulose, fat, and starch) is thermally combusted and yielding three main products (i) biochar (solid fraction), (ii) bio-oil (partly condensed volatile matter), and (iii) non-condensable gases such as carbon monoxide (Co), carbon dioxide (CO_2), methane (CH_4) and hydrogen (H_2) (Suliman et al. 2016). Furthermore, in biochar varying ratio of O/C and H/C is achieved by specific elimination of different elements (C, H, O) into gases and other volatile compounds (Brewer et al. 2012). Essentially, in biochar the ratio of O/C and H/C is directly correlates with aromaticity, biodegradability, and polarity, which are extremely necessary properties for the exclusion of organic pollutants (Crombie et al. 2013). For example, while a biochar formed at higher temperature have lower H/C and O/C ratios as compared to lower temperature, demonstrating a steady increase in aromaticity and lower in polarity with increasing temperature (Suliman et al. 2016; Chen et al. 2016). Van Krevelen diagram is widely used to understand the selective loss of elements (during dehydration and carbonization reactions) by comparing atomic ratios of H/C and O/C. Most of the biochars derived from various sources of feedstock's have decrease ratios of the H/C and O/C due to the removal of H and O atoms during pyrolysis. However, the stability of any biochar depends on high aromaticity and carbon content (Windeatt et al. 2014). Besides, atomic ratios some other factors like pH and temperature also have a major effect on biochar properties. Some researchers had established the relation of biochar high pH with increasing pyrolysis temperature (>500 °C) due to the enrichment of ash content (Table 12.1); greater hydrophobicity and aromaticity, and higher surface area (Windeatt et al. 2014; Keiluweit et al. 2010). Above mentioned all the properties of biochar make a good candidate for highly responsive for removal of organic pollutants. Biochar is act as a zwitterionic which comprises of both positively and negatively charged surfaces (Tan et al. 2017). The negatively charged surface is attracting the cations and influencing the cation exchange capacity (CEC) of soils (Lawrinenko 2014). When pyrolysis was done below 500 °C temperature aids incomplete carbonization resulted smaller pore size biochar formed, lower surface area and greater O-containing functional groups (Lu et al. 2014), which make biochar highly responsive for elimination of inorganic pollutants due to enlarged ionic interactions through interface with O-containing functional groups.

12.3 Effect of Biochar on Soil Microbial Abundance

Microorganisms are present in environment (soil, water, air) and interact with human, plant, animal and neighbouring organism. These microbes also regulate the soil nutrients mobilization, uptake and plant metabolisms. So, it is essential attention to truthfully profiling and also compares the composition of the populations they form. The one of the most important approach for microbial community profiling is by classification of PCR amplicon sequences from the small subunit ribosomal RNA gene (i.e., the 16S rRNA gene of bacteria and archaea). This method is also useful to introduce biases in microbial composition estimation due to variations in 16S rRNA gene copy numbers per genome. The other most common approach for determination of soil microbial abundance and community composition is phospholipid fatty acid (PLFA) analysis (Zhang et al. 2015). The quantitative realtime PCR and Illumina MiSeq sequencing method outcomes revealed that the bacterial abundances and diversity increased with biochar addition (Chen et al. 2013; Yao et al. 2017). A study revealed that the abundance of microbial PLFAs (Gram-positive bacteria, Gram-negative bacteria, actinobacteria and fungi) in biochar amended soil was higher as compared to un-amended soils. The ratios of bacteria/fungi and monounsaturated/branched PLFAs were significantly correlated with the volumetric soil water content, porosity, or computed effective oxygen diffusion coefficients under biochar amended soil.

Several reports are available in which the biochar has both positive and negative impact on microbial community and abundance. Biochar can significantly influencing the soil microbial communities and abundance (Grossman et al. 2010; Jindo et al. 2012; Lehmann et al. (2011), possibly varying the activity of advantageous soil microorganisms and nutrient cycles (Bruun et al. 2014). One of the advantageous aspects that biochar pores are provide habitat for microorganisms such as mycorrhizae and bacteria that also obtain their metabolic needs from these micro-habitats (Lehmann et al. 2011). Biochar surface contained the labile soil organic matters which favours the microbial growth and activity, consequently lead to microbial abundance, microbial activity and mineralization (Wardle et al. 2008; Ameloot et al. 2013). Valuable effects of biochar amendment on crop yield have been documented (Yamato et al. 2006; Jeffery et al. 2011), however broad analysis is needed on soil microbial community and abundance because soil microorganisms play a key role in nutrient cycling and provide thus an important ecosystem service (Costanza et al. 1987). Due to biochar pore size is very small below 5 mm in diameter (Glaser 2007) it protect microorganism from predator such as nematodes, mites, protozoan, collembolans and maintained the microbial diversity (Warnock et al. 2007; Swift et al. 1979; Wright et al. 1995). Contrary to this, some reports are present the negative effect of biochar on microorganism (Graber et al. 2010) due to reduction in reproduction rate. Arbuscular mycorrhizal fungi (AMF) plays a major role in soil aggregation, provide essential nutrient phosphorus to the plant, sequestration of soil carbon and nitrogen under different stress condition like droughts and saline or sodic soil (Wilson et al. 2009). However, some reports are described the AMF abundance decreased with the addition of biochar (Warnock et al. 2010; George et al. 2012) while, others reports reflects had no significant role of AMF and microbial abundance and biomass (Chan et al. 2008; Durenkamp et al. 2010). Zheng et al. (2016) also found that biochar addition increased the bacterial diversity and changes in bacterial community composition in drylands, while under paddy soil it did not alter the microbial community structure (Tian et al. 2016). However, until now, few reports have been available on the changes in the soil microbial community with biochar addition. The possible reason behind no changes in microbial abundance and biomass is biochar were not equally spread across different functional groups and allow to dominate or diminish soil environment might cause some microorganisms to become competitively dominant only specific group of microorganisms (Kuppusamy et al. 2016). Another logic given by Warnock et al. (2010), that organic pyrolytic product (phenolics and polyphenolics) are responsible for the reduction in microbial community and their abundance. Biohar and soil type is also responsible for increasing and/or decreasing the microbial community (Jones et al. 2012; Galvez et al. 2012; Lehmann et al. 2011). One of the important property of biochar is explored by Qui et al. (2009), in which harmful chemical secreted by plant or other organisms (allelochemicals) are detoxify by application of biochar consequently improved the plant growth promoting rhizobacteria (*Paenibacillus* sp., *Rhizobium* sp., *Bradyrhizobium* sp., *Pseudomonas* sp. etc.) and mycorrhizae (mainly arbuscular, ericoid and ectomycorrhiza) growth. Biochar have many O-bearing functional groups which involve in sorption of dissolved organic compounds, simple organic compounds, and ammonium ions, provide favorable microbial habitat (Thies and Rilling 2009; Wardle et al. 2008) and responsible for necessary changes in microbial activity.

12.4 Effect of Biochar on Soil Enzyme Activity

Soil enzyme activity is considered as a most important indicator of soil health. The biochar application had significant long and short term impact on soil enzyme and nutrient cyclic were reported. Biochar is influence the intra and extracellular enzymes activity of the organism in the different soil system like normal and stressed soil. One of the intracellular enzyme i.e. dehydrogenase in a soil have role in respiratory processes and strongly correlated with organic carbon availability (Teutscherova et al. 2018). Though, biochar had pH enhancing property it affects the activity of soil dehydrogenase enzyme by adding labile carbon for neutralizing the acid pH under degraded acid soil. Likewise, β-glucosidase is known for the catalyzing cellulose degradation in the final step of glucose release. On addition of biochar the β-glucosidase enzyme activity observed higher and/or decreased. The other hydrolases enzymes such as β-glucosaminidase, phosphatase and urease are involved in soil organic carbon transformation and nutrient cycling (Teutscherova et al. 2018). Urease enzyme have role in the transformation of soil organic nitrogen into available inorganic nitrogen and had no significant relation with biochar application because the feedstock type, pyrolysis conditions, production method, application rate, and soil types are the governing factors that will influence the nitrogen cycling and urease activity in the soil (Zheng et al. 2019). Invertase enzyme also plays crucial role in improving soluble nutrients in the soil, providing sufficient energy for the soil organisms and increased the activity with addition of biochar. The possible mechanism is biochar increases enzyme activity by enhancing the soil organic matter, microbial activity, and microbial biomass or through co-location of enzymes and their interaction with biochar surface (Zheng et al. 2019).

12.5 Effect of Biochar on Soil Physico-chemical Properties

Soil physico-chemical properties play an important role in plant and microbial growth and development. Biochar amendments can changes soil physico-chemical and biological properties such as reduce bulk density increased water holding capacity (retain plant available water), cation exchange capacity and favour the soil microbial activities. The changes in physical properties of soil are also depends on feedstock type, rate of application, type of biochar and interaction time of biochar with soil (Chaganti et al. 2015). It was well documented that the biochar had high porosity, high inner surface area and large number of micropores, which create a better environment for plant root growth, nutrient capture and air porosity (Zheng et al. 2019). Biochar application efficiently improves the soil fertility and crop productivity and also directly related to improvements in soil characteristics due to the high cation exchangeable capacity, surface area, and nutrient contents of biochar (Major et al. 2010). Under sandy loamy soil, biochar application considerably reduced clay dispersion and aggregate disintegration and increased in filtration rate (Abrol et al. 2016). Additionally, biochar also support the building processes of the soil structure via indirect mechanism, such as providing habitat for soil microorganisms and enzyme activities. The effects of biochar on the growth, nutrient uptake and soil properties are summarized in Table 12.2. On the other hand, under salt affected soil biochar is reduces the Na toxicity to the plant because the accumulation of sodium (Na^+) and also improved the K^+: Na^+ ratio through enhancing potassium (K^+) availability to plant (Saifullah et al. 2018).

Table 12.2 Multifarious role of biochar application under different soil type

Type of biochar application	Soil type	Advantage	References
Corn stalk	Silty clay	Reduce the nitrogen contamination of ground water. Changes in microbial community	Sun et al. (2018)
Holm oak	Acrisol Calcisol	Enhance dehydrogenase & urease activity Enhance aggregates stability	Teutscherova et al. (2018)
Manure compost	Salt affected soil	Increases in nutrient content (especially Ca, Mg, K, N & P)	Lashari et al. (2015)
Peanut shell	Salt affected soil	Improve soil organic C	Bhaduri et al. (2016)
Hardwood	Sodic soil	Reduce sodium uptake by Plants	Akhtar et al. (2015)
Beech, hazel, oak, birch	Saline/ sodic	Reduce Na^+ uptake, & leaching of K^+ and NH_4^+	Di Lonardo et al. (2017)
Peanut shell	Salt affected soil	Improve soil health Enhance nutrient availability Elevated bacterial activities & abundances related to nutrient transformations	Zheng et al. (2017)

12.6 Biochar for Improvement of Soil Nutrient Status

Nutrient retaining in soil for long time become a great interest of researchers because nutrient runoff, erosion and the leaching not only degrade soil quality but also adversely impact the quality of water in streams and reservoirs. Biochar have unique physical property to retain various nutrients in their pores and returns most of nutrients to the soils from which they came. Biochar also have the capability to increases the capacity of soils to adsorb essential plant nutrients (Liang et al. 2006; Cheng et al. 2008) thus reducing runoff or losses of nutrients. During formulation of biochar (pyrolysis) the most of essential plant nutrient such as Ca, Mg, Zn, K, P, and about half of the nitrogen and sulphur in the biomass feedstock are partitioned into the biochar fraction. Indeed, many reports are presented on biochar amendment increased the crop yield simultaneous improve the water holding capacity and nutrient use efficiency (Iswaran et al. 1980; Kishimoto and Sugiura 1985; Marjenah 1994; Yamato et al. 2006). The mechanisms of nutrients immobilization by biochars include (1) physical trapping of nutrients within pores of biochars, (2) direct electrostatic interactions between cationic nutrients and negatively charged carbon surfaces, (3) ionic exchange between nutrients ions and ionisable protons at the surface of acidic carbon, (4) specific binding of nutrients by surface ligands (functional groups) abundant on biochar surfaces, (5) reaction with mineral impurities (ash) and basic nitrogen groups (e.g. pyridine) of carbonaceous materials, (6) forming hydroxides, carbonates and/or various phosphate-involved precipices and (7) redox reactions with biochar along with sportive reactions (Li et al. 2018). In contrast, some reports found a decrease in microbial activity after biochar application (Qin et al.

Table 12.3 Influence of soil pH on addition of different types of biochar

Type of biochar application	Changes in soil pH		Percent increase (%)	References
	Initial	Final		
Bamboo	4.68	4.95	5.7	Li-li et al. 2017)
Mixture of Beech, Hazel, Oak and Birch	5.23	6.76	29.2	Rutigliano et al. (2014)
Oak and Hickory	6.33	7.23	14.2	Laird et al. (2010)
Maize straw	7.89	8.19	3.8	Xiao et al. (2016)

2010). These contrasting results could be related to changes in soil moisture, pH, and nutrient dynamics caused by the chemical components of the different types of biochar used (Table 12.3). Biochar had also good impact on soil respiration and soil microbial biomass (Zheng et al. 2019).

12.7 Conclusions

The problem of the depletion of agricultural land as a result of the pressure caused by the ever-growing population necessitated the sustainable practice of crop production. Biochar application is a unique sustainable approach, which has a significant potential to address number of environmental issues and good way to reduce chemical fertilizer use. Under field condition the biochar is a suitable candidate for the improvement of soil physio-chemical properties, microbial abundance and composition. However, biochar amendment to agricultural land had various effects on soil nutrient composition; changes in soil pH significantly, improving soil fertility, input of organic carbon and nitrogen contents. Several reports available on biochar application significantly reduced the soil bulk density, increased water holding capacity, cation exchange capacity, surface area, and the retention of various essential nutrients like N, P, K, Mg, Ca and several other plant nutrients.

References

Abrol V, Ben-Hur M, Verheijen FG, Keizer JJ, Martins MA, Tenaw H, Tchehansky L, Graber ER (2016) Biochar effects on soil water infiltration and erosion under seal formation conditions: rainfall simulation experiment. J Soils Sediments 16:2709–2719

Ahmad M, Rajapaksha AU, Lim JE, Zhang M, Bolan N, Mohan D, OK YS (2014) Biochar as a sorbent for contaminant management in soil and water: a review. Chemosphere 99:19–33

Akhtar SS, Andersen MN, Liu F (2015) Residual effects of biochar on improving growth, physiology and yield of wheat under salt stress. Agric Water Manag 158:61–68

Ameloot N, De Neve S, Jegajeevagan K, Yildiz G, Buchan D, Funkuin YN, Prins W, Bouckaert L, Sleutel S (2013) Short-term CO_2 and N_2O emissions and microbial properties of biochar amended sandy loam soils. Soil Biol Biochem 57:401–410

Beesley L, Jiménez EM, Eyles JLG (2010) Effects of biochar and green waste compost amendments on mobility, bioavailability and toxicity of inorganic and organic contaminants in a multi-element polluted soil. Environ Pollut 158:2282–2287

Bhaduri D, Saha A, Desai D, Meena HN (2016) Restoration of carbon and microbial activity in salt-induced soil by application of peanut shell biochar during short-term incubation study. Chemosphere 148:86–98

Brewer CE, Hu YY, Schmidt-Rohr K, Loynachan TE, Laird DA, Brown RC (2012) Extent of pyrolysis impacts on fast pyrolysis biochar properties. J Environ Qual 41:1115–1122

Bruun S, Clauson-Kaas S, Bobulska L, Thomsen IK (2014) Carbon dioxide emissions from biochar in soil: role of clay, microorganisms and carbonates. Eur J Soil Sci 65:52–59

Cantrell KB, Hunt PG, Uchimiya M, Novak JM, Ro KS (2012) Impact of pyrolysis temperature and manure source on physicochemical characteristics of biochar. Bioresour Technol 107:419–428

Chaganti VN, Crohn DM, Šimůnek J (2015) Leaching and reclamation of a biochar and compost amended saline–sodic soil with mode rate SAR reclaimed water. Agric Water Manag 158:255–265

Chan K, Van Zwieten L, Meszaros I, Downie A, Joseph S (2008) Using poultry litter biochars as soil amendments. Aust J Soil Res 46:437–444

Chen JH, Liu XY, Zheng JW, Zhang B, Lu HF, Chi ZZ, Pan GX, Li LQ, Zheng JF, Zhang XH, Wang JF, Yu XY (2013) Biochar soil amendment increased bacterial but decreased fungal gene abundance with shifts in community structure in a slightly acid rice paddy from Southwest China. Appl Soil Ecol 71:33–44

Chen T, Liu R, Scott NR (2016) Characterization of energy carriers obtained from the pyrolysis of white ash, switchgrass and corn stover biochar, syngas and bio-oil. Fuel Process Technol 142:124–134

Cheng CH, Lehmann J, Engelhard MH (2008) Natural oxidation of black carbon in soils: changes in molecular form and surface charge along a climosequence. Geochim Cosmochim Acta 72:1598–1610

Costanza R, d'Arge R, de Groot R, Farberk S, Grasso M, Hannon B, Limburg K, Naeem S, O ' Neill RV, Paruelo J, Raskin RG, Suttonkk P, van den Belt M (1987) The value of the world's ecosystem services and natural capital. Nature 387:253–260

Crombie K, Mašek O, Sohi SP, Brownsort P, Cross A (2013) The effect of pyrolysis conditions on biochar stability as determined by three methods. GCB Bioenergy 5(2):122–131

Di Lonardo S, Baronti S, Vaccari FP, Albanese L, Battista P, Miglietta F, Bacci L (2017) Biochar-based nursery substrates: the effect of peat substitution on reduced salinity. Urban For Urban Green 23:27–34

Durenkamp M, Luo Y, Brookes P (2010) Impact of black carbon addition to soil on the determination of soil microbial biomass by fumigation extraction. Soil Biol Biochem 42:2026–2029

Galvez A, Sinicco T, Cayuela ML, Mingorance MD, Fornasier F, Mondini C (2012) Short term effects of bioenergy by-products on soil C and N dynamics, nutrient availability and biochemical properties. Agric Ecosyst Environ 160:3–14

Ge Y, Zhang JB, Zhang LM, Yang M, He JZ (2008) Long-term fertilization regimes affect bacterial community structure and diversity of an agricultural soil in northern China. J Soils Sediments 8:43–50

George C, Wagner M, Kucke M, Rillig MC (2012) Divergent consequences of hydrochar in the plant–soil system: arbuscular mycorrhiza, nodulation, plant growth and soil aggregation effects. Appl Soil Ecol 59:68–72

Glaser B (2007) Prehistorically modified soils of central Amazonia: a model for sustainable agriculture in the twenty-first century. Philos Trans R Soc B 362:187–196

Graber ER, Harel YM, Kolton M, Cytryn E, Silber A, David DR, Tsechansky L, Borenshtein M, Elad Y (2010) Biochar impact on development and productivity of pepper and tomato grown in fertigated soilless media. Plant Soil 337:481–496

Grossman JM, O'Neill BE, Tsai SM, Liang B, Neves E, Lehmann J, Thies JE (2010) Amazonian anthrosols support similar microbial communities that differ distinctly from those extant in adjacent, unmodified soils of the same mineralogy. Microb Ecol 60:192–205
Inyang M, Dickenson E (2015) The potential role of biochar in the removal of organic and microbial contaminants from potable and reuse water: review. Chemosphere 134:232–240
Iswaran V, Jauhri KS, Sen A (1980) Effect of charcoal, coal and peat on the yield of moong, soybean and pea. Soil Biol Biochem 12:191–192
Jeffery S, Verheijena FGA, van der Velde M, Bastos AC (2011) A quantitative re-view of the effects of biochar application to soils on crop productivity using meta-analysis. Agric Ecosyst Environ 144:175–187
Jindo K, Sanchez-Moneder MA, Hernandez T, Garcia C, Furukawa T, Matsumoto K, Sonoki T, Bastida F (2012) Biochar influences the microbial community structure during manure composting with agricultural wastes. Sci Total Environ 416:476–481
Jones DL, Rousk J, Edwards-Jones G, DeLuca TH, Murphy DV (2012) Biochar-mediated changes in soil quality and plant growth in a three year field trial. Soil Biol Biochem 45:113–124
Joseph SD, Camps AM, Lin Y, Munroe P, Chia CH, Hook J, Zwieten LV, Kimber S, Cowie A, Singh BP, Lehmann J, Foidl N, Smernik RJ, Amonette JE (2010) An investigation into the reactions of biochar in soil. Soil Res 48:501–515
Jung C, Oh J, Yoon Y (2015) Removal of acetaminophen and naproxen by combined coagulation and adsorption using biochar: influence of combined sewer over flow components. Environ Sci Pollut Res 22(13):10058–10069
Kambo HS, Dutta A (2015) A comparative review of biochar and hydrochar in terms of production, physico-chemical properties and applications. Renew Sust Energ Rev 45:359–378
Karami N, Clemente R, Moreno JE, Lepp NW, Beesley L (2011) Efficiency of green waste compost and biochar soil amendments for reducing lead and copper mobility and uptake to ryegrass. J Hazard Mater 191:41–48
Keiluweit M, Nico PS, Johnson MG, Kleber M (2010) Dynamic molecular structure of plant biomass-derived black carbon (biochar). Environ Sci Technol 44:1247–1253
Kishimoto S, Sugiura G (1985) Charcoal as a soil conditioner. Int Achieve Future 5:12–23
Kumar A, Singh JS (2017) Cyanoremediation: a green-clean tool for decontamination of synthetic pesticides from agro-and aquatic ecosystems. In: Singh JS, Seneviratne G (eds) Agro-environmental sustainability: vol (2) managing environmental pollution. Springer, Cham, pp 59–83
Kumar A, Kaushal S, Saraf S, Singh JS (2017) Cyanobacterial biotechnology: an opportunity for sustainable industrial production. Clim Change Environ Sustain 5(1):97–110
Kumar A, Kaushal S, Saraf S, Singh JS (2018) Microbial bio-fuels: a solution to carbon emissions and energy crisis. Front Biosci (Landmark) 23:1789–1802
Kuppusamy S, Thavamani P, Megharaj M, Venkateswarlu K, Naidu R (2016) Agronomic and remedial benefits and risks of applying biochar to soil: current knowledge and future research directions. Environ Int 87:1–12
Laird DA, Fleming P, Davis DD, Horton R, Wang B, Karlen DL (2010) Impact of biochar amendments on the quality of a typical Midwestern agricultural soil. Geoderma 158:443–449
Lashari MS, Ye Y, Ji H, Li L, Kibue GW, Lu H, Zheng J, Pan G (2015) Biochar–manure compost in conjunction with pyroligneous solution alleviated salt stress and improved leaf bioactivity of maize in a saline soil from central China: a 2-year field experiment. J Sci Food Agric 95:1321–1327
Lawrinenko M (2014) Anion exchange capacity of biochar. Graduate thesis. Iowa State University, Ames, Iowa, USA. http://lib.dr.iastate.edu/etd
Lehmann J (2007) A handful of carbon. Nature 447:143–144
Lehmann J, Joseph S (2009) Biochar for environmental management: science, technology and implementation. Routledge Publishing, London/New York
Lehmann J, Rillig MC, Thies J, Masiello CA, Hockaday WC, Crowley D (2011) Biochar effects on soil biota – a review. Soil Biol Biochem 43:1812–1836

Li L, Wang S, Li X, Li T, He X, Tao Y (2018) Effects of *Pseudomonas chenduensis* and biochar on cadmium availability and microbial community in the paddy soil. Sci Total Environ 640:1034–1043

Liang B, Lehmann J, Solomon D, Kinyangi J, Grossman J, O'Neill B, Skjemstad JO, Thies J, Luiza FJ, Petersen J, Neves EG (2006) Black carbon increases cation exchange capacity in soils. Soil Sci Soc Am J 70:1719–1730

Li-li H, Zhe-ke Z, Hui-min Y (2017) Effects on soil quality of biochar and straw amendment in conjunction with chemical fertilizers. J Integr Agric 16(3):704–712

Lu K, Yang X, Shen J, Robinson B, Huang H, Liu D, Wang H (2014) Effect of bamboo and rice straw biochars on the bioavailability of Cd, Cu, Pb and Zn to Sedum plumbizincicola. Agric Ecosyst Environ 191:24–132

Luo S, Wang S, Tian L, Li S, Lia X, Shen Y, Tian C (2017) Long-term biochar application influences soil microbial community and its potential roles in semiarid farmland. Appl Soil Ecol 117–118:10–15

Major J, Lehmann J, Rondon M, Goodale C (2010) Fate of soil-applied black carbon: downward migration, leaching and soil respiration. Glob Chang Biol 16(4):1366–1379

Marjenah MS (1994) Effect of charcoaled rice husks on the growth of *Dipterocarpaceae* seedlings in East Kalimantan with special reference to ectomycorrhiza formation. J Jpn For Soc 76:462–464

Oliveira FR, Patel AK, Jaisi DP, Adhikari S, Lu H, Khanal SK (2017) Environmental application of biochar: current status and perspectives. Bioresour Technol 246:110–122

Qin SP, Hu CS, Dong WX (2010) Nitrification results in underestimation of soil urease activity as determined by ammonium production rate. Pedobiologia 53:401–404

Qiu H, Lv L, Pan BC, Zhang QJ, Zhang WM, Zhang QX (2009) Critical review in adsorption kinetic models. J Zhejiang Univ Sci A 10(5):716–724

Rutigliano FA, Romano M, Marzaioli R, Baglivo I, Baronti S, Miglietta F, Castaldi S (2014) Effect of biochar addition on soil microbial community in a wheat crop. Eur J Soil Biol 60:9–15

Saifullah, Dahlawi S, Naeem A, Rengel Z, Naidu R (2018) Biochar application for the remediation of salt-affected soils: challenges and opportunities. Sci Total Environ 625:320–335

Schulz H, Glaser B (2012) Effects of biochar compared to organic and inorganic fertilizers on soil quality and plant growth in a greenhouse experiment. J Plant Nutr Soil Sci 175:410–422

Singh JS (2013) Anticipated effects of climate change on methanotrophic methane oxidation. Clim Change Environ Sustain 1(1):20–24

Singh JS (2014) Cyanobacteria: a vital bio-agent in eco-restoration of degraded lands and sustainable agriculture. Clim Change Environ Sustain 2:133–137

Singh JS (2015) Biodiversity: current perspective. Clim Change Environ Sustain 3(1):71–72

Singh JS (2016) Microbes play major roles in ecosystem services. Clim Change Environ Sustain 3:163–167

Singh JS (ed) (2019) New and future developments in microbial biotechnology and bioengineering: microbes in soil, crop and environmental sustainability. Elsevier, San Diego

Singh JS, Boudh S (2016) Climate change resilient crops to sustain Indian agriculture. Clim Change Environ Sustain 5:97–110

Singh JS, Singh DP (eds) (2019) New and future developments in microbial biotechnology and bioengineering: microbial biotechnology in agro-environmental sustainability. Elsevier, San Diego

Singh C, Tiwari S, Boudh S, Singh JS (2017a) Biochar application in management of paddy crop production and methane mitigation. In: Singh JS, Seneviratne G (eds) Agro-environmental sustainability: vol (2) managing environmental pollution. Springer, Cham, pp 123–145

Singh JS, Koushal S, Kumar A, Vimal SR, Gupta VK (2017b) Book review: microbial inoculants in sustainable agricultural productivity- Vol. II: functional application. Front Microbiol 7:2015

Singh JS, Kumar A, Singh M (2019a) Cyanobacteria: a sustainable and commercial bioresource in production of bio-fertilizer and bio-fuel from waste waters. Environ Sustain Indic 3-4:100008

Singh MK, Rai PK, Rai A, Singh S, Singh JS (2019b) Poly-β-hydroxybutyrate production by the cyanobacterium Scytonema geitleri Bharadwaja under varying environmental conditions. Biomol Ther 9(198):1–10

Suliman W, Harsh JB, Abu-Lail NI, Fortuna AM, Dallmeyer I, Garcia-Perez M (2016) Influence of feedstock source and pyrolysis temperature on biochar bulk and surface properties. Biomass Bioenergy 84:37–48

Sun J, Chen L, Rene ER, Hu Q, Ma W, Shen Z (2018) Biological nitrogen removal using soil columns for the reuse of reclaimed water: performance and microbial community analysis. J Environ Manag 217:100–109

Swift MJ, Heal OW, Anderson JM (1979) Decomposition in terrestrial ecosystems. University of California Press, Berkeley

Tan XF, Liu SB, Liu YG, Gu YL, Zeng GM, Hu XJ, Wang X, Liu SH, Jiang LH (2017) Biochar as potential sustainable precursors for activated carbon production: multiple applications in environmental protection and energy storage. Bioresour Technol 227:359–372

Teixidó M, Pignatello JJ, Beltrán JL, Granados M, Peccia J (2011) Speciation of the ionizable antibiotic sulfamethazine on black carbon (biochar). Environ Sci Technol 45:10020–10027

Teutscherova N, Lojka B, Houška J, Masaguer A, Benito M, Vazquez E (2018) Application of holm oak biochar alters dynamics of enzymatic and microbial activity in two contrasting Mediterranean soils. Eur J Soil Biol 88:15–26

Thies E, Rilling MC (2009) Characteristics of biochar: biological properties. In: Lehmann J, Joseph S (eds) Biochar for environmental management: science and technology. Earthscan, London, pp 85–106

Tian J, Wang J, Dippold M, Gao Y, Blagodatskaya E, Kuzyakov Y (2016) Biochar affects soil organic matter cycling and microbial functions but does not alter microbial community structure in a paddy soil. Sci Total Environ 556:89–97

Tiwari AP, Singh JS (2017) Plant growth promoting rhizospheric *Pseudomonas aeruginosa* strain inhibits seed germination in *Triticum aestivum* (L) and *Zea mays* (L). Microbiol Res 8(7233):73–79

Uras U, Carrier M, Hardie AG, Knoetze JH (2012) Physico-chemical characterization of biochars from vacuum pyrolysis of South African agricultural wastes for application as soil amendments. J Anal Appl Pyrolysis 98:207–213

Vimal SR, Singh JS (2019) Salt tolerant pgpr and fym application in saline soil paddy agriculture sustainability. Clim Change Environ Sustain 7(1):23–33

Vimal SR, Gupta J, Singh JS (2018) Effect of salt tolerant *Bacillus* sp. and *Pseudomonas* sp. on wheat (*Triticum aestivum* L.) growth under soil salinity: a comparative study. Microbiol Res 9(1):1–14

Wardle DA, Nilsson MC, Zackrisson O (2008) Response to comment on fire-derived charcoal causes loss of forest humus. Science 321:1295d

Warnock DD, Lehman J, Kuype TW, Rillig MC (2007) Mycorrhizal responses to biochar in soil–concepts and mechanisms. Plant Soil 300:9–20

Warnock DD, Mummey DL, McBride B, Major J, Lehmann J, Rillig MC (2010) Influences of non-herbaceous biochar on Arbuscular Mycorrhizal fungal abundances in roots and soils: results from growth-chamber and field experiments. Appl Soil Ecol 46:450–456

Wilson GW, Rice CW, Rillig MC, Springer A, Hartnett DC (2009) Soil aggregation and carbon sequestration are tightly correlated with the abundance of arbuscular mycorrhizal fungi: results from long-term field experiments. Ecol Lett 12:452–461

Windeatt JH, Ross AB, Williams PT, Forster PM, Nahil MA, Singh S (2014) Characteristics of biochars from crop residues: potential for carbon sequestration and soil amendment. J Environ Manag 146:189–197

Wright DA, Killham K, Glover LA, Prosser JI (1995) Role of pore size location in determining bacterial activity during predation by protozoa in soil. Appl Environ Microbiol 61:3537–3543

Xiao Q, Zhu LX, Zhang HP, Li XY, Shen YF, Li SQ (2016) Soil amendment with biochar increases maize yields in a semi-arid region by improving soil quality and root growth. Crop Pasture Sci 67:495–507

Xu T, Lou L, Luo L, Cao R, Duan D, Chen Y (2012) Effect of bamboo biochar on pentachlorophenol leachability and bioavailability in agricultural soil. Sci Total Environ 414:727–731

Yamato M, Okimori Y, Wibowo I, Anshiori S, Ogawa M (2006) Effects of the application of charred bark of *Acacia mangium* on the yield of maize, cowpea and peanut, and soil chemical properties in South Sumatra, Indonesia. Soil Sci Plant Nutr 52:489–495

Yao Q, Liu J, Yu Z, Li Y, Jin J, Liu X, Wang G (2017) Changes of bacterial community compositions after three years of biochar application in a black soil of northeast China. Appl Soil Ecol 113:11–21

Zama EF, Zhu YG, Reid BJ, Sun GX (2017) The role of biochar properties in influencing the sorption and desorption of Pb (II), Cd (II) and As (III) in aqueous solution. J Clean Prod 148:127–136

Zhang XK, Li Q, Liang WJ, Zhang M, Bao XL, Xie ZB (2013) Soil nematode response to biochar addition in a Chinese wheat field. Pedosphere 23:98–103

Zhang H, Ding W, Yu H, He X (2015) Linking organic carbon accumulation to microbial community dynamics in a sandy loam soil: result of 20 years compost and inorganic fertilizers repeated application experiment. Biol Fertil Soils 51:137–150

Zheng W, Guo M, Chow T, Bennett DN, Rajagopalan N (2010) Sorption properties of green waste biochar for two triazine pesticides. J Hazard Mater 181:121–126

Zheng JF, Chen JH, Pan GX, Liu XY, Zhang XH, Li LQ, Bian RJ, Cheng K, Zheng JW (2016) Biochar decreased microbial metabolic quotient and shifted community composition four years after a single incorporation in a slightly acid rice paddy from Southwest China. Sci Total Environ 571:206–217

Zheng H, Wang X, Chen L, Wang Z, Xia Y, Zhang Y, Wang H, Luo X, Xing B (2017) Enhanced growth of halophyte plants in biochar-amended coastal soil: roles of nutrient availability and rhizosphere microbial modulation. Plant Cell Environ 41(3):517–532. https://doi.org/10.1111/pce.12944

Zheng Y, Han X, Li Y, Yang J, Li N, An N (2019) Effects of biochar and straw application on the physicochemical and biological properties of Paddy soils in northeast China. Sci Rep 9:16531. https://doi.org/10.1038/s41598-019-52978-w

Zwieten LV, Kimber S, Morris S, Chan KY, Downie A, Rust J, Joseph S, Cowie A (2010) Effects of biochar from slow pyrolysis of papermill waste on agronomic performance and soil fertility. Plant Soil 327:235–246

The manufacturer's authorised representative in the EU is Springer Nature Customer Service Centre GmbH, Europaplatz 3, 69115 Heidelberg, Germany. If you have any concerns regarding our products, please contact ProductSafety@springernature.com

Printed and bound by CPI Group (UK) Ltd, Croydon, CR0 4YY
22/09/2026
02221065-0003